Edexcel Chemistry for AS

Graham Hill
Andrew Hunt

HODDER EDUCATION
AN HACHETTE UK COMPANY

The Publishers would like to thank the following for permission to reproduce copyright material:

Photo credits: p.1 Photodisc; **p.2** © lookGaleria/Alamy (inset: AJ Photo/Science Photo Library); **p.3** *t* Gregory Ochocki/Science Photo Library, *c* © IBM; **p.4** *t* Colin Cuthbert/Science Photo Library, *c* Geoff Tompkinson/Science Photo Library, *b* © Scott Camazine; **p.7** © Christina Kennedy/Brand X/Corbis; **p.8** Tim Graham/Getty Images; **p.9** Bill Barksdale/AGStockUSA/Science Photo Library; **p.10** Adam Hart-Davis/Science Photo Library; **p.11** *t* © JUPITERIMAGES/BananaStock/Alamy, *c* Andrew Lambert/Science Photo Library; **p.16** both Martyn Chillmaid; **p.21** © Bettmann/Corbis; **p.26** Saturn Stills/Science Photo Library; **p.29** © Hodder Education; **p.30** Saturn Stills/Science Photo Library; **p.32** www.ctpimaging.co.uk, **p.45** Science Photo Library; **p.50** Geoff Tompkinson/Science Photo Library; **p.51** © Gilbert Iundt, Jean-Yves Ruszniewski/TempSport/Corbis; **p.63** Philippe Plailly/Eurelios/Science Photo Library; **p.64** *t* © Peter Arnold, Inc./Alamy, *cl* © Alan Goldsmith/Corbis, *cr* © David Samuel Robbins/Corbis; **p.65** Andrew Lambert/Science Photo Library; **p.66** © Owaki-Kulla/Corbis; **p.68** Andrew Lambert/Science Photo Library; **p.77** © David Noton Photography/Alamy; **p.78** Andrew Lambert/Science Photo Library; **p.79** © PjrFoto/Studio/Alamy; **p.83** *c* © Bob Krist/Corbis, *r* © Nordicphotos/Alamy; **p.86** Geoff Tompkinson/Science Photo Library; **p.92** © Car Culture/Corbis; **p.96** Ingram; **p.97** © Justin Kase zfourz/Alamy; **p.99** Andrew Lambert/Science Photo Library; **p.104** Sue Cunningham Photographic; **p.108** reproduced by kind permission of Unilever PLC (from an original in Unilever Archives); **p.113** *t* Emma Lee/Life File, *b* Andrew Lambert/Science Photo Library; **p.114** © Worldwide Picture Library/Alamy; **p.115** © allOver Photography/Alamy; **p.117** Photodisc; **p.118** © AGStockUSA, Inc./Alamy; **p.123** © Eye Ubiquitous; **p.124** © Reuters/Corbis; **p. 125** courtesy Zettl Research Group, Lawrence Berkeley National Laboratory and University of California at Berkeley; **p.130** Charles D. Winters/Science Photo Library; **p.131** Volker Steger/Science Photo Library; **p.135** Peter Scoones/Science Photo Library; **p.147** all Andrew Lambert/Science Photo Library; **p.148** Andrew Lambert/Science Photo Library; **p.150** *tl* Russ Lappa/Science Photo Library, *c* Bob Gibbons/Science Photo Library, *bl* Andrew Lambert/Science Photo Library; **p.153** *t* Martin Land/Science Photo Library, *c* Natural History Museum, London, *b* Science Photo Library; **p.154** Martin Bond/Science Photo Library; **p.156** both Andrew Lambert/Science Photo Library; **p.157** Andrew Lambert/Science Photo Library; **p.159** Andrew Lambert/Science Photo Library; **p.162** © Randy Faris/Corbis; **p.164** Ria Novosti/Science Photo Library; **p.165** Andrew Lambert Photography/Science Photo Library; **p.166** Andrew Lambert Photography/Science Photo Library; **p.167** Geoff Tompkinson/Science Photo Library; **p.178** *t* Geoff Tompkinson/Science Photo Library, *c* © Israel Sanchez/epa/Corbis; **p.184** Science Photo Library; **p.190** *t* Leonard Lessin/Science Photo Library, *c* © Corbis, All Rights Reserved, *b* Andrew Lambert/Science Photo Library; **p.196** *t* Dr Peter Linke/Leibniz-Institut für Meereswissenschaften (IFM-Geomar), *c* Dr Tim Evans/Science Photo Library; **p.199** © Amit Bhargava/Corbis; **p.202** Andrew Lambert Photography/Science Photo Library; **p.204** Bill Barksdale/AGStockUSA/Science Photo Library; **p.207** © Jeff Morgan Recycling & Waste Management/Alamy; **p.211** Colin Cuthbert/Science Photo Library; **p.212** Volker Steger/Science Photo Library; **p.215** US Dept of Energy/Science Photo Library; **p.220** *t* © SHOUT/Alamy, *c* © Gareth Price; **p.222** PA Photos/AP Photo/Amy Sinisterra; **p.225** Maximilian Stock Ltd/Science Photo Library; **p.226** David Zutler, Founder and Developer of World's First Planet Friendly Bottle; **p.227** courtesy of Monsanto; **p.228** © Bill Stormont/Corbis; **p.233** Tek Image/Science Photo Library; **p.235** E. Wolff/BAS; **p.239** Alex Bartel/Science Photo Library; **p.241** © Reuters/Corbis; **p.242** NASA.

b = bottom, *c* = centre, *l* = left, *r* = right, *t* = top

Artwork credits: Philip Allan Updates: Dr Eric Wolff, graph of estimated temperature difference from today and carbon dioxide and methane in ppmv against year before, from ***Chemistry Review,*** volume 15, number 1 (September, 2005); **Carbon Footprint Ltd**: Redrawn pie chart showing components of a typical footprint from (www.carbonfootprint.com/carbon_footprint.html); **Chemicals Industries Association**: Pie chart of products produced by the chemical industry in the UK from **www.cia.org.uk** (Chemicals Industries Association, 2006); **www.greener.industry.org**: Diagram of manufacture of ethanoic acid from carbon monoxide and ethanol from diagram of Reaction mechanism on screen 6 of the series of web pages on ethanoic acid; **IPCC** : Graph of temperature differences from 1961-1990 average against year (redrawn) from Presentation ***of*** the Working Group 1 report (www.ipcc.ch/present/presentations.htm); **LGC**: Vicki Barwick and Elizabeth Pritchard, extracts from *Introducing Measurement Uncertainty* (LCG, 2003), © LGC; **Royal Society of Chemistry**: Graph of carbon dioxide in ppmv against year AD from *Education in Chemistry*, volume 41, number 5 (September, 2004);**Transport for London**: Flow diagram of hydrogen-fuelled London bus and supply of hydrogen (slightly modified) from www.tfl.gov.uk.

Acknowledgements Every effort has been made to trace all copyright holders, but if any have been inadvertently overlooked the Publishers will be pleased to make the necessary arrangements at the first opportunity.

Orders: please contact Bookpoint Ltd, 130 Milton Park, Abingdon, Oxon OX14 4SB. Telephone: (44) 01235 827720. Fax: (44) 01235 400454. Lines are open 9.00 – 5.00, Monday to Saturday, with a 24-hour message answering service. Visit our website at www.hoddereducation.co.uk

First published in 2008 by
Hodder Education,
An Hachette UK Company
338 Euston Road
London NW1 3BH

Impression number 10 9 8
Year 2014, 2013

Cover photo © Steve Gschmeissner/Science Photo Library
Illustrations by Barking Dog Art
Typeset in Goudy 10.5pt by Fakenham Photosetting Ltd, Fakenham Norfolk
Printed in Dubai

A catalogue record for this title is available from the British Library

ISBN: 978 0340 94908 5

Contents

Introduction

Welcome to *Edexcel Chemistry for AS*. This book covers everything in the Edexcel specification with 6 topics for Unit 1 and 13 topics for Unit 2. Test yourself questions throughout the book will help you to think about what you are studying while the Activities give you the chance to apply what you have learnt in a range of modern contexts. At the end of each topic you will find exam-style Review questions to help you check your progress.

Student support for this book can be found on the **Dynamic Learning Student** website which contains an interactive copy of the book. Students can access a range of free digital resources by visiting **www.dynamic-learning-student.co.uk** and using the code printed on the inside front cover of this book to gain access to relevant resources. These free digital resources include:

- Data tables, for use when answering questions
- Tutorials, which work through selected problems and concepts using a voiceover and animated diagrams
- Practical guidance to support experimental and investigative skills.

All diagrams and photographs can be launched and enlarged. There are also Learning outcomes available at the beginning of every topic, and answer files to all Test yourself and Activity questions. Extension questions, covering some ideas in greater depth, are available at the end of each topic. The Student Online icon, shown on the left, indicates where a resource such as a Data Sheet, Tutorial, Practical Guidance or Extension questions is provided on the Dynamic Learning Student website. All resources can be saved to your local hard drive.

With the powerful Search tool, key words can be found in an instant, leading you to the relevant page or alternatively to resources associated with each key word.

Acknowledgements

We are grateful to John Apsey, Senior Examiner with Edexcel, for reading and commenting on drafts of the book and CD-ROM resources. We would also like to acknowledge the suggestions from teachers who commented on our initial plans: Ian Davis, Neil Dixon and Tim Joliffe.

The team at Hodder Education, led by Gillian Lindsey, has made an extremely valuable contribution to the development of the book and the CD-ROM resources. In particular, we would like to thank Anne Trevillion, Anne Wanjie, Deborah Sanderson, Anne Russell, Marion Edsall and Tony Clappison for their skilful work on the print and electronic resources.

Graham Hill and Andrew Hunt

A Note for Teachers

Edexcel Chemistry for AS Network disc

The *Edexcel Chemistry for AS Network disc* which accompanies the Student's Book and Dynamic Learning Student website provides a complete bank of resources for teachers and technicians following the Edexcel specification. Powered by Dynamic Learning, the Network disc contains the same interactive version of the Student's Book that is available on the Dynamic Learning Student website, plus every resource that teachers might wish to use in activities, discussions, practical work and assessment.

These additional resources on the Network disc include:

- a synoptic Topic overview for each of the nineteen topics showing how the text and resources cover and follow the Edexcel specification, including coverage of 'How Science Works'. These synopses also indicate how the resources on the Network disc can be used to create a teaching programme and lesson plans
- Introductory PowerPoints for each topic
- Practical worksheets for students covering the entire course
- Teacher's and technician's notes for all practicals showing the intentions of each practical, a suggested approach to it, health and safety considerations, the materials and apparatus required by each group and answers to the questions on the worksheets
- additional Weblinks
- additional Activities involving data analysis, application and evaluation
- 3D Rotatable models of molecules of interest
- answers to all Review questions in the Student's Book, as well as answers to all Extension questions available to students from the Dynamic Learning Student website and to questions in the additional Activities on the Network disc
- Interactive objective tests for each topic with answers.

All these resources are launched from interactive pages of the Student's Book on the Network disc. Teachers can also search for resources by resource type or key words to allow greater flexibility in using the resource material.

A Lesson Builder allows teachers to build lessons by dragging and dropping resources they want to use into a lesson group that can be saved and launched later from a single screen.

The tools provided also enable teachers to import their own resources and weblinks into the lesson group and to populate a VLE at the click of a button.

Risk assessment

As a service to users, a risk assessment for this text and associated resources has been carried out by CLEAPSS and is available on request to the Publishers. However, the Publishers accept no legal responsibility on any issue arising from this risk assessment; whilst every effort has been made to check the instructions for practical work in this book and associated resources, it is still the duty and legal obligation of schools to carry out their own risk assessment.

Unit 1

The Core Principles of Chemistry

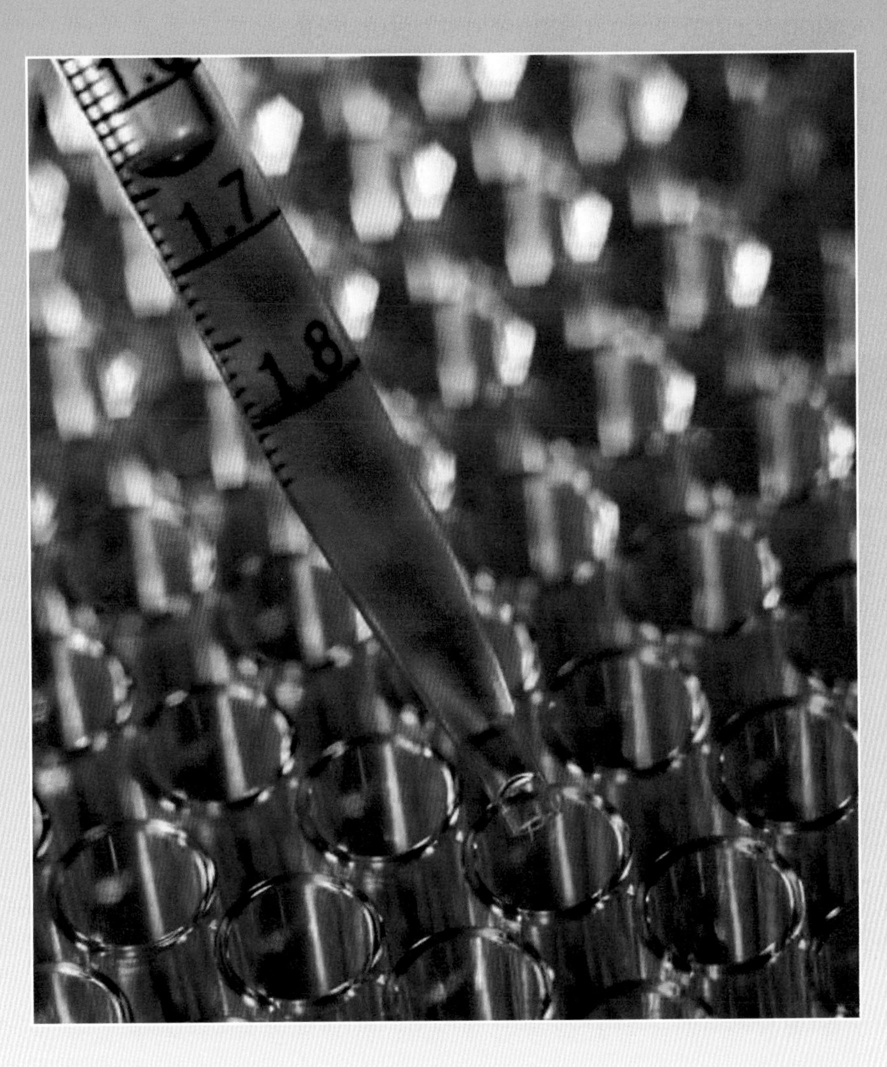

1 Chemical quantities

Modern chemistry began when scientists understood the difference between elements and compounds. Soon afterwards, chemists were able to explain chemical changes in terms of atoms and molecules and write chemical formulae. This led to the use of chemical equations and experiments to determine the quantities of chemicals which react and which are produced in chemical reactions.

1.1 Why study chemistry?

Chemistry is one of the sciences which helps us to develop our knowledge, and to understand ourselves and all the materials and living things in the Universe.

Some of the reasons for studying chemistry are outlined below. How well do these reasons match your own expectations as you start this advanced chemistry course?

Looking for patterns in chemical behaviour

Part of being a chemist involves getting a feel for the way in which chemicals behave. Chemists get to know chemicals just as people get to know their friends and family. They look for patterns in behaviour and recognise that some of the patterns are familiar. For example, the elements sodium and potassium are both soft and stored under oil because they react so readily with air and water; copper sulfate is blue, like other copper compounds. By understanding patterns, chemists can identify and make plastics like polythene and medicines like aspirin.

Figure 1.1 ▶
Aspirin is probably the commonest medicine we use. The bark of willow trees was used to ease pain for more than 2000 years. Early in the twentieth century, chemists extracted the active ingredient from willow bark. Their understanding of patterns in the behaviour of similar compounds enabled them to synthesise aspirin.

Working out the composition and structure of materials

New materials exist only because chemists understand how atoms, ions and molecules are arranged in different materials and the forces which hold these particles together. Thanks to this knowledge, we can enjoy fibres that breathe but are waterproof, plastic ropes that are 20 times stronger than similar ropes of steel (Figure 1.2) and metal alloys which can remember their shape. Understanding the structure and bonding of materials is a central theme in modern chemistry.

Controlling chemical changes

Four simple questions are at the heart of most chemical investigations.

- *How much?* – how much of the reactants do we need to make a product and how much of the product will we produce?
- *How fast?* – how can we make sure that a reaction goes at the right speed: not too fast and not too slow? How can we control the speed of reactions?
- *How far?* – will the chemicals we use react completely and make the product, or will the reaction stop before we get all we want? If it does, what can we do to get as big a yield as possible?
- *How do reactions occur?* – which bonds between atoms are breaking and which new bonds are forming during a reaction?

Figure 1.2 ▲
Plastic ropes made of polymers twenty times stronger than steel have made rock and ice climbing safer and easier.

Developing new skills

Chemistry involves *doing* things as well as gaining knowledge and understanding about materials. Chemists use their thinking skills and practical skills to solve practical problems. One of the frontiers of today's chemistry involves nanotechnology, in which chemists work with particles as small as individual atoms (Figure 1.3).

Increasingly, chemists rely on modern instruments to explore structures and chemical changes. They also use information technology to store data, search for information and to publish their findings.

Figure 1.3 ▲
In the 1990s, two scientists working for IBM cooled a nickel surface to −269 °C in a vacuum chamber. Then they introduced a tiny amount of xenon so that some of the xenon atoms stuck to the nickel surface. Using a special instrument called a scanning tunnelling microscope, the scientists were able to move individual xenon atoms around on the nickel surface and construct the IBM logo. Each blue blob is the image of a single xenon atom.

Recognising the value of chemistry to society

Chemists study materials and try to change raw materials into more useful substances. On a large scale, the chemical industry converts raw materials from the earth, the sea and the air into valuable new products for society. A good example of this is the Haber process, which turns natural gas and air into ammonia – the chemical used to make fertilisers, dyes and explosives.

A vital task for chemists is to analyse materials and find out what they are made of. When chemists have analysed a substance, they use symbols and formulae to show the elements it contains. Symbols are used to represent the atoms in elements; formulae are used to represent the ions and molecules in compounds. Analysis is involved in checking that our drinking water is pure and our food is safe to eat. People worry about pollution of the environment, but without chemical analysis we would not understand the causes or the scale of this pollution.

Linking theories and experiments

Scientists test their theories by doing experiments. In chemistry, experiments often begin with careful observation of what happens as chemicals react and change. Theories are more likely to be accepted if predictions made from them turn out to be correct when tested by experiment.

One of the reasons why Mendeléev's periodic table was so successful was because his predictions about the properties of missing elements turned out to be so accurate (Table 1.1).

Mendeléev's predictions in 1871	Actual properties in 1886
Grey metal Density 5.5 g cm^{-3} Relative atomic mass 73.4 Melting point 800 °C	Pale grey metal Density 5.35 g cm^{-3} Relative atomic mass 72.6 Melting point 937 °C
Formula of oxide GeO_2	Ge forms GeO_2

Table 1.1 ▲
Mendeléev's predictions for germanium in 1871 and the properties it was found to have after its discovery in 1886

Definitions

Elements are the simplest chemicals – they cannot be broken down into simpler chemicals.

Compounds are chemicals containing two or more elements chemically combined together.

An **atom** is the smallest particle of an element.

A **molecule** is a particle containing two or more atoms joined together chemically.

An **ion** is an atom, or a group of atoms, which has become electrically charged by the loss or gain of one or more electrons.

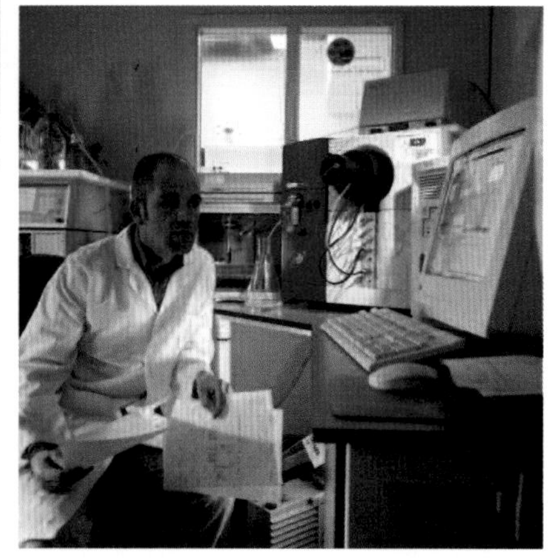

Figure 1.4 ▲
This scientist is using a machine which uses liquid chromatography followed by mass spectrometry to research anti-cancer (chemotherapy) drugs.

Studying chemistry is more than about 'what we know' – it is also about 'how we know'. For example, the study of atomic structure has provided evidence about the nature and properties of electrons, and this has led to an explanation of the properties of elements and the patterns in the periodic table in terms of the electron structures of atoms.

Learning to enjoy and take an interest in chemistry

As a schoolgirl, Dorothy Hodgkin became intensely interested in crystals and their structures. In 1964, she won a Nobel Prize for her use of X-rays to determine the structures of complex molecules such as penicillin and vitamin B12. In 1996, Harry Kroto shared the Nobel Chemistry Prize for his part in the discovery of a new form of carbon called 'buckyballs'. According to Sir Harry, 'science is to do with fun and solving puzzles.'

Figure 1.5 ▲
Professor Harry Kroto with models of his buckyballs

Definitions

Symbols are used to represent elements, e.g. Fe for iron and C for carbon.

Formulae are used to represent compounds, e.g. H_2O for water and NaCl for sodium chloride.

A **molecular formula** shows the numbers of atoms of the different elements in one molecule of a compound, e.g. CH_4 for methane and C_2H_6 for ethane.

An **empirical formula** shows the simplest whole number ratio of the atoms of different elements in a compound, e.g. CH_4 for methane and CH_3 for ethane.

The practical side of chemistry can really inspire some people. They get pleasure from working with chemicals, producing high yields of products and obtaining accurate results. Others are fascinated by the theory of chemistry and the use of models to explain how materials react and change. Yet others are interested in chemistry because of the ways in which it can improve our lives – especially in medicine, nutrition, pharmacy, dentistry and the science of materials.

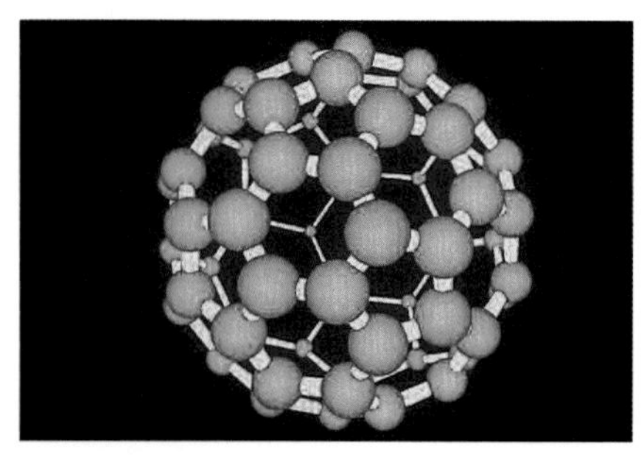

Figure 1.6 ▲
A computer model of a C_{60} buckyball – 60 carbon atoms arranged like a cage

Test yourself

1 Remind yourself of some patterns in the ways that chemicals behave.
  a) What happens when a more reactive metal (such as zinc) is added to a solution in water of a compound of a less reactive metal (such as copper sulfate)?
  b) What forms at the negative electrode (cathode) during the electrolysis of a metal compound?
  c) What happens on adding an acid (such as hydrochloric acid) to a carbonate (such as calcium carbonate)?
  d) What do sodium chloride, sodium bromide and sodium iodide look like?

2 This question concerns the substances ice, salt, sugar, copper, steel and limestone.

Which of these substances contain:

a) uncombined atoms

b) ions

c) molecules?

3 The structure of the main constituent in antifreeze is:

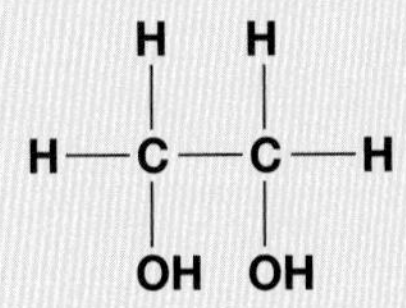

What is:

a) its molecular formula

b) its empirical formula?

1.2 From elements to compounds

Compounds form when two or more elements combine. Apart from the atoms of helium and neon, all atoms combine with other atoms.

Compounds of non-metals with non-metals

Water, carbon dioxide, methane in natural gas, sugar and ethanol ('alcohol') are examples of compounds of two or more non-metals. Most of these non-metal compounds melt and vaporise easily. They may be gases, liquids or solids at room temperature and they do not conduct electricity.

In these compounds of non-metals, the atoms combine in small groups to form molecules. For example, methane contains one carbon atom bonded to four hydrogen atoms. The formula of the molecule is CH_4. Figure 1.7 shows three ways of representing a methane molecule. Models like that in the first diagram represent atoms as little spheres.

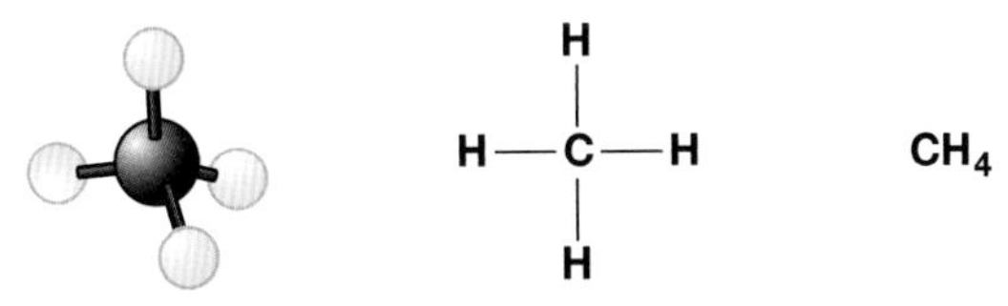

Figure 1.7 ▲
Ways of representing a molecule of methane

It is possible to work out the formula of most non-metal compounds if you know how many bonds the atoms normally form (Table 1.2).

Water is a compound of oxygen and hydrogen. Oxygen atoms form two bonds and hydrogen atoms form one bond – so two hydrogen atoms can bond to one oxygen atom (Figure 1.8) and the formula of water is H_2O.

Element	Symbol	Number of bonds formed	Colour in molecular models
Carbon	C	4	Black
Nitrogen	N	3	Blue
Oxygen	O	2	Red
Sulfur	S	2	Yellow
Hydrogen	H	1	White
Chlorine	Cl	1	Green

Table 1.2 ▲
Symbols, number of bonds and colour codes of some non-metals

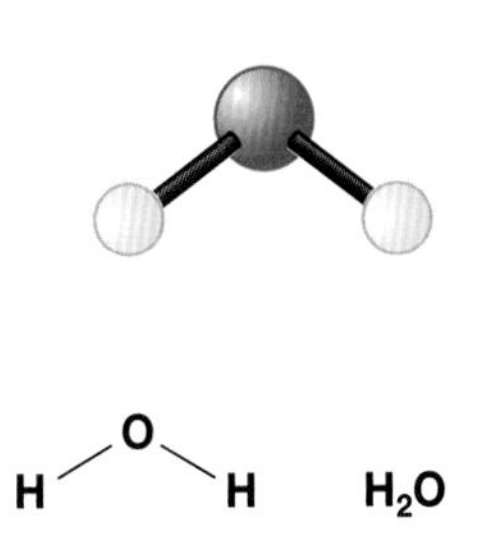

Figure 1.8 ▲
Ways of representing a molecule of water

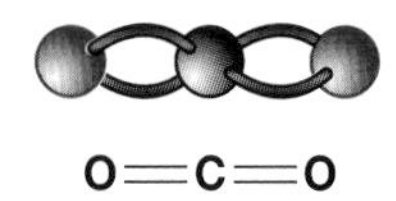

Figure 1.9 ▲
Bonding in carbon dioxide showing the double bonds between atoms

There are double and even triple bonds between the atoms in some non-metal compounds (Figure 1.9). Notice also that there is a strict colour code for the atoms of different elements in molecular models – these colours are shown in Table 1.2.

In practice, it is not possible to predict the formulae of all non-metal compounds. For example, the bonding rules in Table 1.2 cannot account for the formulae of carbon monoxide, CO, or sulfur dioxide, SO_2.

Test yourself

4 Draw the various ways – like those for methane in Figure 1.7 – of representing the following compounds:
   a) hydrogen chloride
   b) carbon disulfide.

5 Name the elements present and work out the formula of the following compounds:
   a) hydrogen sulfide
   b) dichlorine oxide
   c) hydrogen nitride (ammonia).

Compounds of metals with non-metals

Common salt (sodium chloride), limestone (calcium carbonate) and copper sulfate are all examples of compounds of metals with non-metals. These metal/non-metal compounds melt at much higher temperatures than compounds of non-metals – they are solids at room temperature. They conduct electricity as molten liquids but not as solids. Metal/non-metal compounds conduct electricity as liquids because they consist of ions. For example, sodium chloride consists of sodium ions, Na^+, and chloride ions, Cl^-.

The formula of sodium chloride is NaCl (or Na^+Cl^-) because the positive charge on one Na^+ ion is balanced by the negative charge on one Cl^- ion. In a crystal of sodium chloride there are equal numbers of sodium ions and chloride ions (Figure 1.10).

The formulae of all metal/non-metal (ionic) compounds can be worked out by balancing the charges on positive and negative ions. For example, the formula of potassium oxide is K_2O (or $(K^+)_2O^{2-}$). Here, two K^+ ions balance the charge on one O^{2-} ion.

Elements such as iron, which have two different ions (Fe^{2+} and Fe^{3+}), have two sets of compounds – iron(II) compounds such as iron(II) chloride, $FeCl_2$, and iron(III) compounds such as iron(III) chloride, $FeCl_3$.

Table 1.3 shows the names and formulae of some ionic compounds. Notice that the formula of magnesium nitrate is $Mg(NO_3)_2$ (or $Mg^{2+}(NO_3^-)_2$). The brackets round NO_3^- show that it is a single unit containing one nitrogen and

Figure 1.10 ▶
A space-filling model and a ball-and-stick model showing the structure of sodium chloride

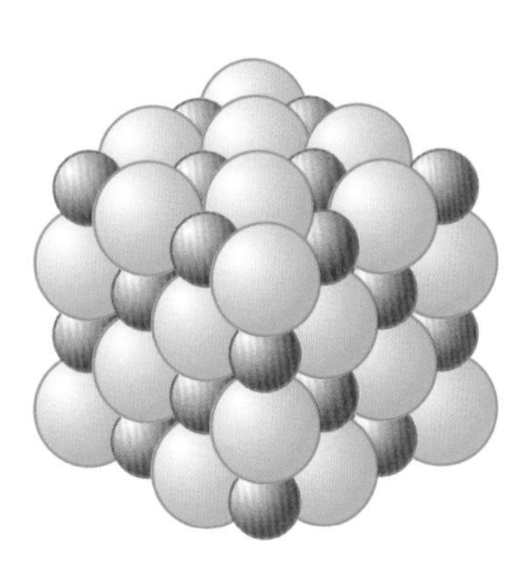

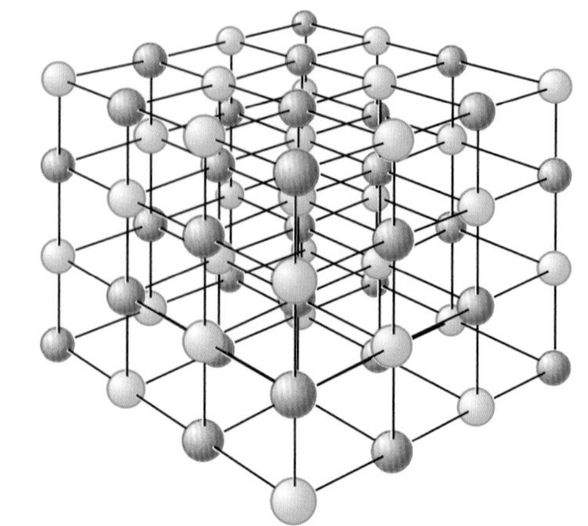

three oxygen atoms bonded together with a 1− charge. Other ions, like OH^-, SO_4^{2-} and CO_3^{2-}, must also be treated as single units and put in brackets when there are two or three of them in a formula.

Name of compound	Formula
Magnesium nitrate	$Mg^{2+}(NO_3^-)_2$, $Mg(NO_3)_2$
Aluminium hydroxide	$Al^{3+}(OH^-)_3$, $Al(OH)_3$
Zinc bromide	$Zn^{2+}(Br^-)_2$, $ZnBr_2$
Lead nitrate	$Pb^{2+}(NO_3^-)_2$, $Pb(NO_3)_2$
Calcium iodide	$Ca^{2+}(I^-)_2$, CaI_2
Copper(II) carbonate	$Cu^{2+}CO_3^{2-}$, $CuCO_3$
Silver sulfate	$(Ag^+)_2SO_4^{2-}$, Ag_2SO_4

Table 1.3 ▲
The names and formulae of some ionic compounds

Data

Test yourself

6 The formula of aluminium hydroxide must be written as $Al(OH)_3$. Why is $AlOH_3$ wrong?

7 Write the formulae of the following ionic compounds:
- a) potassium sulfate
- b) aluminium oxide
- c) lead carbonate
- d) zinc hydroxide
- e) iron(III) sulfate

8 Which of the following compounds consist of molecules and which consist of ions?
- a) octane (C_8H_{18}) in petrol
- b) copper(I) oxide
- c) pure sulfuric acid
- d) lithium fluoride
- e) phosphorus trichloride

9 Compare non-metal (molecular) compounds with metal/non-metal (ionic) compounds in
- a) melting temperatures and boiling temperatures
- b) conduction of electricity as liquids.

Data

1.3 Writing chemical equations

Burning and rusting are good examples of chemical reactions. When they occur, chemical bonds in the reactants break and then new bonds form in the products. The photo in Figure 1.11 shows sparks from a sparkler – these sparks are bits of burning magnesium. When magnesium burns in air, it reacts with oxygen to form white magnesium oxide. A word equation for the reaction is:

magnesium + oxygen → magnesium oxide

Figure 1.11 ▲
When sparklers burn, bits of magnesium react with oxygen in the air.

Balanced chemical equations

When chemists write equations, they use symbols and formulae rather than words. However, writing a word equation is a useful first step towards getting a balanced chemical equation with symbols. There are four key steps in writing an equation.

Step 1: Write a word equation for the reaction.

magnesium + oxygen → magnesium oxide

Step 2: Write symbols for the elements and formulae for the compounds in the word equation.

$$Mg + O_2 \rightarrow MgO$$

Remember that oxygen, hydrogen, nitrogen and the halogens are diatomic molecules with two atoms in their molecules – so they are written as O_2, H_2, N_2, F_2, Cl_2, Br_2 and I_2. All other elements are shown as single atoms.

Step 3: Balance the equation by putting numbers in front of the symbols and formulae, so that the numbers of each type of atom are the same on both sides of the equation.

$$2Mg + O_2 \rightarrow 2MgO$$

Never change a formula to make an equation balance. The formula of magnesium oxide is always MgO and never MgO_2 or Mg_2O; or anything else for that matter.

Step 4: Add state symbols to show the state of each substance in the equation. Use (s) for solid, (l) for liquid, (g) for gas and (aq) for an aqueous solution (a substance dissolved in water).

$$2Mg(s) + O_2(g) \rightarrow 2MgO(s)$$

Definition

A **balanced chemical equation** describes a chemical reaction using symbols for the reactants and products. The numbers of each kind of atom are the same on both sides of the equation.

Balanced chemical equations are more useful than word equations because they show:

- the symbols and formulae of the reactants and products
- the relative numbers of atoms and molecules in the reaction.

Modelling equations

Chemists often use models to understand what is happening to the different atoms, molecules and ions during a reaction. Figure 1.12 shows how molecular models can give a picture of the reaction between hydrogen and oxygen at an atomic level.

Figure 1.12 ▶

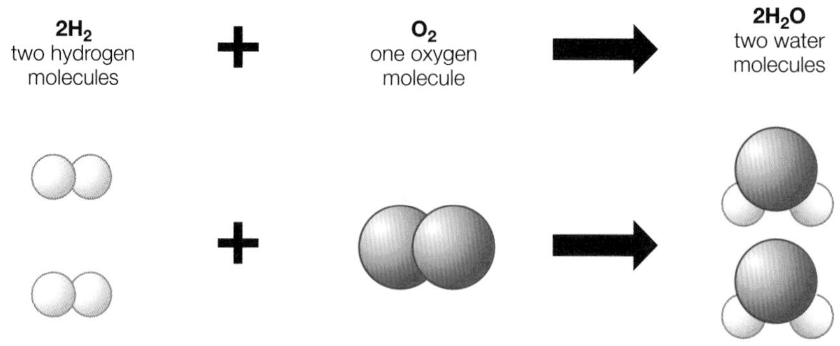

$2H_2$ two hydrogen molecules + O_2 one oxygen molecule → $2H_2O$ two water molecules

Using molecular models like those in Figure 1.12, it is possible to see which bonds break in the reactants and which new bonds form in the products. The models also confirm that the atoms have simply been rearranged during the reaction, as there are exactly the same atoms on both sides of the equation.

Figure 1.13 ▲
When methane (natural gas) burns on a hob, it reacts with oxygen in the air to form carbon dioxide and water.

Test yourself

10 a) Use molecular models to show what happens when methane burns in oxygen (Figure 1.13).
b) Draw a diagram for this reaction, similar to that shown in Figure 1.12.

11 Write balanced equations for the following word equations:
a) sodium + oxygen → sodium oxide
b) sodium oxide + water → sodium hydroxide
c) hydrogen + chlorine → hydrogen chloride
d) zinc + hydrochloric acid (HCl) → zinc chloride + hydrogen
e) methane (CH_4) + oxygen → carbon dioxide + water
f) iron + chlorine → iron(III) chloride.

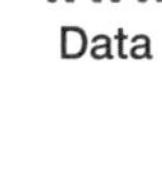

Data

1.4 Types of chemical change

Chemists classify chemical reactions in order to make sense of the thousands of reactions which they study. By putting similar reactions into groups or sets, chemists can predict how different substances will behave.

Thermal decomposition

Calcium carbonate breaks up (decomposes) on heating to form calcium oxide and carbon dioxide (Figure 1.14):

$$CaCO_3(s) \rightarrow CaO(s) + CO_2(g)$$

This is an example of thermal decomposition. Calcium carbonate is stable at room temperature, but it becomes unstable when heated with a hot Bunsen flame.

Other metal carbonates, except those of group 1 metals, also decompose on heating. Carbonates of metals below copper in the reactivity series are so unstable that they cannot exist at room temperature.

Note

The decomposition of calcium carbonate (limestone) to calcium oxide (quicklime) is important for agriculture and industry. Quicklime is used to neutralise acid soils and it also reacts with water to make calcium hydroxide (slaked lime), which is an important industrial alkali.

Figure 1.14 ▲
Quicklime is used to neutralise acid soils, making them more fertile.

Hydrated compounds like hydrated copper sulfate, $CuSO_4.5H_2O$, which contain water as part of their structure, also decompose on heating. Fairly gentle heating causes most hydrates to give off water vapour which often condenses to water on the cooler parts of the apparatus.

Definition

Thermal decomposition is the breaking up of substances into simpler substances by heating.

$$\underset{\text{blue hydrated copper sulfate}}{CuSO_4.5H_2O(s)} \xrightarrow{\text{heat}} \underset{\text{white anhydrous copper sulfate}}{CuSO_4(s)} + 5H_2O(g)$$

Reactions of acids

Acids are compounds with characteristic properties:

- they form solutions in water with a pH below 7
- they change the colour of indicators
- they react with metals above hydrogen in the reactivity series forming hydrogen plus an ionic metal compound called a salt

$$Mg(s) + 2HCl(aq) \rightarrow MgCl_2(aq) + H_2(g)$$

- they react with metal oxides and metal hydroxides to form salts and water

$$CuO(s) + H_2SO_4(aq) \rightarrow CuSO_4(aq) + H_2O(l)$$

- they react with carbonates to form salts, carbon dioxide and water (Figure 1.15)

$$CaCO_3(s) + 2HCl(aq) \rightarrow CaCl_2(aq) + CO_2(g) + H_2O(l)$$

Figure 1.15 ▲
This statue is made from limestone (calcium carbonate). Its worn and pitted appearance has been caused over many years by the action of acids in 'acid rain'.

What makes an acid an acid?

Acids do not simply mix with water when they dissolve in it – they react with it to produce aqueous hydrogen ions, $H^+(aq)$.

$$HCl(g) \xrightarrow{\text{water}} H^+(aq) + Cl^-(aq)$$

All acids produce $H^+(aq)$ ions with water and this is why they all react in a similar way. The typical reactions of dilute acids in water are the reactions of aqueous hydrogen ions. So, we can rewrite the equations for acids showing the ions in acids and in metal/non-metal (ionic) compounds.

With metals

$$Mg(s) + 2H^+(aq) + 2Cl^-(aq) \rightarrow Mg^{2+}(aq) + 2Cl^-(aq) + H_2(g)$$

Notice that the chloride ions are the same before and after the reaction – because of this they are called spectator ions. For clarity, chemists often omit spectator ions from ionic equations like the one above. So, the equation becomes:

$$Mg(s) + 2H^+(aq) \rightarrow Mg^{2+}(aq) + H_2(g)$$

Definition

Spectator ions are ions which take no part in a reaction.

With metal oxides and hydroxides

$$Cu^{2+}O^{2-}(s) + 2H^+(aq) + SO_4^{2-}(aq) \rightarrow Cu^{2+}(aq) + SO_4^{2-}(aq) + H_2O(l)$$

This time, the Cu^{2+} and SO_4^{2-} ions take no part in the reaction. So, we can simplify the equation to:

$$O^{2-}(s) + 2H^+(aq) \rightarrow H_2O(l)$$

With carbonates

$$Ca^{2+}CO_3^{2-}(s) + 2H^+(aq) + 2Cl^- \rightarrow Ca^{2+}(aq) + 2Cl^-(aq) + CO_2(g) + H_2O(l)$$

In this case, the Ca^{2+} and $2Cl^-$ are spectator ions, so the equation becomes:

$$CO_3^{2-}(s) + 2H^+(aq) \rightarrow CO_2(g) + H_2O(l)$$

Reactions of bases and alkalis

Bases are 'anti-acids' – they are the chemical opposites of acids. Acids donate (give up) hydrogen ions; bases (such as oxides, hydroxides and carbonates) accept hydrogen ions.

Alkalis are bases which dissolve in water. The common laboratory alkalis are sodium hydroxide, potassium hydroxide, calcium hydroxide and ammonia. Alkalis form solutions with a pH above 7, so they change the colours of acid–base indicators.

We use alkalis in our homes to neutralise acids and to remove grease. Magnesium carbonate and calcium carbonate in antacid tablets, such as Rennies®, neutralise stomach acid which causes indigestion.

Manufacturers produce powerful oven and drain cleaners containing sodium hydroxide or potassium hydroxide. These strong alkalis are highly 'caustic' and they can even remove grease. They attack skin, producing a chemical burn. Even dilute solutions of these alkalis can be hazardous, especially if they get into your eyes.

All alkalis dissolve in water to produce hydroxide ions, OH^-. Sodium hydroxide (Na^+OH^-) and potassium hydroxide (K^+OH^-) contain hydroxide ions in the solid as well as in solution. Ammonia produces hydroxide ions by reacting with water, taking hydrogen ions from water molecules.

$$NH_3(g) + H_2O(aq) \rightarrow NH_4^+(aq) + OH^-(aq)$$

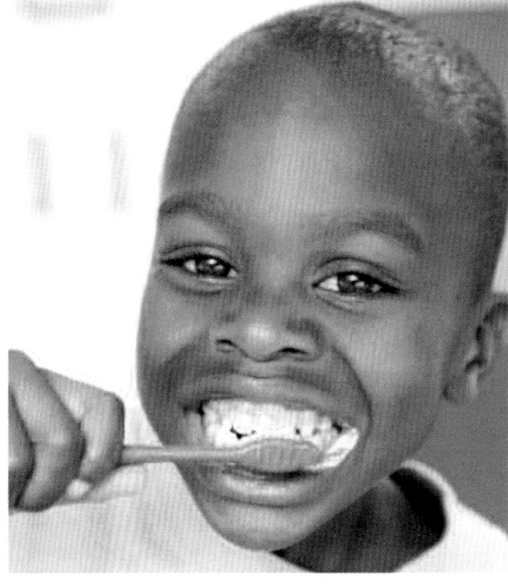

Figure 1.16 ▲
Toothpaste is mildly alkaline. It neutralises the acids present in the mouth which attack teeth and cause decay.

Neutralisation

During a neutralisation reaction, an acid reacts with a base to form a salt. Mixing the right amounts of hydrochloric acid and sodium hydroxide (a base) produces a neutral solution of sodium chloride (common salt).

$$HCl(aq) + NaOH(aq) \rightarrow NaCl(aq) + H_2O(l)$$

If we rewrite this equation showing the ions in the acid and in the metal/non-metal (ionic) compounds, we get:

$$H^+(aq) + Cl^-(aq) + Na^+(aq) + OH^-(aq) \rightarrow Na^+(aq) + Cl^-(aq) + H_2O(l)$$

By cancelling the spectator ions ($Na^+(aq)$ and $Cl^-(aq)$), we get:

$$H^+(aq) + OH^-(aq) \rightarrow H_2O(l)$$

This shows that acids and alkalis neutralise each other because hydrogen ions react with hydroxide ions to form water, which is neutral.

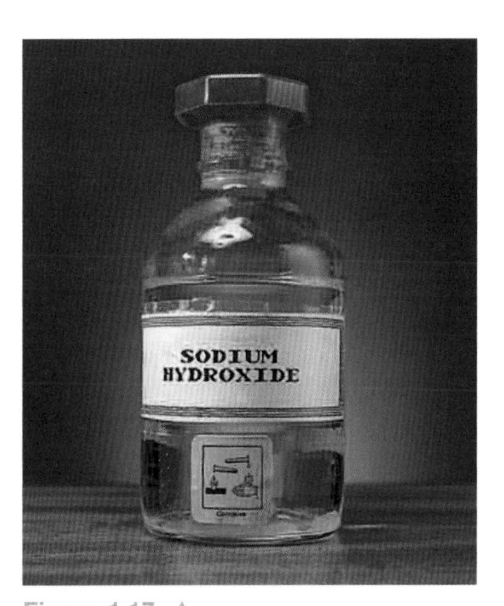

Figure 1.17 ▲
The label on a bottle of sodium hydroxide showing the hazard warning. Sodium hydroxide is highly caustic – hence its older name, caustic soda.

Test yourself

12 Give the names and symbols of the ions formed when these acids dissolve in water:
   a) nitric acid, HNO_3
   b) sulfuric acid, H_2SO_4.

13 Write equations for the reactions of:
   a) zinc with sulfuric acid
   b) calcium oxide with nitric acid
   c) sodium carbonate with hydrochloric acid.

14 Rewrite the full equations in question 13 as ionic equations. Then cancel the spectator ions to get equations showing only the reacting atoms and ions.

Precipitation

A precipitate is an insoluble solid which separates from a solution during a reaction. A precipitate of greasy 'scum' forms whenever we wash our hands in hard water. The whitish 'scum' is a precipitate of calcium stearate. This forms because calcium ions, Ca^{2+}, in the hard water react with stearate ions (shortened to X^- below) in the soap.

$$Ca^{2+}(aq) + 2X^-(aq) \rightarrow \underset{\text{'scum'}}{Ca^{2+}(X^-)_2(s)}$$

Precipitates form if two solutions can produce an insoluble substance when they mix. For example, when silver nitrate solution is added to sodium chloride solution, an insoluble white precipitate of silver chloride is produced.

Definitions

Acids are substances which donate hydrogen ions, H^+ (protons).

Bases are substances which accept hydrogen ions, H^+ (protons).

Alkalis are soluble bases.

Neutralisation occurs when H^+ ions react with OH^- or O^{2-} ions forming water.

Salts are ionic compounds formed when an acid reacts with a base. Most salts are metal/non-metal compounds.

On mixing the two solutions, there are two possible new combinations of ions – silver ions with chloride ions, and sodium ions with nitrate ions. Silver chloride is insoluble, so it precipitates. Sodium nitrate is soluble so the sodium and nitrate ions stay in solution.

$$AgNO_3(aq) + NaCl(aq) \rightarrow AgCl(s) + NaNO_3(aq)$$

Rewriting this equation in terms of ions, we get:

$$Ag^+(aq) + NO_3^-(aq) + Na^+(aq) + Cl^-(aq) \rightarrow Ag^+Cl^-(s) + Na^+(aq) + NO_3^-(aq)$$

Cancelling the spectator ions ($Na^+(aq)$ and $NO_3^-(aq)$) gives a simpler and clearer equation for the reaction.

$$Ag^+(aq) + Cl^-(aq) \rightarrow Ag^+Cl^-(s)$$

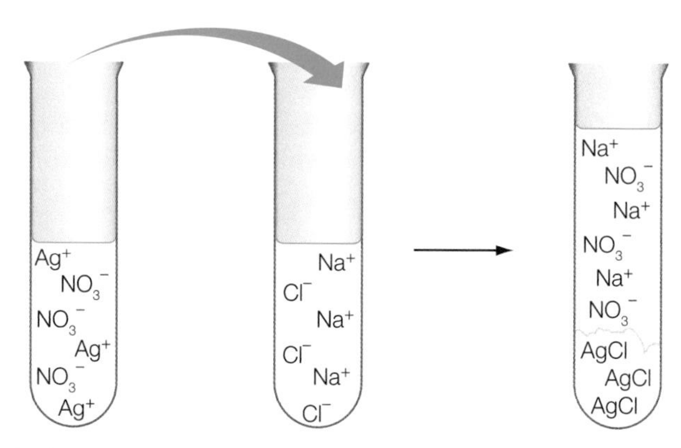

Figure 1.18 ▶ Precipitation of silver chloride

Lead chloride ($PbCl_2$) is also insoluble like silver chloride. So, lead chloride precipitates when lead nitrate solution (containing $Pb^{2+}(aq)$ and $NO_3^-(aq)$ ions) is mixed with sodium chloride solution (containing $Na^+(aq)$ and $Cl^-(aq)$ ions).

$$Pb^{2+}(aq) + 2Cl^-(aq) \rightarrow PbCl_2(s)$$

Activity

Testing for metal ions with sodium hydroxide solution

Many tests for ions involve the formation of precipitates that can be recognised by their colour. Most metal hydroxides are insoluble in water and this is the basis of a scheme for identifying common metal ions. Table 1.4 shows what happens when:

- a little sodium hydroxide solution is added and then
- excess sodium hydroxide solution is added

to solutions of some common cations.

Look carefully at Table 1.4 and answer the following questions.

1 Write a balanced equation for the reaction of sodium hydroxide solution with magnesium sulfate solution to form a precipitate of magnesium hydroxide.

2 Use your equation from question **1** to write a summary equation for the precipitation of magnesium hydroxide, which could go in Table 1.4.

3 a) Which cations in Table 1.4 give no precipitate with sodium hydroxide solution?
b) How can you identify these cations using flame tests?

Cation in solution	3 drops of NaOH(aq) added to 3 cm^3 of solution of cation	Excess NaOH(aq) added
Potassium, K^+	No precipitate	No precipitate
Sodium, Na^+	No precipitate	No precipitate
Calcium, Ca^{2+}	A white precipitate of $Ca(OH)_2$ forms $Ca^{2+}(aq) + 2OH^-(aq) \rightarrow Ca(OH)_2(s)$	White precipitate remains
Magnesium, Mg^{2+}	A white precipitate of $Mg(OH)_2$ forms	White precipitate remains
Aluminium, Al^{3+}	A white precipitate of $Al(OH)_3$ forms $Al^{3+}(aq) + 3OH^-(aq) \rightarrow Al(OH)_3(s)$	White precipitate dissolves to give a colourless solution
Iron(II), Fe^{2+}	A green precipitate of $Fe(OH)_2$ forms $Fe^{2+}(aq) + 2OH^-(aq) \rightarrow Fe(OH)_2(s)$	Green precipitate remains
Iron(III), Fe^{3+}	A brown precipitate of $Fe(OH)_3$ forms $Fe^{3+}(aq) + 3OH^-(aq) \rightarrow Fe(OH)_3(s)$	Brown precipitate remains
Zinc, Zn^{2+}	A white precipitate of $Zn(OH)_2$ forms $Zn^{2+}(aq) + 2OH^-(aq) \rightarrow Zn(OH)_2(s)$	White precipitate dissolves to give a colourless solution
Copper(II), Cu^{2+}	A blue precipitate of $Cu(OH)_2$ forms $Cu^{2+}(aq) + 2OH^-(aq) \rightarrow Cu(OH)_2(s)$	Blue precipitate remains

Table 1.4 ▲
Testing for metal ions with sodium hydroxide solution

4 a) Which cations give a white precipitate that does not dissolve with excess sodium hydroxide?

b) How can you identify these cations using flame tests?

5 Which cations give a white precipitate that dissolves in excess sodium hydroxide? Further tests are needed to identify the metal ions in these hydroxides.

6 a) Which cations give a coloured (non-white) precipitate with sodium hydroxide solution?

b) How can you tell the difference between these cations from the colour of their hydroxides?

7 The ammonium cation, NH_4^+, can also be identified using sodium hydroxide solution. If the ammonium compound containing NH_4^+ ions is warmed with sodium hydroxide solution, NH_4^+ ions react with hydroxide ions (OH^-) producing water and ammonia (NH_3). Ammonia has a very pungent smell and is the only common alkaline gas.

a) How can damp litmus paper be used to identify ammonia and ammonium compounds?

b) Copy and complete the following equation for the reaction of ammonium ions with hydroxide ions:

$$NH_4^+(aq) + OH^-(aq) \rightarrow \_\_\_\_\_\_\_\_ + \_\_\_\_\_\_\_\_$$

1.5 Understanding chemical quantities

Chemists often need to measure how much of a particular chemical there is in a sample of it. Analysts in pharmaceutical companies test samples of tablets and medicines to check that they contain the right amount of a drug. Food manufacturers check the purity of the raw materials they buy and the foods they produce. Industrial chemists measure the amounts of substances they need for chemical processes.

Chemical amounts

When chemists are determining formulae or working with equations, they need to measure amounts containing equal numbers of atoms, molecules or ions. Chemists have balances to measure masses and graduated containers to measure volumes, but there is no instrument for measuring chemical amounts directly. Instead, chemists must first measure the masses or volumes of substances and then calculate the chemical amounts.

Chemists use the term 'mole' for an amount of substance containing a standard number of atoms, molecules or ions. The word 'mole' entered the language of chemistry at the end of the nineteenth century. It is based on the Latin word for a heap or a pile.

Relative atomic masses

The key to working with chemical amounts in moles is to know the relative masses of different atoms. The accurate method for determining relative atomic masses involves using a mass spectrometer (Section 3.2).

Chemists originally measured atomic masses relative to hydrogen. Then, because of the existence of isotopes, it became necessary to choose one particular isotope as the standard. Today, the isotope carbon-12 ($^{12}_{6}C$) is chosen as the standard and given a relative mass of exactly 12. The symbol for relative atomic mass is A_r, where 'r' stands for relative. So, the relative atomic mass of an element is the average mass of an atom of the element relative to one twelfth the mass of an atom of the isotope carbon-12.

$$\text{Relative atomic mass} = \frac{\text{average mass of an atom of the element}}{\frac{1}{12} \times \text{the mass of one atom of carbon-12}}$$

Using this scale, the relative atomic mass of hydrogen is 1.0 and that of chlorine is 35.5. So, we can write $A_r(H) = 1.0$ and $A_r(Cl) = 35.5$, or simply H = 1 and Cl = 35.5 for short.

Notice that the values of relative atomic masses have no units because they are relative. Table 1.5 lists a few relative atomic masses. The relative atomic masses of all elements are shown in the periodic table on page 244.

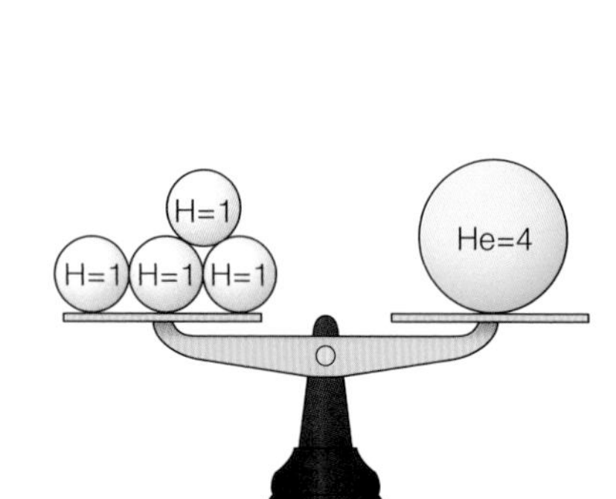

Figure 1.19 ▲
If atoms could be weighed, the scales would show that helium atoms are four times as heavy as hydrogen atoms.

Relative molecular mass and relative formula mass, M_r

Relative atomic masses can also be used to compare the masses of different molecules. The relative masses of molecules are called relative molecular masses (symbol M_r).

Definition

The **relative molecular mass** of an element or compound is the sum of the relative atomic masses of all the atoms in its molecular formula.

Element	Symbol	Relative atomic mass
Carbon	C	12.0
Hydrogen	H	1.0
Helium	He	4.0
Nitrogen	N	14.0
Oxygen	O	16.0
Magnesium	Mg	24.3
Sulfur	S	32.1
Iron	Fe	55.8
Copper	Cu	63.5

Table 1.5 ▲
Relative atomic masses of some common elements

For oxygen, O_2, $M_r(O_2) = 2 \times A_r(O) = 2 \times 16.0 = 32.0$

and for sulfuric acid, $M_r(H_2SO_4) = 2 \times A_r(H) + A_r(S) + 4 \times A_r(O)$

$= (2 \times 1.0) + 32.1 + (4 \times 16.0) = 98.1$

Metal compounds consist of giant structures of ions and not molecules. To avoid the suggestion that their formulae represent molecules, chemists use the term 'relative formula mass', not relative molecular mass, for ionic compounds and for other compounds with giant structures such as silicon dioxide, SiO_2.

For magnesium nitrate,

$$M_r(Mg(NO_3)_2) = A_r(Mg) + 2 \times A_r(N) + 6 \times A_r(O)$$
$$= 24.3 + (2 \times 14.0) + (6 \times 16.0) = 148.3$$

Definition

The **relative formula mass** of a compound is the sum of the relative atomic masses of all the atoms in its formula.

Test yourself

15 Look carefully at Table 1.5. How many times heavier (to the nearest whole number) are:
- a) C atoms than H atoms
- b) Mg atoms than C atoms
- c) S atoms than He atoms
- d) C atoms than He atoms
- e) Fe atoms than N atoms?

16 What is the relative molecular mass of:
- a) chlorine, Cl_2
- b) sulfur, S_8
- c) ethanol, C_2H_5OH
- d) tetrachloromethane, CCl_4?

17 What is the relative formula mass of:
- a) magnesium chloride, $MgCl_2$
- b) iron(III) oxide, Fe_2O_3
- c) hydrated copper(II) sulfate, $CuSO_4.5H_2O$?

Molar mass

Tutorial

In chemistry, the amount of a substance is measured in moles, and 1 mole of an element is equal to its relative atomic mass in grams. So, 1 mole of carbon is 12 g and 1 mole of copper is 63.5 g. These masses of 1 mole are usually called 'molar masses' (symbol M). So, the molar mass of carbon, $M(C) = 12\,g\,mol^{-1}$ and the molar mass of copper, $M(Cu) = 63.5\,g\,mol^{-1}$.

Similarly the molar mass of the molecules of an element or a compound is numerically equal to its relative molecular mass. So, the molar mass of oxygen molecules, $M(O_2) = 32\,g\,mol^{-1}$ and the molar mass of sulfuric acid, $M(H_2SO_4) = 98.1\,g\,mol^{-1}$. Likewise, the molar mass of an ionic compound is numerically equal to its relative formula mass. The molar mass of magnesium nitrate is therefore $148.3\,g\,mol^{-1}$.

Definition

Molar mass is the mass of one mole of a chemical – the unit is $g\,mol^{-1}$. As always with molar amounts, the symbol or formula of the chemical must be specified.

Amount in moles

The mole is the SI unit for amount of substance. So, the amount of any substance is measured in moles, which is normally abbreviated to 'mol'. So,

12 g of carbon contains 1 mol of carbon atoms

24 g of carbon contains 2 mol of carbon atoms

240 g of carbon contains 20 mol of carbon atoms.

Notice that: $$\text{amount of substance/mol} = \frac{\text{mass of substance/g}}{\text{molar mass/g mol}^{-1}}$$

Figure 1.20 ▶ One mole amounts of copper, carbon, iron, aluminium, mercury and sulfur

Tutorial

The Avogadro constant

Relative atomic masses show that one atom of carbon is 12 times heavier than one atom of hydrogen. This means that 12 g of carbon will contain the same number of atoms as 1 g of hydrogen. Similarly, one atom of oxygen is 16 times as heavy as one atom of hydrogen, so 16 g of oxygen will also contain the same number of atoms as 1 g of hydrogen.

In fact, the relative atomic mass in grams (i.e. 1 mole) of every element (1 g of hydrogen, 12 g carbon, 16 g oxygen, etc.) will contain the same number of atoms. This number is called the Avogadro constant – in honour of the Italian scientist Amedeo Avogadro. Experiments show that the Avogadro constant, L, is $6 \times 10^{23}\,\text{mol}^{-1}$.

Written out in full this is 600 000 000 000 000 000 000 000 per mole.

Figure 1.21 ▶ One mole amounts of some ionic compounds

The Avogadro constant is the number of atoms, molecules or formula units in one mole of any substance. So, 1 mole of oxygen (O_2) contains 6×10^{23} O_2 molecules, and 2 moles of oxygen (O_2) contains $2 \times 6 \times 10^{23}$ O_2 molecules.

∴ The number of atoms, molecules or formula units
= amount of chemical in moles × the Avogadro constant

It is, of course, vital to specify the particles concerned in calculating the amount of a substance or the number of particles of a substance. For example, 2 g of hydrogen will contain 2 mol of hydrogen (H) atoms (12×10^{23} atoms) but only 1 mol of hydrogen (H_2) molecules (6×10^{23} molecules).

Practical guidance

Test yourself

18 What is the amount, in moles, of:
- a) 20.05 g of calcium atoms
- b) 3.995 g of bromine atoms
- c) 159.8 g of bromine molecules
- d) 6.41 g of sulfur dioxide molecules
- e) 10 g of sodium hydroxide, NaOH?

19 What is the mass of:
- a) 0.1 mol of iodine atoms
- b) 0.25 mol of chlorine molecules
- c) 2 mol of water molecules
- d) 0.01 mol of ammonium chloride, NH_4Cl
- e) 0.125 mol of sulfate ions, SO_4^{2-}?

20 How many moles of:
- a) sodium ions are there in 1 mol of sodium carbonate, Na_2CO_3
- b) bromide ions are there in 0.5 mol of barium bromide, $BaBr_2$
- c) nitrogen atoms are there in 2 mol of ammonium nitrate, NH_4NO_3?

21 Use the Avogadro constant to calculate:
- a) the number of chloride ions in 0.5 mol of sodium chloride, NaCl
- b) the number of oxygen atoms in 2 mol of oxygen molecules, O_2
- c) the number of sulfate ions in 3 mol of aluminium sulfate, $Al_2(SO_4)_3$

22 One cubic decimetre of tap water was found to contain 0.1116 mg of iron(III) ions (Fe^{3+}) and 12.40 mg of nitrate ions (NO_3^-).
- a) What are these masses of Fe^{3+} and NO_3^- in grams?
- b) What are the amounts in moles of Fe^{3+} and NO_3^-?
- c) What are the numbers of Fe^{3+} and NO_3^- ions?

1.6 Finding formulae

Although the formulae of most compounds can be predicted, the only sure way of knowing a formula is by experiment. This has been done for all common compounds and their formulae can be checked in tables of data.

Experimental (empirical) formulae

'Empirical' evidence is information based on experience or experiment – so chemists use the term 'empirical formulae' for formulae calculated from the results of experiments.

An experiment to find an empirical formula involves measuring the masses of elements which combine in the compound. From these masses, it is possible to calculate the number of moles of atoms which react, and hence the ratio of atoms which react. This gives an empirical formula which shows the simplest whole number ratio for the atoms of different elements in a compound.

Worked example

Analysis of 20.1 g of iron bromide showed that it contained 3.8 g of iron and 16.3 g of bromine. What is its formula?

Answer

	Iron	Bromine
Combined masses	3.8 g	16.3 g
Molar mass	$55.8\ g\ mol^{-1}$	$79.9\ g\ mol^{-1}$
Combined moles of atoms	$\frac{3.8\ g}{55.8\ g\ mol^{-1}} = 0.068\ mol$	$\frac{16.3\ g}{79.9\ g\ mol^{-1}} = 0.204\ mol$
Ratio of combined atoms	$\frac{0.068}{0.068} = 1$	$\frac{0.204}{0.068} = 3$
Simplest ratio of atoms	1 :	3

So, the empirical formula is $FeBr_3$

Test yourself

23 What is the formula of the compound in which:
   a) 0.6 g carbon combines with 0.2 g hydrogen
   b) 1.02 g vanadium combines with 2.84 g chlorine
   c) 1.38 g sodium combines with 0.96 g sulfur and 1.92 g oxygen?

24 What is the formula of the compound in which the percentages of the elements present are:
   a) 34.6% copper, 30.5% iron and 34.9% sulfur
   b) 2.04% hydrogen, 32.65% sulfur and 65.31% oxygen
   c) 52.18% carbon, 13.04% hydrogen and 34.78% oxygen?

Sometimes, the analysis of a compound shows the percentages of the different elements, rather than their masses. These percentages can, of course, be regarded as combining masses and the empirical formula can be worked out in the same way. An example of this is shown in Table 1.6. Notice how the calculation can be summarised in a table.

	Ca	N
Combined masses	81.08 g	18.92 g
Molar mass		
Combined moles of atoms	$\frac{81.08\ g}{} = 2.02\ mol$	$\frac{18.92\ g}{} = 1.35\ mol$
Ratio of combined atoms	$\frac{2.02}{1.35} = 1.50$	$\frac{1.35}{1.35} = 1.00$
Simplest ratio of atoms	3 :	2
Formula = Ca_3N_2		

Table 1.6 ▲
Finding the empirical formula of a compound of calcium and nitrogen

Activity

Finding the formula of red copper oxide

A student investigated the formula of red copper oxide by reducing it to copper using natural gas as shown in Figure 1.22.

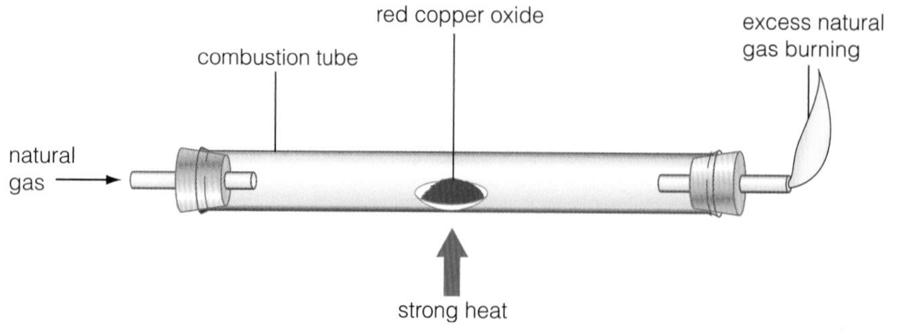

Figure 1.22 ▲
Reducing red copper oxide by heating in natural gas

The experiment was carried out five times, starting with different amounts of red copper oxide. The results are shown in Table 1.7.

Table 1.7

Experiment number	Mass of red copper oxide/g	Mass of copper in the oxide/g
1	1.43	1.27
2	2.14	1.90
3	2.86	2.54
4	3.55	3.27
5	4.29	3.81

1 Look at Figure 1.22. What safety precautions should the student take during the experiments?

2 What steps should the student take to ensure that all the copper oxide is reduced to copper?

3 Start a spreadsheet program on a computer and open up a new spreadsheet for your results. Enter the experiment numbers and the masses of copper oxide and copper in the first three columns of your spreadsheet, as in Table 1.7.

4 **a)** Enter a formula in column 4 to work out the mass of oxygen in the red copper oxide used.
b) Enter a formula in column 5 to find the amount of copper in moles in the oxide.
c) Enter a formula in column 6 to find the amount of oxygen in moles in the oxide.

5 From the spreadsheet, plot a line graph of amount of copper (*y*-axis) against amount of oxygen (*x*-axis). Print out your graph. If you can't plot graphs directly from the spreadsheet, draw the graph by hand.

6 Which of the points should be disregarded in drawing the line of best fit?

7 **a)** What is the average value of $\frac{\text{amount of copper/mol}}{\text{amount of oxygen/mol}}$ from your graph?
b) How many moles of copper combine with 1 mole of oxygen in red copper oxide?
c) What is the formula of red copper oxide?

8 How did the student improve the reliability of the results?

9 Is the result for the formula of red copper oxide valid? Explain your answer. (Hint: Valid results are reliable, obtained by fair tests in which only one variable changes and which are free of observer or experimenter bias.)

10 Write a word equation, and then a balanced equation, for the reduction of red copper oxide to copper using methane (CH_4) in natural gas. (Hint: The only solid product is copper.)

11 Look at the nutritional data on the packaging of a chocolate bar.
a) Is there any evidence that the data is **i)** accurate, **ii)** reliable, **iii)** valid?
b) What could producers do to provide evidence that the data about their product is **i)** accurate, **ii)** reliable, **iii)** valid?

1.7 Calculations from equations

An equation is not just a useful shorthand for describing what happens during a reaction. In industry, in medicine and anywhere that chemists make products from reactants, it is vitally important to know the amounts of reactants that are needed for a chemical process and the amount of product that can be obtained. Chemists can calculate these amounts using equations.

Calculating the masses of reactants and products

There are four key steps in solving problems using equations.

Step 1: Write the balanced equation for the reaction.

Step 2: Write down the amounts in moles of the relevant reactants and products.

Step 3: Convert the amounts in moles of the relevant reactants and products to masses.

Step 4: Scale the masses to the quantities required.

Worked example

What mass of iron can be obtained from 1 kg of iron(III) oxide (iron ore)?

Answer

Step 1: $Fe_2O_3(s) + 3CO(g) \rightarrow 2Fe\,(s) + 3CO_2(g)$

Step 2: $1\,mol\ Fe_2O_3 \rightarrow 2\,mol\ Fe$

Step 3: $M(Fe_2O_3) = (2 \times 55.8\,g\,mol^{-1}) + (3 \times 16.0\,g\,mol^{-1})$
$= 111.6 + 48.0 = 159.6\,g\,mol^{-1}$

So, $159.6\,g\ Fe_2O_3 \rightarrow 2 \times 55.8\,g\ Fe = 111.6\,g\ Fe$

Step 4: $159.6\,g\ Fe_2O_3 \rightarrow 111.6\,g\ Fe$

$$\therefore 1\,g\ Fe_2O_3 \rightarrow \frac{111.6}{159.6}\,g\ Fe = 0.7\,g\ Fe$$

$\therefore$ 1 kg of iron(III) oxide produces 0.7 kg of iron.

Test yourself

25 What mass of calcium oxide, CaO, forms when 25 g calcium carbonate, $CaCO_3$, decomposes on heating?

26 What mass of sulfur combines with 8.0 g copper to form copper(I) sulfide, Cu_2S?

27 What mass of sulfur is needed to produce 1 kg of sulfuric acid, H_2SO_4?

Calculating the volumes of gases in reactions

The volume of a gas depends on three things:

- the amount of gas in moles
- the temperature
- the pressure.

At a fixed temperature and fixed pressure, the volume of a gas depends only on the amount of gas in moles – the type or the formula of the gas does not matter.

After thousands of measurements, chemists have found that the volume of 1 mole of every gas occupies about 24 dm^3 (24 000 cm^3) at room temperature (25 °C) and atmospheric pressure. This volume of one mole of gas is called the molar volume.

At 0 °C (273 K) and 1 atmosphere pressure, the molar volume is 22.4 dm^3 mol^{-1} (22 400 cm^3 mol^{-1}).

So, 1 mole of oxygen (O_2) and 1 mole of carbon dioxide (CO_2) each occupy 24 dm^3 at room temperature. Therefore 2 moles of O_2 will occupy 48 dm^3 and 0.5 moles of O_2 will occupy 12 dm^3 at room temperature. Notice from these simple calculations that:

$$\text{Volume of gas/cm}^3 = \text{amount of gas/mol} \times \text{molar volume/cm}^3\,\text{mol}^{-1}$$

So, under laboratory conditions:

$$\text{Volume of gas/cm}^3 = \text{amount of gas/mol} \times 24\,000/\text{cm}^3\,\text{mol}^{-1}$$

Definition

The **molar volume** of a gas is the volume of 1 mole. At room temperature and atmospheric pressure, the molar volume of all gases is 24 dm^3 mol^{-1}.

Figure 1.23 ▲
The Italian scientist Amedeo Avogadro (1776–1856). Avogadro was the first to suggest that equal volumes of all gases, at the same temperature and pressure, contain the same number of molecules. Later, scientists found that 22.4 dm^3 of all gases, at 273 K and 1 atmosphere pressure, contain 6 × 10^{23} molecules. This number is now known as the Avogadro constant.

Measuring the volumes of gases in reactions

The apparatus in Figure 1.24 can be used to measure the reacting volumes of dry ammonia (NH_3) and dry hydrogen chloride (HCl).

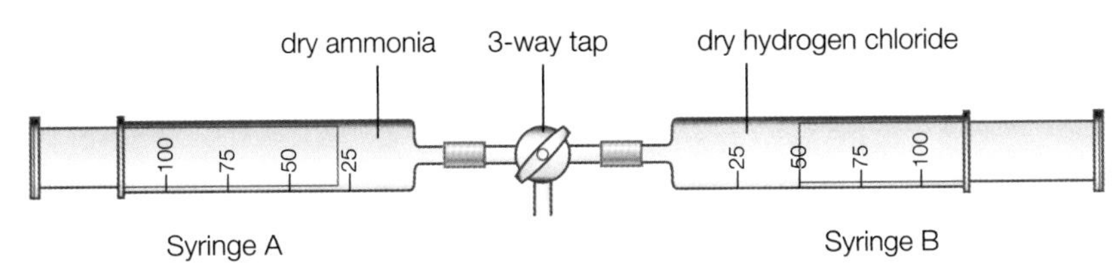

Figure 1.24 ▲
Measuring the reacting volumes of ammonia and hydrogen chloride

When 30 cm^3 of ammonia and 50 cm^3 of hydrogen chloride are mixed, ammonium chloride (NH_4Cl) forms as a white solid – the volume of this solid is insignificant. The volume of gas remaining is 20 cm^3, which turns out to be excess hydrogen chloride.

So, 30 cm^3 of NH_3 reacts with 30 cm^3 of HCl

∴ 1 cm^3 of NH_3 reacts with 1 cm^3 of HCl

and 24 dm^3 of NH_3 reacts with 24 dm^3 of HCl.

⇒ 1 mole of NH_3 reacts with 1 mole of HCl.

Notice that the ratio of the reacting volumes of these gases is the same as the ratio of the reacting amounts in moles. This is always the case when gases react.

$NH_3(g)$ +	$HCl(g) \rightarrow NH_4Cl(s)$
1 mol	1 mol
1 volume	1 volume

Note

Temperature scales

Two temperature scales are in common use – the Kelvin scale and the Celsius scale. On the Kelvin scale, temperatures are measured in units called kelvin, symbol K. On the Celsius scale, temperatures are measured in degrees Celsius (or centigrade), symbol °C. Strictly speaking, the standard (SI) unit of temperature is the kelvin.

Fortunately, temperatures on both scales use the same size units. Therefore,

absolute zero,	−273 °C is	0 K
	0 °C is	273 K
	25 °C is	298 K
	100 °C is	373 K
and in general,	*t* °C is	(273 + *t*) K

Test yourself

28 What is the amount, in moles, of gas at room temperature and pressure in:
- a) 240 000 cm^3 chlorine
- b) 48 cm^3 hydrogen
- c) 3 dm^3 ammonia?

29 What are the volumes of the following amounts of gas at room temperature and pressure:
- a) 2 mol nitrogen
- b) 0.0002 mol neon
- c) 0.125 mol carbon dioxide?

Gas volume calculations

Gas volume calculations are straightforward when all the relevant substances are gases. In these cases, the ratio of the gas volumes in the reaction is the same as the ratio of the numbers of moles in the equation.

Worked example

What volume of oxygen reacts with 60 cm^3 methane and what volume of carbon dioxide is produced if all volumes are measured at the same temperature and pressure?

Answer

The equation for the reaction is:

	$CH_4(g)$ +	$2O_2(g)$ →	$CO_2(g)$ +	$2H_2O(l)$
	1 mol	2 mol	1 mol	insignificant volume
∴	1 volume	2 volumes	1 volume	of liquid below 100 °C

So, 60 cm^3 methane reacts with 120 cm^3 oxygen to produce 60 cm^3 carbon dioxide.

Practical guidance

By measuring the volumes of gases involved in reactions, and then converting these volumes to amounts of gas in moles, it is possible to deduce equations for reactions. This is illustrated in the activity on page 23.

Test yourself

30 Assuming that all gas volumes are measured under the same conditions of temperature and pressure, what volume of:

a) nitrogen forms when 2 dm^3 ammonia, NH_3, decomposes into its elements

b) oxygen is needed to react with 50 cm^3 ethane, C_2H_6, when it burns, and what volume of carbon dioxide forms?

31 a) Copy and balance this equation for the complete combustion of propane (Calor gas)

$$C_3H_8(g) + \_\_\_\_ O_2(g) \rightarrow \_\_\_\_ CO_2(g) + \_\_\_\_ H_2O(g)$$

b) What volume of oxygen reacts with 200 cm^3 of propane and what volume of carbon dioxide is produced? (Assume that all volumes are measured at room temperature and pressure.)

c) What is the mass of the carbon dioxide produced?

32 What volume of gas forms at room temperature and pressure when:

a) 0.654 g of zinc reacts with excess dilute hydrochloric acid

b) 2.022 g of potassium nitrate, KNO_3, decomposes on heating to potassium nitrite, KNO_2, and oxygen?

Calculating the concentration of solutions

The concentration of a solution tells us how much solute is dissolved in a certain volume of solution. It is usually measured in grams per cubic decimetre ($g\,dm^{-3}$) or moles per cubic decimetre ($mol\,dm^{-3}$). For example, a solution of sodium hydroxide containing 1.0 $mol\,dm^{-3}$ has 1 mole of sodium hydroxide (40 g of NaOH) in 1 dm^3 (1000 cm^3) of solution. Figure 1.25 shows how to prepare a solution with a specified concentration.

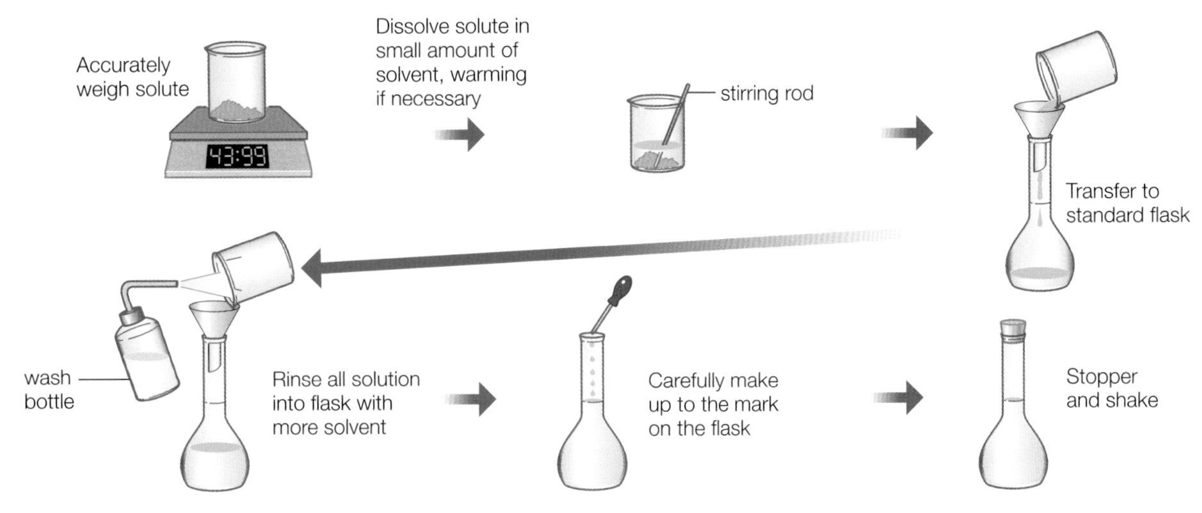

Figure 1.25 ▲
Using a standard flask to prepare a solution with a specified concentration

Activity

Finding an equation for the reaction of magnesium with hydrochloric acid

A small piece of magnesium was weighed and then added to excess dilute hydrochloric acid (HCl(aq)) in the apparatus shown in Figure 1.26. A vigorous reaction occurred and hydrogen gas (H_2) was produced. Eventually all the magnesium reacted and the reaction stopped.

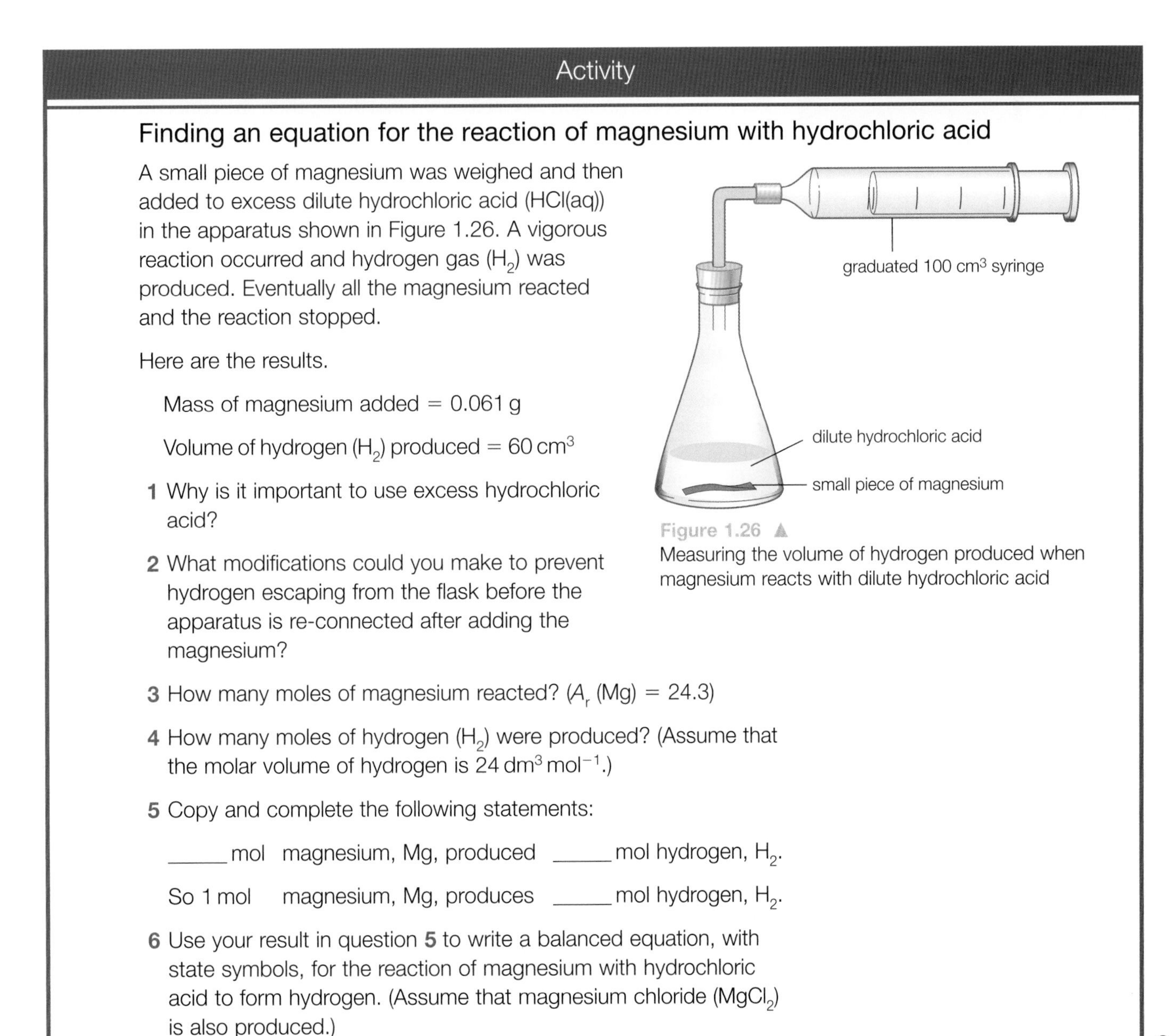

Figure 1.26 ▲
Measuring the volume of hydrogen produced when magnesium reacts with dilute hydrochloric acid

Here are the results.

Mass of magnesium added = 0.061 g

Volume of hydrogen (H_2) produced = 60 cm^3

1 Why is it important to use excess hydrochloric acid?

2 What modifications could you make to prevent hydrogen escaping from the flask before the apparatus is re-connected after adding the magnesium?

3 How many moles of magnesium reacted? (A_r (Mg) = 24.3)

4 How many moles of hydrogen (H_2) were produced? (Assume that the molar volume of hydrogen is 24 $dm^3\,mol^{-1}$.)

5 Copy and complete the following statements:

_____ mol magnesium, Mg, produced _____ mol hydrogen, H_2.

So 1 mol magnesium, Mg, produces _____ mol hydrogen, H_2.

6 Use your result in question **5** to write a balanced equation, with state symbols, for the reaction of magnesium with hydrochloric acid to form hydrogen. (Assume that magnesium chloride ($MgCl_2$) is also produced.)

Worked example

A car battery contains 2354.4 g of sulfuric acid (H_2SO_4) in 6 dm^3 of the battery liquid. What is the concentration of sulfuric acid in **a)** g dm^{-3}, **b)** mol dm^{-3}?

Answer

a) $\text{Concentration/g dm}^{-3} = \frac{\text{mass of solute/g}}{\text{volume of solution/dm}^3} = \frac{2354.4\text{ g}}{6\text{ dm}^3} = 392.4\text{ g dm}^{-3}$

b) $M(H_2SO_4) = 98.1\text{ g mol}^{-1}$

$$\text{So, amount of } H_2SO_4 \text{ in the battery} = \frac{2354.4\text{ g}}{\quad} = 24\text{ mol.}$$

$$\therefore \text{Concentration} = \frac{\text{amount of solute/mol}}{\text{volume of solution/dm}^3} = \frac{24\text{ mol}}{6\text{ dm}^3} = 4\text{ mol dm}^{-3}$$

When ionic compounds dissolve, the ions separate in the solution.

$$\text{For example, } CaCl_2(s) \xrightarrow{(aq)} Ca^{2+}(aq) + 2Cl^-(aq)$$

So, if the concentration of $CaCl_2$ is 0.1 mol dm^{-3}, then the concentration of Ca^{2+} is also 0.1 mol dm^{-3}, but the concentration of Cl^- is 0.2 mol dm^{-3}.

Definition

Concentration/g dm^{-3} = $\frac{\text{mass of solute/g}}{\text{volume of solution/dm}^3}$

Concentration/mol dm^{-3} = $\frac{\text{amount of solute/mol}}{\text{volume of solution/dm}^3}$

Very dilute solutions

Concentrations expressed in mol dm^{-3} are not so useful when dealing with very low concentrations. The concentrations of some dilute solutions, such as those of various substances in the blood, are so small that they are measured in millimoles per cubic decimetre (mmol dm^{-3}).

$$1\text{ mmol dm}^{-3} = \frac{1\text{ mol dm}^{-3}}{1000} = 10^{-3}\text{ mol dm}^{-3}$$

In some cases, such as pollutants and toxic chemicals in the atmosphere or in water, it is convenient to measure the very low concentrations in parts per million (ppm).

The number of parts of solute to one part of the solution

$$= \frac{\text{mass of solute/g}}{\text{mass of solution/g}}$$

∴ the concentration of solute in parts per million of solution,

$$\text{ppm} = \frac{\text{mass of solute/g}}{\text{mass of solution/g}} \times 10^6$$

Worked example

A sample of tap water contains 0.15 mg of Fe^{3+} per dm^3. What is its concentration in ppm? (Assume that 1 cm^3 of tap water has a mass of 1.0 g.)

Answer

$$0.15\text{ mg of } Fe^{3+} \text{ in } 1\text{ dm}^3 \text{ solution} = \frac{0.15\text{ g}}{1000}\ Fe^{3+} \text{ in } 1000\text{ cm}^3 \text{ solution}$$

$$= 0.000\,15\text{ g } Fe^{3+} \text{ in } 1000\text{ g solution}$$

$$\text{Concentration/ppm} = \frac{0.000\,15\text{ g}}{1000\text{ g}} \times 10^6 = 0.15$$

Tutorial

Test yourself

33 What is the concentration, in mol dm^{-3}, of a solution containing:
- **a)** 4.25 g silver nitrate, $AgNO_3$, in 500 cm^3 solution
- **b)** 4.0 g sodium hydroxide, NaOH, in 250 cm^3 of solution
- **c)** 20.75 g potassium iodide, KI, in 200 cm^3 of solution?

34 What mass of solute is present in:
- **a)** 50 cm^3 of 2.0 mol dm^{-3} sulfuric acid
- **b)** 100 cm^3 of 0.01 mol dm^{-3} potassium manganate(VII), $KMnO_4$
- **c)** 250 cm^3 of 0.2 mol dm^{-3} sodium carbonate, Na_2CO_3?

35 Highland Spring mineral water contains 0.133 g of hydrogencarbonate per cubic decimetre. What is the concentration of hydrogencarbonate in parts per million?

1.8 Theoretical and percentage yields

If a chemical reaction is totally efficient, all the starting reactant is converted to the required product. This gives a 100% yield. But few reactions are completely efficient and many give low yields. There are various reasons why yields are not 100%:

- the reactants may not be totally pure
- some of the product may be lost during transfer of the chemicals from one container to another, when the product is separated and purified
- there may be side reactions in which the reactants form different products
- some of the reactants may not react because the reaction is so slow (Topic 13) or because it comes to equilibrium (Topic 14).

Definitions

The **actual yield** is the mass of product obtained.

The **theoretical yield** is the mass of product obtained if the reaction goes according to the equation.

The **percentage yield** = $\frac{\text{actual yield}}{\text{theoretical yield}} \times 100$

A **limiting reactant** is a substance which is present in an amount which limits the theoretical yield. Often, some reactants are added in excess to ensure that the most valuable reactant is converted to as much product as possible.

Worked example

A modern gas-fuelled lime kiln produces 500 kg of calcium oxide, CaO (quicklime), from 1000 kg of crushed calcium carbonate, $CaCO_3$ (limestone). What is the percentage yield of calcium oxide?

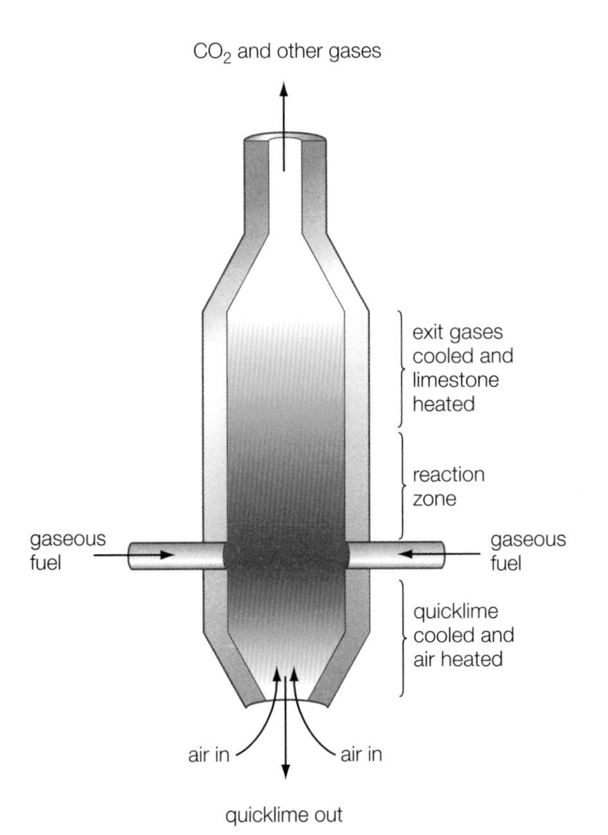

Figure 1.27 ◀ A gas-fuelled lime kiln

Figure 1.28 ▲
The production of ibuprofen is an excellent example of atom economy. Ibuprofen is an important medicine which reduces swelling and pain. In the 1960s, Boots made ibuprofen in five steps with an atom economy of only 40%. When the patent expired, another company developed a new process requiring just two steps with an atom economy of 100%.

Answer

The equation for the reaction involved is:

$$CaCO_3(s) \rightarrow CaO(s) + CO_2(g)$$

From the equation, 1 mole $CaCO_3 \rightarrow$ 1 mole CaO

$$\therefore [40.1 + 12 + (3 \times 16)]\,g\ CaCO_3 \rightarrow (40.1 + 16)\,g\ CaO$$

$$100.1\,g\ CaCO_3 \rightarrow 56.1\,g\ CaO$$

$$\text{So, } 1\,g\ CaCO_3 \rightarrow \frac{56.1}{100.1}\,g\ CaO$$

$$\therefore \text{Theoretical yield from } 1000\,kg\ CaCO_3 = \frac{56.1}{100.1} \times 1000\,kg\ CaO$$

$$= 560.4\,kg$$

The actual yield of CaO = 500 kg CaO

$$\therefore \text{Percentage yield} = \frac{500}{560.4} \times 100 = 89\%$$

Test yourself

36 1000 kg of pure iron(III) oxide, Fe_2O_3, was reduced to iron and 630 kg of iron was obtained. What are the theoretical and percentage yields of iron?

37 500 kg of calcium oxide (quicklime) was reacted with water to produce calcium hydroxide, $Ca(OH)_2$ (slaked lime). 620 kg of calcium hydroxide was produced. Calculate the theoretical and percentage yields.

Atom economy

The yield in a laboratory or industrial process focuses on the desired product. But many atoms in the reactants don't end up in the desired product, which leads to a huge waste of material. For example, when calcium carbonate (limestone) is decomposed to produce calcium oxide (quicklime), part of the calcium carbonate is lost as carbon dioxide.

The waste in many reactions has led scientists and industrialists to use the term 'atom economy' in calculating the overall efficiency of a chemical process. The atom economy of a reaction is the mass of the atoms in the desired product expressed as a percentage of the mass of the atoms in all the reactants as shown in the equation.

Definition

Atom economy

$$= \frac{\text{mass of atoms in desired product}}{\text{mass of atoms in all reactants}} \times 100$$

Worked example

Titanium is manufactured by heating titanium(IV) chloride with magnesium. The equation for the reaction is:

$$TiCl_4(g) + 2Mg(l) \rightarrow Ti(s) + 2MgCl_2(s)$$

What is the atom economy of this process?

Answer

$$\text{Relative mass of reactants} = M_r(TiCl_4) + 2A_r(Mg)$$

$$= A_r(Ti) + 4A_r(Cl) + 2A_r(Mg)$$

$$= 47.9 + 142 + 48.6 = 238.5$$

$$\text{Relative mass of desired product} = A_r(Ti) = 47.9$$

$$\therefore \text{Atom economy} = \frac{47.9}{238.5} \times 100 = 20.1\%$$

Almost 80% of the reactants are 'wasted' in the manufacture of titanium by the process described above because magnesium and chlorine atoms are lost as magnesium chloride. If we are to use raw materials as efficiently as possible, chemists must look for high atom economies as well as high percentage yields, particularly in industrial processes.

Test yourself

38 Calculate the atom economy for:

a) the conversion of nitrogen (N_2) to ammonia (NH_3) in the Haber process:

$$N_2(g) + 3H_2(g) \rightarrow 2NH_3(g)$$

b) the fermentation of glucose ($C_6H_{12}O_6$) to ethanol (C_2H_5OH)

$$C_6H_{12}O_6(aq) \rightarrow 2C_2H_5OH(aq) + 2CO_2(g)$$

c) the manufacture of tin (Sn) from tinstone (tin(IV) oxide, SnO_2)

$$SnO_2(s) + 2C(s) \rightarrow Sn(s) + 2CO(g)$$

REVIEW QUESTIONS

Extension questions

1 a) Which reactant in the equation below is acting as an acid?

$NH_3(g) + H_2O(l) \rightarrow NH_4^+(aq) + OH^-(aq)$ **(1)**

b) Concentrated ammonia solution contains 17.9 mol dm^{-3} of ammonia. What volume of this solution is required to make 1 dm^3 of a 2.0 mol dm^{-3} solution? **(2)**

2 For each of the following equations, state the type of reaction which it represents.

a) $Ca(NO_3)_2(aq) + K_2CO_3(aq) \rightarrow CaCO_3(s) + 2KNO_3(aq)$

b) $Mg(s) + 2HCl(aq) \rightarrow MgCl_2(aq) + H_2(g)$

c) $2Zn(NO_3)_2(s) \rightarrow 2ZnO(s) + 4NO_2(g) + O_2(g)$

d) $H_2SO_4(aq) + 2NaOH(aq) \rightarrow Na_2SO_4(aq) + 2H_2O(l)$ **(4)**

3 Excess calcium reacted vigorously with 25 cm^3 of 1.0 mol dm^{-3} hydrochloric acid producing calcium chloride solution and hydrogen.

a) Write a balanced equation, with state symbols, for the reaction. **(3)**

b) Draw a labelled diagram showing how the hydrogen gas could be collected during the reaction. **(2)**

c) Describe how you would obtain clean, dry crystals of hydrated calcium chloride, $CaCl_2.6H_2O$, from the solution formed. **(4)**

d) What is the maximum possible yield of hydrated calcium chloride, $CaCl_2.6H_2O$, from the reaction? (Ca = 40, Cl = 35.5, O = 16) **(4)**

e) Suggest two reasons why the actual yield of hydrated calcium chloride is much less than the mass calculated in part **d)**. **(2)**

4 a) How many molecules are present in 4.0 g of oxygen, O_2? (O = 16) **(3)**

b) How many ions are present in 9.4 g of potassium oxide, K_2O? (K = 39, O = 16) (Avogadro constant = 6.0×10^{23} mol^{-1}) **(3)**

5 a) Ammonium sulfate was prepared by adding ammonia solution to 25 cm^3 of 2.0 mol dm^{-3} sulfuric acid.

$$2NH_3(aq) + H_2SO_4(aq) \rightarrow (NH_4)_2SO_4(aq)$$

i) What volume of 2.0 mol dm^{-3} ammonia solution will just neutralise the sulfuric acid? **(1)**

ii) How could you test to check that enough ammonia has been added to neutralise all the acid? **(2)**

iii) What is the atom economy of the reaction? **(2)**

b) Iron(II) sulfate, $FeSO_4$, was added to the ammonium sulfate solution to produce ammonium iron(II) sulfate hexahydrate, $(NH_4)_2SO_4.FeSO_4.6H_2O$.

i) What mass of iron(II) sulfate should be added to the ammonium sulfate solution? **(3)**

ii) What mass of ammonium iron(II) sulfate hexahydrate will be obtained if the percentage yield is 50%? (H = 1, N = 14, Fe = 56, S = 32, O = 16) **(4)**

6 Balance the following equations:

a) $Cu_2S(s) + O_2(g) \rightarrow CuO(s) + SO_2(g)$ (1)

b) $FeS(s) + O_2(g) + SiO_2(s) \rightarrow FeSiO_3(s) + SO_2(g)$ (3)

c) $Fe(NO_3)_3(s) \rightarrow Fe_2O_3(s) + NO_2(g) + O_2(g)$ (3)

7 A sample of limestone (impure calcium carbonate) weighing 0.20 g was reacted with 50 cm^3 of 0.10 mol dm^{-3} hydrochloric acid.

$$CaCO_3(s) + 2HCl(aq) \rightarrow CaCl_2(aq) + CO_2(g) + H_2O(l)$$

The hydrochloric acid was in excess, and this excess just reacted with 14 cm^3 of 0.10 mol dm^{-3} sodium hydroxide.

a) Write an equation, including state symbols, for the reaction of hydrochloric acid with sodium hydroxide. (2)

b) What excess volume of 0.10 mol dm^{-3} hydrochloric acid reacted with the 14 cm^3 of 0.10 mol dm^{-3} sodium hydroxide? (1)

c) i) What volume of 0.10 mol dm^{-3} hydrochloric acid reacted with the calcium carbonate in the limestone? (1)

ii) How many moles of hydrochloric acid reacted with the calcium carbonate? (1)

iii) How many moles of calcium carbonate reacted with the hydrochloric acid? (1)

iv) What mass of calcium carbonate reacted? (Ca = 40, C = 12, O = 16) (1)

d) What is the percentage by mass of calcium carbonate in the limestone? (1)

8 a) Explain why hydrogen chloride, HCl, acts as an acid and why ammonia, NH_3, acts as a base in the reaction

$$HCl(g) + NH_3(g) \rightarrow NH_4^+Cl^-(s)$$ (4)

b) Which of the reactants acts as an acid in the following equation?

$$HCO_3^-(aq) + H_2O(l) \rightarrow CO_3^{2-}(aq) + H_3O^+(aq)$$ (1)

c) Which of the reactants acts as a base in the following equation?

$$HCO_3^-(aq) + H_3O^+(aq) \rightarrow H_2CO_3(aq) + H_2O(l)$$ (1)

9 The concentration of cholesterol ($C_{27}H_{46}O$) in a patient's blood was found to be 6.0 mmol dm^{-3}.

a) What is the concentration of cholesterol in mol dm^{-3}? (1000 mmol = 1 mol) (1)

b) Calculate the concentration of cholesterol in g dm^{-3}. (2)

c) What is the mass of cholesterol in 10 cm^3 of the patient's blood? (1)

2 Energetics

In the natural environment, energy changes from the Sun create differences in temperature which stir the winds and vaporise water. The Sun also provides the energy for photosynthesis in plants which produces the concentrated sources of chemical energy in foods. In many mechanical systems, the energy changes from burning fuels create high temperatures in engines to keep machines turning and vehicles moving. There is a close connection between these energy changes and chemical reactions.

2.1 Energy changes

The study of energy changes during chemical reactions is called thermochemistry. Thermochemistry is important because it helps chemists to measure the energy changes in chemical reactions and explain the stability of compounds. With the help of thermochemistry, chemists can also decide whether or not reactions are likely to occur.

The term 'system' is important in thermochemistry and it has a precise meaning. It describes just the material or the mixture of chemicals being studied. Everything around the system is called the surroundings (Figure 2.1). The surroundings include the apparatus, the air in the laboratory – in theory everything else in the Universe.

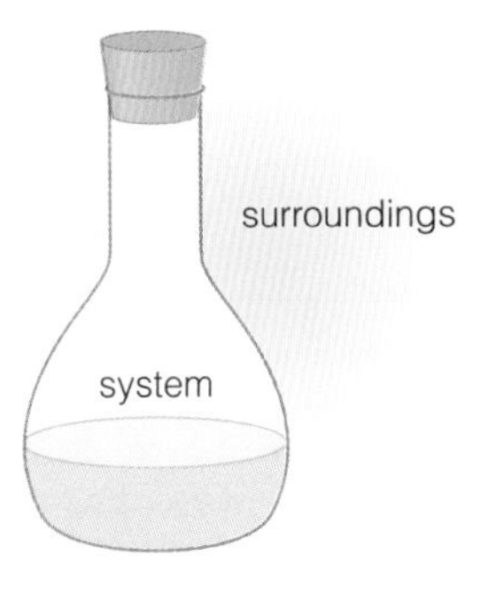

Figure 2.1 ▲
A system and its surroundings

In a closed system like that in Figure 2.1, the system cannot exchange matter with its surroundings because the flask is closed with a bung. It can, however, exchange energy, such as heat, with the surroundings. If the bung is removed, the system is described as 'open'. An open system can exchange both energy and matter with its surroundings.

Definition

An **enthalpy change**, ΔH, is the overall energy exchanged with the surroundings when a change happens at constant pressure and the final temperature is the same as the starting temperature.

Note

Scientists use the capital Greek letter 'delta', Δ, for a change or difference in a physical quantity. So, ΔH means change in enthalpy and ΔT means change in temperature.

2.2 Enthalpy changes

Whenever a change occurs in a system, there is almost always an energy change involving transfer of energy between the system and its surroundings.

The energy transferred between a system and its surroundings is described as an enthalpy change when the change happens at constant pressure. The symbol for an enthalpy change is ΔH and its units are kJ mol^{-1}.

Exothermic and endothermic changes

Exothermic changes give out energy to the surroundings. In an **ex**othermic change, energy leaves the system, just as people leave a building by the **ex**it.

Burning is an obvious exothermic chemical reaction. Respiration is another exothermic reaction in which foods are oxidised to provide energy for living things to grow and move. Hot packs used in self-warming drinks and in treating painful rheumatic conditions also involve exothermic reactions (Figure 2.2).

Figure 2.2 ▶
A self-warming can of coffee – pressing the bulb on the bottom of the can starts an exothermic reaction in a sealed compartment. The energy released heats the coffee.

Figure 2.3 shows what happens in the exothermic reaction between calcium oxide and water. This reaction is used in hot packs such as the self-warming can of coffee in Figure 2.2.

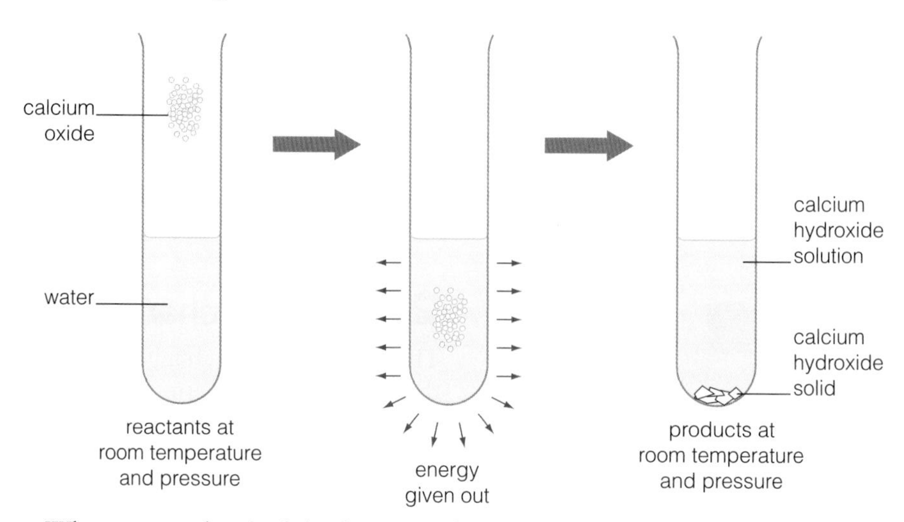

Figure 2.3 ▶
The exothermic reaction that takes place in some hot packs

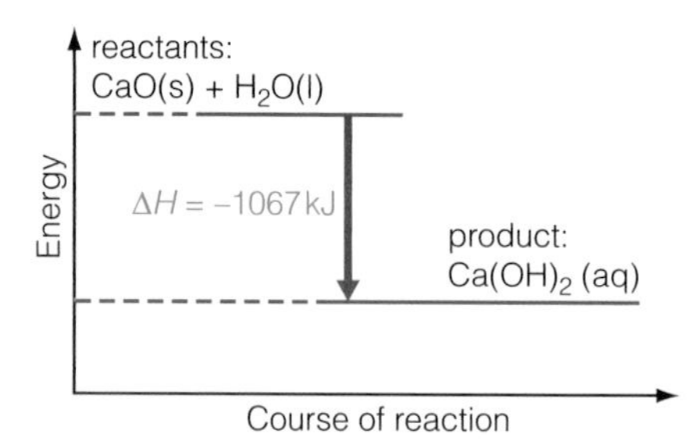

Figure 2.4 ▲
An enthalpy level diagram for the reaction of calcium oxide with water

When one mole of solid calcium oxide reacts with water to form calcium hydroxide solution, 1067 kJ of heat are given out. The system has lost energy (heat) to the surroundings. This loss of energy from the system means that ΔH is negative and we can write:

$$CaO(s) + H_2O(l) \rightarrow Ca(OH)_2(aq) \quad \Delta H = -1067\,kJ\,mol^{-1}$$

The energy changes in chemical reactions can be summarised in enthalpy level diagrams.

Figure 2.4 shows the enthalpy level diagram for the reaction of calcium oxide with water. Energy is lost to the surroundings as heat and therefore the products are at a lower energy level than the reactants. For this and all other exothermic reactions, ΔH is negative.

Endothermic changes take in energy from the surroundings. They are the opposite of exothermic changes. Melting and vaporisation are endothermic changes of state. Photosynthesis is an endothermic chemical change. During photosynthesis, plants take in energy from the Sun in order to convert carbon dioxide and water to glucose. Figure 2.5 illustrates the use of an endothermic reaction in a cold pack.

An enthalpy level diagram shows that the system has more energy after an endothermic reaction than it had at the start. So, for endothermic reactions, the enthalpy change, ΔH, is positive and the products are at a higher energy level than the reactants (Figure 2.6).

Figure 2.5 ▲
Twist a cold pack and it gets cold enough to reduce the pain of a sports injury. When chemicals in the cold pack react, they take in energy as heat and the pack gets cold. This, in turn, cools the sprained or bruised area and helps to reduce painful swelling.

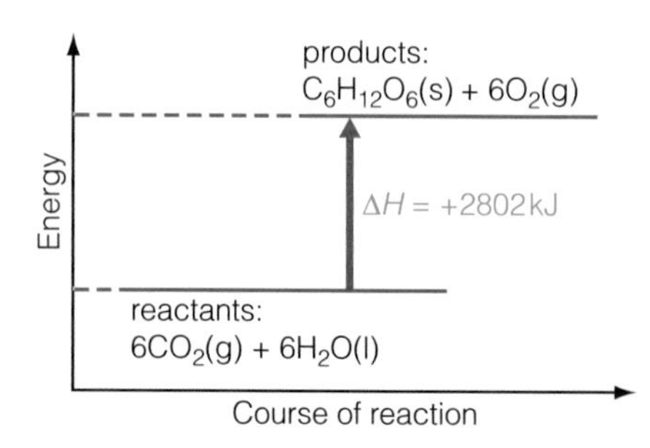

Figure 2.6 ▲
An enthalpy level diagram for photosynthesis

Test yourself

1 Which of the following changes are exothermic and which are endothermic?
  a) melting ice b) burning wood c) condensing steam
  d) metabolising sugar e) subliming iodine.

2 When 1 mole of carbon burns completely, 394 kJ of heat is given out.
  a) Write an equation for the reaction including state symbols and show the value of the enthalpy change.
  b) Draw an enthalpy level diagram for the reaction including the enthalpy change.

3 When 0.2 g of methane, CH_4 (natural gas), was burned completely, it produced 11 000 J.
  a) Write an equation for the reaction when methane burns completely.
  b) Calculate the molar mass of methane.
  c) How much heat is produced when 1 mole of methane burns completely?
  d) Draw an enthalpy level diagram for the reaction showing the value of the enthalpy change.

2.3 Measuring enthalpy changes

The heat given out or taken in during many chemical reactions can be measured fairly conveniently and this allows us to determine enthalpy changes.

Enthalpy changes from burning fuels

Figure 2.7 shows the simple apparatus that could be used to measure the heat given out from a liquid fuel like meths. Wear eye protection if you try this experiment and remember that liquid fuels are highly flammable.

In the experiment we will assume that *all* the heat produced from the burning meths heats up the water. The results from one experiment are shown in Table 2.1.

- From the mass of water in the can and its temperature rise, we can work out the heat produced.
- From the loss in mass of the liquid burner, we can find the mass of meths which has burned.
- We can then calculate the heat produced when 1 g of the fuel burns.

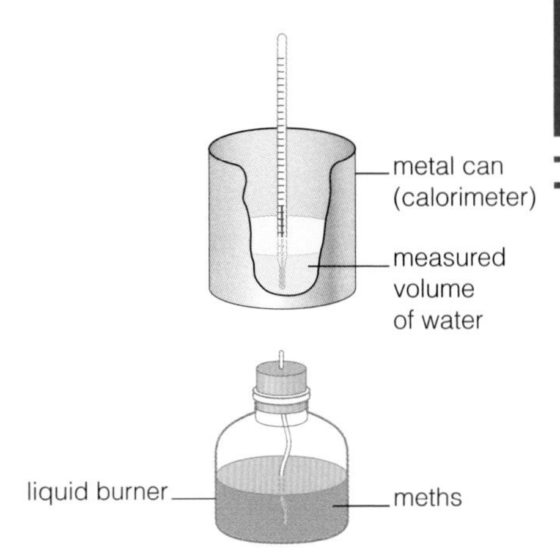

Figure 2.7 ▲
Measuring the enthalpy change when meths is burned

Mass of burner + meths at start of experiment	= 271.8 g
Mass of burner + meths at end of experiment	= 271.3 g
∴ Mass of meths burned	= 0.5 g
Volume of water in can	= 250 cm^3
∴ Mass of water in can	= 250 g
Rise in temperature of water	= 10 °C = 10 K

Table 2.1 ▲

The specific heat capacity of water is $4.2\,J\,g^{-1}\,K^{-1}$. This means that:

$4.2\,J$ will raise the temperature of 1 g of water by 1 K (1 °C)
∴ $m \times 4.2\,J$ will raise the temperature of m g of water by 1 K (1 °C)
and $m \times 4.2 \times \Delta T\,J$ will raise the temperature of m g of water by ΔT K (ΔT °C)

In general, if $\boldsymbol{m}$ g of a substance with a specific heat capacity of $\mathbf{c}\,J\,g^{-1}\,K^{-1}$ changes in temperature by $\boldsymbol{\Delta T}$ K, then

$$\textbf{Energy transferred/J} = \boldsymbol{m}/g \times \mathbf{c}/J\,g^{-1}\,K^{-1} \times \boldsymbol{\Delta T}/K$$

Using this equation with the results of our experiment:

0.5 g of burning meths transfers $250\,g \times 4.2\,J\,g^{-1}\,K^{-1} \times 10\,K$
$= 10\,500\,J$ to the water

∴ 1 g of meths produces 21 000 J = 21 kJ

Meths is mainly ethanol (alcohol), C_2H_6O. If we assume that meths is pure ethanol ($M_r = 46$), then:

Heat produced when 1 mole of ethanol burns = 46 × 21 kJ = 966 kJ

So we can write:

$$C_2H_6O(l) + 3O_2(g) \rightarrow 2CO_2(g) + 3H_2O(l) \quad \Delta H = -966\,kJ\,mol^{-1}$$

The heat transferred when one mole of a fuel burns completely is usually called the enthalpy change of combustion of the fuel and given the symbol ΔH_c. Therefore:

$$\Delta H_c[\text{ethanol}] = -966\,kJ\,mol^{-1}$$

Definition

The **specific heat capacity** of a material, c, is the energy needed to raise the temperature of 1 g of the material by 1 K.
For water $c = 4.2\,J\,g^{-1}\,K^{-1}$.

Assumptions and errors in thermochemical experiments

In the experiment just described, we assumed that all the heat (energy) from the flame heats the water. In practice a proportion of the energy heats the metal can and the surrounding air. In addition, the flame is affected by draughts and sometimes the fuel burns incompletely, leaving soot on the bottom of the metal can. Our initial assumption is clearly flawed and our result is certainly inaccurate. The major sources of error (loss of heat, flame disturbance and incomplete combustion) all reduce the heat transferred to the water. This leads to a result that is lower than the true value.

Figure 2.8 ▶
A bomb calorimeter

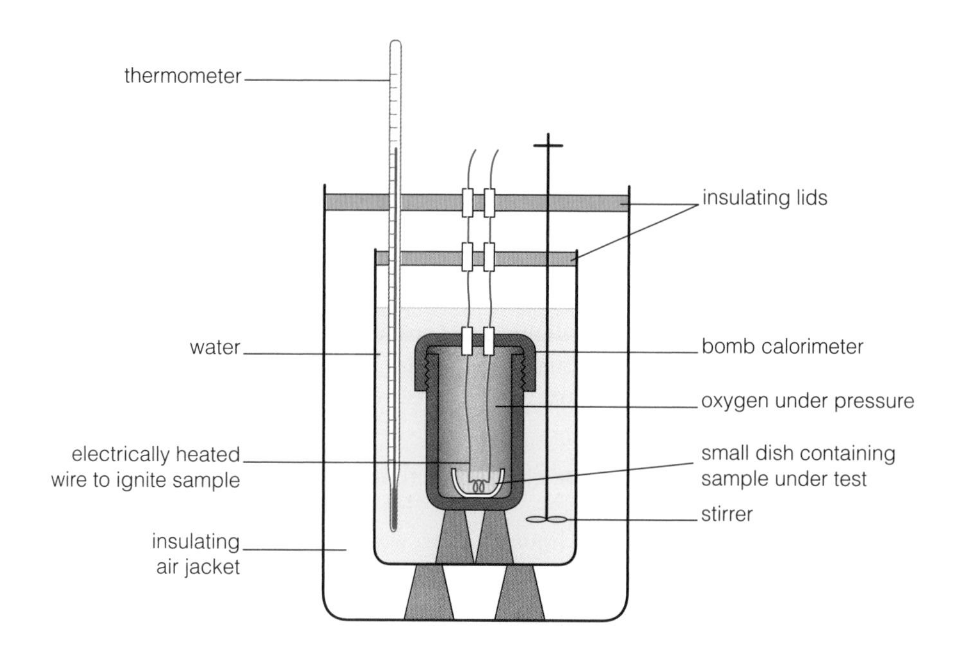

Accurate values for enthalpy changes of combustion are obtained using a bomb calorimeter (Figure 2.8). The apparatus is specially designed to ensure that the sample burns completely and that heat losses are avoided. A measured amount of the sample burns in excess oxygen under pressure. There must be enough oxygen to ensure that all carbon in the compound is fully oxidised to carbon dioxide and no carbon monoxide or soot are produced.

Heat losses are eliminated altogether by coupling the thermochemical experiment with an electrical calibration. After the chemical reaction has been carried out, the experiment is repeated, but this time an electrical heating coil replaces the reactants. The current in the coil is continually adjusted to give the same temperature rise in the same time as the chemical reaction. By recording the current during the time of this electrical calibration, it is possible to calculate the electrical energy supplied very accurately. This is the same as the energy transferred in the chemical reaction.

Enthalpy changes of combustion are important in the fuel and food industries (Figure 2.9). Scientists use bomb calorimeters like that in Figure 2.8 to determine the energy values of fuels and foods. The prices of fuels are closely related to their energy values and dieticians give advice related to their knowledge of energy-providing foods.

Figure 2.9 ▶
Scientists use bomb calorimeters like the one in this photo to measure the energy produced by different fuels and foods.

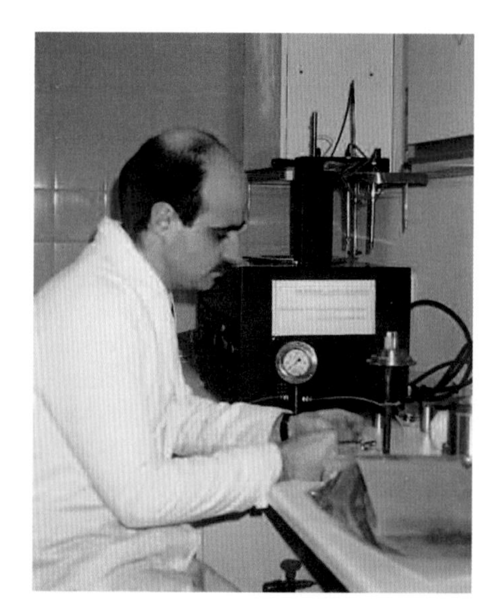

Enthalpy changes in solution

Enthalpy changes for reactions in solution can be measured using insulated plastic containers such as polystyrene cups (Figure 2.10). Polystyrene is an excellent insulator and it has a negligible specific heat capacity.

If the reaction is exothermic, the energy released cannot escape to the surroundings so it heats up the solution. If the reaction is endothermic, no energy can enter from the surroundings so the solution cools. If the solutions are dilute, it is sufficiently accurate to calculate the enthalpy changes by assuming that the solutions have the same density and specific heat capacity as water.

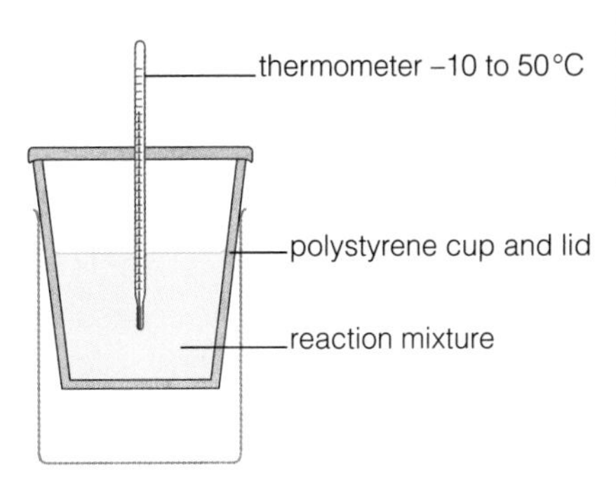

Figure 2.10 ▲
Measuring the enthalpy change of a reaction in solution

Worked example

When 4.0 g of ammonium nitrate (NH_4NO_3) dissolves in 100 cm^3 of water, the temperature falls by 3.0 °C. Calculate the enthalpy change per mole when NH_4NO_3 dissolves in water under these conditions.

Answer

$$\text{Amount of } NH_4NO_3 \text{ used} = \frac{\text{mass}}{M(NH_4NO_3)} = \frac{4\,\text{g}}{80\,\text{g mol}^{-1}} = 0.05\,\text{mol}$$

Assuming that the solution has the same specific heat capacity and density as water:

Energy taken in from the solution

= mass × specific heat capacity × temperature change
= $100\,\text{g} \times 4.2\,\text{J g}^{-1}\text{K}^{-1} \times 3\,\text{K}$
= 1260 J

$$\therefore \text{Energy taken in per mole of } NH_4NO_3 = \frac{1260\,\text{J}}{0.05\,\text{mol}} = 25\,200\,\text{J mol}^{-1}$$

$$= 25\,\text{kJ mol}^{-1}$$

The reaction is endothermic, so the enthalpy change for the system is positive.

$$NH_4NO_3(s) \xrightarrow{(aq)} NH_4NO_3(aq) \qquad \Delta H = +25\,\text{kJ mol}^{-1}$$

Test yourself

4 Burning butane, C_4H_{10}, from a Camping Gaz® container raised the temperature of 200 g water from 18 °C to 28 °C. The Gaz container was weighed before and after and the loss in mass was 0.29 g. Estimate the molar enthalpy change of combustion of butane.

5 On adding 25 cm^3 of 1.0 mol dm^{-3} nitric acid to 25 cm^3 1.0 mol dm^{-3} potassium hydroxide in a plastic cup, the temperature rise is 6.5 °C.
   a) Write an equation for the reaction.
   b) Calculate the enthalpy change for the neutralisation reaction per mole of nitric acid.

6 On adding excess powdered zinc to 25.0 cm^3 of 0.2 mol dm^{-3} copper(II) sulfate solution, the temperature rises by 9.5 °C.
   a) Write an equation for the reaction.
   b) Calculate the enthalpy change of the reaction for the molar amounts in the equation.

Activity

Measuring and evaluating the enthalpy change for the reaction of zinc with copper(II) sulfate solution

Two students decided to measure the enthalpy change for the reaction between zinc and copper(II) sulfate solution.

$$Zn(s) + CuSO_4(aq) \rightarrow ZnSO_4(aq) + Cu(s)$$

The method which they used is shown in Figure 2.11 and their results are shown in Table 2.2. After adding the zinc, it took a little while for the temperature to reach a peak and then the mixture began to cool.

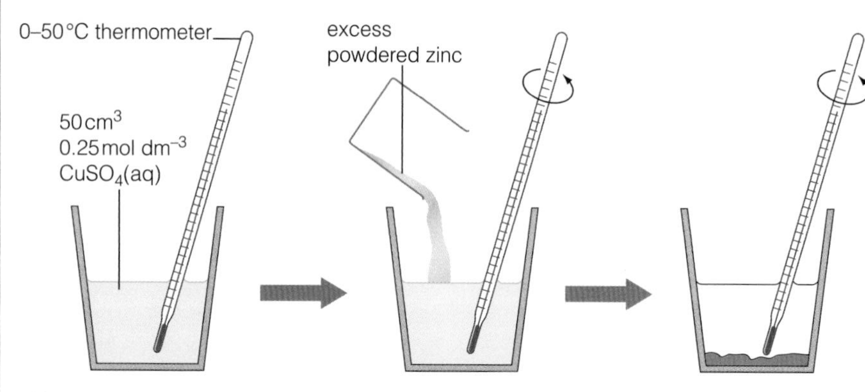

Figure 2.11 ◀ Measuring the enthalpy change for the reaction of zinc with copper(II) sulfate solution

Table 2.2 ◀

Time/min	Temperature /°C	Time/min	Temperature /°C	Time/min	Temperature /°C
0	24.1	3.5	34.2	6.5	33.7
0.5	24.0	4.0	34.8	7.0	33.6
1.0	24.1	4.5	35.0	7.5	33.5
1.5	24.1	5.0	34.6	8.0	33.4
2.0	24.2	5.5	34.2	8.5	33.2
2.5	24.1	6.0	33.9	9.0	33.1
3.0	–				

1 Plot a graph of temperature (vertically) against time (horizontally) using the results in Table 2.2.

2 Extrapolate the graph backwards from 9 minutes to 3 minutes, as in Figure 2.12, in order to estimate the maximum temperature. This assumes that all the zinc reacted at once and there was no loss of heat to the surroundings.

a) What is the estimated maximum temperature at 3 minutes?

b) What is the temperature rise, ΔT, for the reaction?

3 Calculate the energy given out during the reaction using the equation:

Energy transferred = mass × specific heat capacity × temperature change

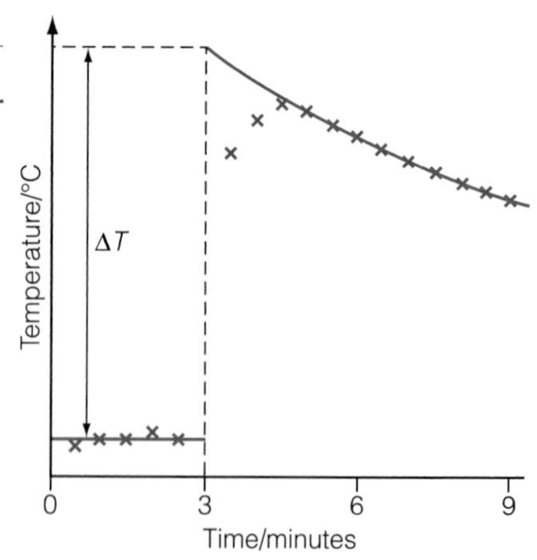

Figure 2.12 ▲ Estimating the maximum temperature of the mixture when zinc reacts with copper(II) sulfate solution.

Assume that:

- all the heat is transferred to the solution in the polystyrene cup
- the density of the solution is 1 g cm^{-3}
- the specific heat capacity of the solution is 4.2 J g^{-1} K^{-1}.

4 How many moles of each chemical reacted?

a) $CuSO_4$ **b)** Zn

5 What is the enthalpy change of the reaction, $\Delta H_{reaction}$, for the amounts of Zn and $CuSO_4$ in the equation?
(State the value of $\Delta H_{reaction}$ in kJ mol^{-1} with the correct sign.)

6 Use the Practical guidance: 'Errors and uncertainty' on the Dynamic Learning Student website to copy and complete Table 2.3 for the various measurements in the experiment.

Practical guidance

Table 2.3 The values, uncertainties and percentage uncertainties of measurements in the experiment

Measurement	Value	Uncertainty	Percentage uncertainty
Concentration of copper(II) sulfate solution			
Volume of copper(II) sulfate solution measured from a 100 cm^3 measuring cylinder			
Temperature rise, ΔT, estimated from the difference in two readings taken with a 0–50 °C thermometer			

7 What is the total percentage uncertainty in the experiment?

8 What is the total uncertainty in the value you have calculated for the enthalpy change?

9 Write a value for the enthalpy change showing the uncertainty as a ± amount.

10 What are the main sources of error in the measurements and procedure for the experiment?

11 Look critically at the procedures in the experiment and suggest improvements to minimise errors and increase the reliability of the result.

2.4 Standard enthalpy changes

Our studies in this topic have already shown that it is only energy changes and enthalpy changes that can be measured and not absolute levels of energy or enthalpy in a system. It is, however, very useful to compare enthalpy changes.

In order to compare different enthalpy changes accurately, it is important that the conditions under which measurements and experiments are carried out are the same in all cases. These conditions for the accurate comparison of enthalpy changes and other thermochemistry measurements are called **standard conditions**.

Standard conditions are:

- a pressure of 1 atmosphere (1.013×10^5 Pa $\simeq$ 100 kPa)
- a temperature of 298 K (25 °C)
- substances in their standard (most stable) state at 1 atm pressure and 298 K
- solutions with a concentration of 1 mol dm^{-3}.

Definition

The **standard enthalpy change of a reaction**, $\Delta H^{\ominus}_{reaction}$, is the energy transferred when the molar quantities of reactants as stated in the equation react under standard conditions.

Any enthalpy change measured under these conditions is described as a standard enthalpy change and given the symbol $\Delta H^{\ominus}_{298}$ or simply $\Delta H^{\ominus}$, pronounced 'delta *H* standard'.

In thermochemistry, it is important to specify the states of the substances and therefore to include state symbols in equations. So, $\Delta H^{\ominus}$ for the reaction:

$$2H_2(g) + O_2(g) \rightarrow 2H_2O(l)$$

must relate to hydrogen gas, oxygen gas and liquid water (not steam). The states of elements and compounds must also be the most stable at 298 K and 1 atmosphere. Thus, $\Delta H^{\ominus}$ measurements involving carbon should use graphite, which is energetically more stable than diamond.

Worked example

When 50 cm^3 of 2.0 mol dm^{-3} hydrochloric acid is mixed with 50 cm^3 of 2.0 mol dm^{-3} sodium hydroxide in a bomb calorimeter at 25 °C and 1 atmosphere pressure, the temperature rises by 13.7 °C. What is the enthalpy change for the neutralisation reaction?

Answer

The equation for the reaction is:

$$HCl(aq) + NaOH(aq) \rightarrow NaCl(aq) + H_2O(l)$$

Amount of HCl used = amount of NaOH used

$$= \frac{50}{1000}\ dm^3 \times 2.0\ mol\ dm^{-3} = 0.1\ mol$$

Energy given out and used to heat the solution

$$= 100\ g \times 4.2\ J\ g^{-1} K^{-1} \times 13.7\ K = 5754\ J$$

$$\therefore \text{Energy given out per mole of acid} = \frac{5754\ J}{0.1\ mol} = 57\,540\ J\ mol^{-1}$$

$$\therefore \Delta H_{neutralisation} = -57.5\ kJ\ mol^{-1}$$

Definition

The **standard enthalpy change of neutralisation** is the enthalpy change when the amounts of acid and alkali in the equation for the reaction neutralise each other under standard conditions.

Now, as the concentrations of the two solutions were effectively 1.0 mol dm^{-3} immediately after mixing, the temperature was 25 °C and the pressure 1 atmosphere, the value of $\Delta H_{neutralisation}$ has been obtained under standard conditions.

$$\therefore HCl(aq) + NaOH(aq) \rightarrow NaCl(aq) + H_2O(l)$$

$$\Delta H^{\ominus}_{neutralisation} = -57.5\ kJ\ mol^{-1}$$

This is called a standard enthalpy change of neutralisation.

Standard enthalpy changes of combustion

The standard enthalpy change of combustion of an element or compound, $\Delta H^{\ominus}_c$, is the enthalpy change when one mole of the substance burns completely in oxygen under standard conditions. The substance and the products of burning must be in their stable (standard) states. For a carbon compound, complete combustion means that all the carbon burns to carbon dioxide and there is no soot or carbon monoxide. If the substance contains hydrogen, the water formed must end up as liquid and not as a gas.

Values of enthalpies of combustion are much easier to measure than many other enthalpy changes. They can be calculated from measurements taken with a bomb calorimeter (Section 2.3).

Chemists use two ways to summarise standard enthalpy changes of combustion. One way is to write the equation with the enthalpy change alongside it. So, for the standard enthalpy change of combustion of carbon, they write:

$$C_{(graphite)} + O_2(g) \rightarrow CO_2(g) \quad \Delta H^{\ominus}_c = -393.5\,\text{kJ mol}^{-1}$$

The other way is to use a shorthand form. For the standard enthalpy change of combustion of methane, this is written as:

$$\Delta H^{\ominus}_c[CH_4(g)] = -890\,\text{kJ mol}^{-1}$$

Remember that all combustion reactions are exothermic, so $\Delta H^{\ominus}_c$ values are always negative.

Definition

The **standard enthalpy change of combustion** of a substance, $\Delta H^{\ominus}_c$ is the enthalpy change when one mole of the substance burns completely in oxygen under standard conditions.

Standard enthalpy changes of formation

The standard enthalpy change of formation of a compound, $\Delta H^{\ominus}_f$, is the enthalpy change when one mole of the compound forms from its elements. The elements and the compound formed must be in their stable standard states. The more stable state of an element is chosen where there are allotropes (different forms in the same state) such as graphite and diamond.

As with standard enthalpies of combustion, there are two ways of representing standard enthalpy changes of formation.

One way is to write the equation with the enthalpy change alongside it. For the standard enthalpy change of formation of water this is:

$$H_2(g) + \tfrac{1}{2}O_2(g) \rightarrow H_2O(l) \quad \Delta H^{\ominus}_f = -286\,\text{kJ mol}^{-1}$$

The other way is to use shorthand. For the standard enthalpy change of formation of ethanol this is:

$$\Delta H^{\ominus}_f[C_2H_5OH(l)] = -277\,\text{kJ mol}^{-1}$$

Like all thermochemical quantities, the precise definition of the standard enthalpy change of formation is important. Books of data tabulate values for standard enthalpies of formation. These tables are very useful because they make it possible to calculate the enthalpy changes for many reactions (Section 2.5).

Unfortunately it is difficult to measure some enthalpy changes of formation directly. For example, it is impossible to convert carbon, hydrogen and oxygen straight to ethanol under any conditions. Because of this, chemists have had to find an indirect method of measuring the standard enthalpy change of formation of ethanol and other compounds (Section 2.5).

Definition

The **standard enthalpy change of formation** of a compound, $\Delta H^{\ominus}_f$, is the enthalpy change when one mole of the compound forms from its elements under standard conditions with the elements and the compound in their standard (stable) states.

One important consequence of the definition of standard enthalpy changes of formation is that, for an element, $\Delta H^{\ominus}_f = 0\,\text{kJ mol}^{-1}$ because there is no change, and therefore no enthalpy change, when an element forms from itself. In other words, the standard enthalpy change of formation of an element is zero. So:

$$\Delta H^{\ominus}_f[Cu(s)] = 0 \quad \text{and} \quad \Delta H^{\ominus}_f[O_2(g)] = 0$$

Obviously, $\Delta H^{\ominus}_f$ for the process $H_2(g) \rightarrow H_2(g)$ is zero, but this is not so for the process $H_2(g) \rightarrow 2H(g)$, i.e. the conversion of H_2 molecules to single H atoms. The latter process involves atomisation and the standard enthalpy change of atomisation of hydrogen ($\Delta H^{\ominus}_{at}[\tfrac{1}{2}H_2(g)]$) involves the change:

$$\tfrac{1}{2}H_2(g) \rightarrow H(g) \quad \Delta H^{\ominus}_{at} = +218\,\text{kJ mol}^{-1}$$

Definition

The **standard enthalpy change of atomisation** of an element, $\Delta H^{\ominus}_{at}$, is the enthalpy change when one mole of gaseous atoms is formed from the element under standard conditions.

2.5 Hess's Law and the indirect determination of enthalpy changes

The enthalpy change of a reaction is the same whether the reaction happens in one step or in a series of steps. As long as the reactants and products are the same, the overall enthalpy change is the same whether the reactants are converted to products directly or through two or more intermediates. This is Hess's Law. In Figure 2.13 the enthalpy change for Route 1 and the overall enthalpy change for Route 2 are the same.

Hess's Law is an example of a mathematical model. This is shown by the precise quantitative relationship between ΔH_1, ΔH_2, ΔH_3 and ΔH_4 in Figure 2.13. Using Hess's Law it is possible to bring together data and calculate enthalpy changes which cannot be measured directly by experiment. So, Hess's Law can be used to calculate:

- standard enthalpy changes of formation from standard enthalpy changes of combustion and
- standard enthalpy changes of reaction from standard enthalpy changes of formation.

Hess's Law is a chemical version of the law of conservation of energy. Suppose the enthalpy change for Route 1 in Figure 2.13 was more exothermic than the total enthalpy change for Route 2. It would be possible to go round the cycle in Figure 2.13 from A to D direct and back to A via C and B, ending up with the same starting chemical but with a net release of energy. This would contravene the law of conservation of energy.

Definition

Hess's Law says that the enthalpy change in converting reactants to products is the same regardless of the route taken, provided the initial and final conditions are the same.

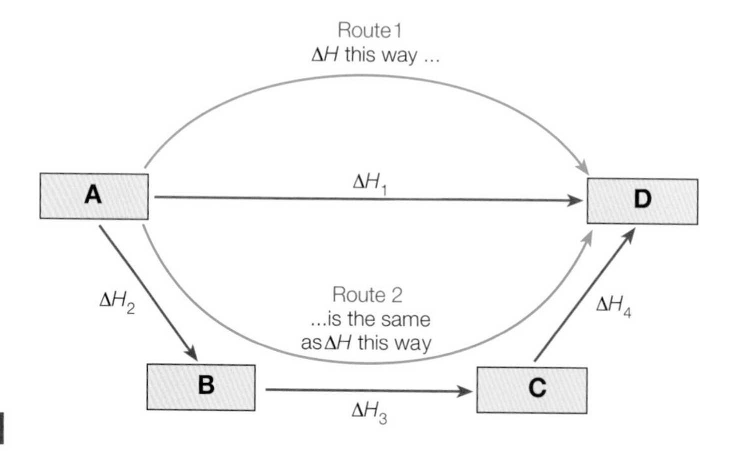

Figure 2.13 ▶ A diagram to illustrate Hess's Law:

$$\Delta H_1 = \Delta H_2 + \Delta H_3 + \Delta H_4$$

Note

Reversing the direction of a reaction in an energy cycle reverses the sign of ΔH.

Enthalpies of formation from enthalpies of combustion

Figure 2.14 shows the form of the energy cycle which we can use to calculate enthalpy changes of formation from enthalpy changes of combustion.

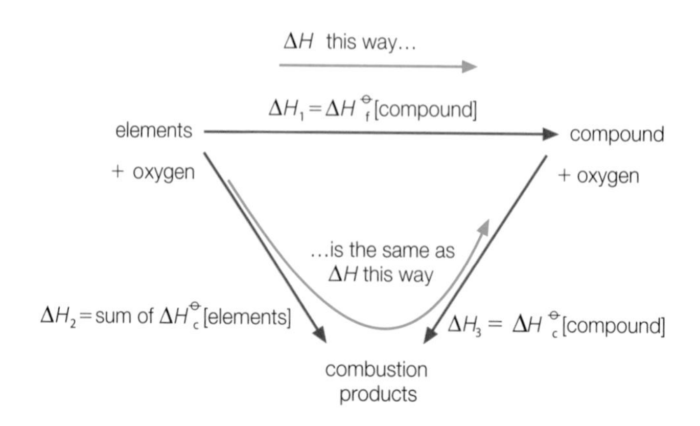

Figure 2.14 ▶ An energy cycle for calculating standard enthalpies of formation from standard enthalpies of combustion:

$$\Delta H_1 = \Delta H_2 - \Delta H_3$$

Worked example

Calculate the enthalpy change of formation of propane, C_3H_8, at 298 K given the following standard enthalpies of combustion.

propane, $\Delta H^{\ominus}_{c}[C_3H_8(g)] = -2220\,kJ\,mol^{-1}$

carbon, $\Delta H^{\ominus}_{c}[C_{(graphite)}] = -393\,kJ\,mol^{-1}$

hydrogen $\Delta H^{\ominus}_{c}[H_2(g)] = -286\,kJ\,mol^{-1}$

Notes on the method

Draw up an energy cycle using the model in Figure 2.14. Use Hess's Law to produce an equation linking the relevant enthalpy changes. Pay careful attention to the signs. Put the value and sign for a quantity in brackets when adding or subtracting enthalpy values.

Answer

An energy cycle linking the formation of propane with its combustion and the combustion of its constituent elements is shown in Figure 2.15. Sometimes, energy cycles like the one in Figure 2.15 are called Hess cycles.

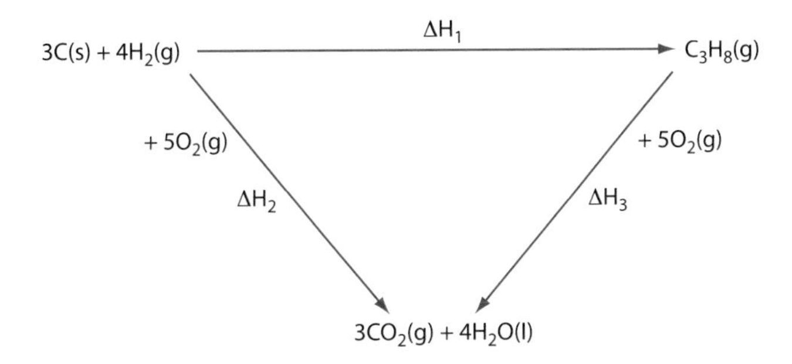

Figure 2.15 ◀ An energy cycle for the combustion of propane and its constituent elements

According to Hess's Law, $\Delta H_1 = \Delta H_2 - \Delta H_3$

$$\Delta H_1 = \Delta H^{\ominus}_{f}[C_3H_8(g)]$$

$$\Delta H_2 = 3 \times \Delta H^{\ominus}_{c}[C_{(graphite)}] + 4 \times \Delta H^{\ominus}_{c}[H_2(g)]$$

$$= 3 \times (-393\,kJ\,mol^{-1}) + 4 \times (-286\,kJ\,mol^{-1}) = -2323\,kJ\,mol^{-1}$$

$$\Delta H_3 = \Delta H^{\ominus}_{c}[C_3H_8(g)] = -2220\,kJ\,mol^{-1}$$

Hence:

$$\Delta H^{\ominus}_{f}[C_3H_8(g)] = (-2323\,kJ\,mol^{-1}) - (-2220\,kJ\,mol^{-1})$$

$$= -2323 + 2220$$

$$= -103\,kJ\,mol^{-1}$$

Test yourself

7 a) By writing a balanced equation, show that the standard enthalpy change of formation of carbon dioxide is the same as the standard enthalpy change of combustion of carbon (graphite).
b) Write equations for the standard enthalpy change of formation of:
i) aluminium oxide, Al_2O_3
ii) hydrogen chloride, HCl
iii) propane, C_3H_8.

8 Use the values of standard enthalpies of combustion below to calculate the standard enthalpy change of formation of methanol, CH_3OH.
$\Delta H^{\ominus}_{c}[CH_3OH(l)] = -726\,kJ\,mol^{-1}$
$\Delta H^{\ominus}_{c}[C_{(graphite)}] = -393\,kJ\,mol^{-1}$
$\Delta H^{\ominus}_{c}[H_2(g)] = -286\,kJ\,mol^{-1}$

9 Why is it useful to have standard enthalpies of combustion which can be used to calculate standard enthalpies of formation?

Enthalpies of reaction from enthalpies of formation

Data books contain tables of standard enthalpies of formation for both inorganic and organic compounds. The great value of this data is that it allows us to calculate the standard enthalpy change for any reaction involving the substances listed in the tables.

The standard enthalpy change of a reaction is the enthalpy change when the amounts shown in the chemical equation react. Like other standard quantities in thermochemistry, the standard enthalpy change of reaction is defined at 298 K and 1 atmosphere pressure with the reactants and products in their normal stable states. The concentration of any solution is 1 mol dm^{-3}.

Thanks to Hess's Law it is easy to calculate the standard enthalpy change of a reaction from tabulated values of standard enthalpy changes of formation (Figure 2.16).

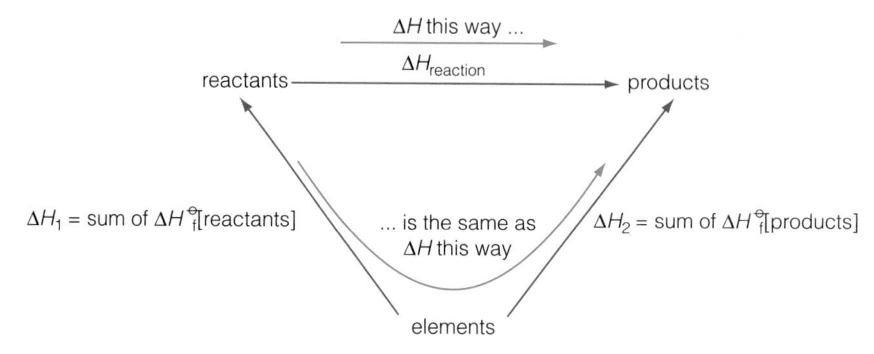

Figure 2.16 ▶ An energy cycle for calculating standard enthalpies of reaction from standard enthalpies of formation

According to Hess's Law:

$$\Delta H_{reaction} = -\Delta H_1 + \Delta H_2 = \Delta H_2 - \Delta H_1$$

Notice the negative sign in front of ΔH_1. This is because the direction of ΔH_1 for the equation is opposite to the direction of ΔH_1 in Figure 2.16.

So $\Delta H_{reaction}$ = sum of $\Delta H_f^{\ominus}$[products] − sum of $\Delta H_f^{\ominus}$[reactants]

Worked example

Calculate the enthalpy change for the reduction of iron(III) oxide by carbon monoxide.

$\Delta H_f^{\ominus}[Fe_2O_3] = -824\,kJ\,mol^{-1}$
$\Delta H_f^{\ominus}[CO] = -110\,kJ\,mol^{-1}$
$\Delta H_f^{\ominus}[CO_2] = -393\,kJ\,mol^{-1}$

Notes on the method

Write the balanced equation for the reaction and then draw an energy cycle (Hess cycle) using the model in Figure 2.16.

Remember that, by definition, $\Delta H_f^{\ominus}[element] = 0\,kJ\,mol^{-1}$.

Pay careful attention to the signs. Put the value and sign for a quantity in brackets when adding or subtracting enthalpy values.

Answer

$$Fe_2O_3(s) + 3CO(g) \rightarrow 2Fe(s) + 3CO_2(g)$$

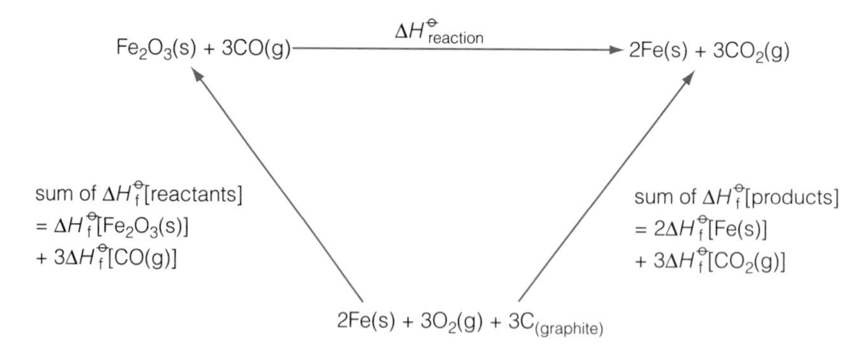

Figure 2.17 ▶ An energy cycle (Hess cycle) for calculating the enthalpy change of reaction between iron(III) oxide and carbon monoxide

Applying Hess's Law to Figure 2.17:

$\Delta H^{\ominus}_{reaction}$ = − sum of $\Delta H_f^{\ominus}$[reactants] + sum of $\Delta H_f^{\ominus}$[products]

= sum of $\Delta H_f^{\ominus}$[products] − sum of $\Delta H_f^{\ominus}$[reactants]

www Tutorial

$\therefore \Delta H^{\ominus}_{reaction} = 2 \times \Delta H_f^{\ominus}[Fe] + 3 \times \Delta H_f^{\ominus}[CO_2] - \Delta H_f^{\ominus}[Fe_2O_3] - 3 \times \Delta H_f^{\ominus}[CO]$

$= 0 + (3 \times -393\,kJ\,mol^{-1}) - (-824\,kJ\,mol^{-1}) - (3 \times -110\,kJ\,mol^{-1})$

$= -1179 + 824 + 330$

$\Delta H^{\ominus}_{reaction} = -25\,kJ\,mol^{-1}$

2.6 Enthalpy changes and the direction of change

Strike a match and it catches fire and burns. Put a spark to petrol and it burns furiously. These are two exothermic reactions which, once started, tend to 'go'. They are examples of the many exothermic reactions which just keep going once they have started. In general, chemists expect that a reaction will go if it is exothermic.

What this means is that reactions which give out energy to their surroundings are the ones which happen. This ties in with the common experience that change happens in the direction in which energy is spread around and dissipated in the surroundings. So the sign of ΔH is a guide to the likely direction of change, but it is not a totally reliable guide for three main reasons.

- The direction of change may depend on the conditions of temperature and pressure. One example is the condensation of a vapour such as steam. Steam condenses to water below 100 °C and heat is given out. This is an exothermic change.

  $$H_2O(g) \rightarrow H_2O(l) \quad \Delta H = -44\,kJ\,mol^{-1}$$

  At temperatures above 100 °C, the change goes in the opposite direction and this process is endothermic.
- There are some examples of endothermic reactions which occur readily under normal conditions. So some reactions for which ΔH is positive can happen. One example of this is the reaction of citric acid solution with sodium hydrogencarbonate. The mixture fizzes vigorously and cools rapidly.
- Some exothermic reactions never occur because the rate of reaction is so slow and the mixture of reactants is effectively inert. For example, the change from diamond to graphite is exothermic, but diamonds don't suddenly turn into black flakes.

Test yourself

10 The standard enthalpy change of formation of sucrose (sugar), $C_{12}H_{22}O_{11}$, is $-2226\,kJ\,mol^{-1}$. Write the balanced equation for which the standard enthalpy change of reaction is $-2226\,kJ\,mol^{-1}$.

11 When calculating standard enthalpy changes for reactions involving water, why is it important to specify that the H_2O is present as water and not as steam?

12 Calculate the standard enthalpy of reaction of hydrazine, $N_2H_4(l)$ with oxygen, O_2, to form nitrogen, N_2, and water, H_2O.

$\Delta H_f^{\ominus}[N_2H_4(l)] = +51\,kJ\,mol^{-1}$

$\Delta H_f^{\ominus}[H_2O(l)] = -286\,kJ\,mol^{-1}$

2.7 Enthalpy changes and bonding

During reactions, the bonds in reactants break and then new bonds form in the products. For example, when hydrogen reacts with oxygen:

$$2H_2(g) + O_2(g) \rightarrow 2H_2O(g)$$

Bonds in the H_2 and O_2 molecules first break to form H and O atoms (Figure 2.18). New bonds then form between the H and O atoms to produce water, H_2O.

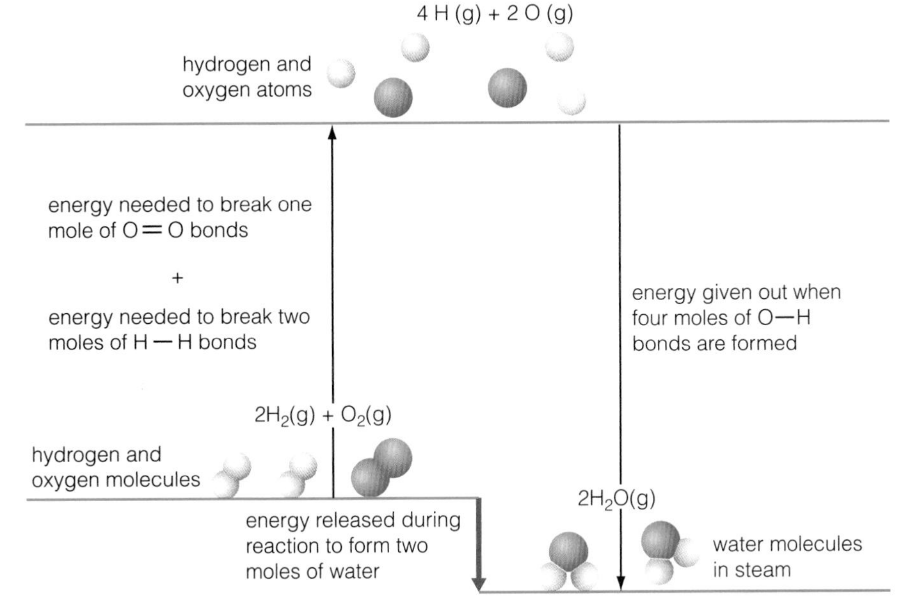

Figure 2.18 ◀ An energy level diagram for the reaction between hydrogen and oxygen

Note

Bond breaking is endothermic.
Bond making is exothermic.

Obviously, energy is needed to break the bonds between atoms. So, energy must be released when the reverse occurs and a bond forms.

Chemical reactions involve bond breaking followed by bond making. This means that the enthalpy change of a reaction is the energy difference between bond-breaking and bond-making processes.

When hydrogen and oxygen react, more energy is released in making four new O–H bonds in the two H_2O molecules than in breaking the bonds in two H_2 molecules and one O_2 molecule, so the overall reaction is exothermic (Figure 2.18).

A definite quantity of energy known as the bond enthalpy or bond energy can be associated with each type of bond. This energy is absorbed when the bond is broken and given out when the bond is formed. In measuring and using bond enthalpies, chemists distinguish between the terms bond enthalpy and mean (average) bond enthalpy.

Bond enthalpies are precise values for specific bonds in compounds (e.g. for the C–Cl bond in CH_3Cl).

Mean bond enthalpies are average values for one kind of bond in different compounds (e.g. an average value for the C–Cl bond in all compounds). Mean bond enthalpies take into account the fact that:

Definition

The **bond enthalpy** of a particular bond is the energy required to break one mole of the bonds in a substance in the gaseous state.

- the bond enthalpy for a specific covalent bond varies slightly from one compound to another (i.e. the O–H bond will have a slightly different bond enthalpy in H_2O and C_2H_5OH)
- successive bond enthalpies are not the same in compounds such as water and methane. (The energy needed to break the first O–H bond in H–O–H(g) is 498 kJ mol^{-1}, but the energy needed to break the second O–H bond in OH(g) is 428 kJ mol^{-1}.)

The mean bond enthalpies for various bonds are provided on a data sheet on your CD-ROM. The symbol for bond enthalpy is E, so the C–H bond enthalpy is written as E(C–H) = 413 kJ mol^{-1}.

Data

Using bond enthalpies

The most important use of mean bond enthalpies is in estimating the enthalpy changes in chemical reactions involving molecular substances with covalent bonds. These estimates are particularly helpful when experimental measurements cannot be made, as in the following worked example.

Worked example

Use mean bond enthalpies to estimate the enthalpy of formation of hydrazine, N_2H_4.

Note on the method

Write out the equation for the reaction showing all the atoms and bonds in the molecules. This makes it easier to count the number of bonds broken and formed.

Answer

N≡N + 2H—H ⟶ H₂N—NH₂ (structural formula: each N bonded to the other N and to two H atoms)

Bonds broken	*kJ mol^{-1}*	*Bonds formed*	*kJ mol^{-1}*
1 N≡N	+945	1 N–N	−158
2 H–H	+(2 × 436)	4 N–H	−(4 × 391)

$$\therefore \Delta H_{reaction} = +945 + (2 \times 436) - 158 - (4 \times 391)$$
$$= +1817 - 1722 = +95\ \text{kJ mol}^{-1}$$

For many reactions, the values of ΔH estimated from mean bond enthalpies agree closely with experimental values. However, there are limitations to the use of bond energy data in this way, and significant differences between the values of ΔH estimated from bond enthalpies and those obtained by experiment do occur. These differences usually arise:

- either from variations in the strength of one kind of bond in different molecules and mean bond enthalpies should not be used,
- or when one of the reactants or products is not in the gaseous state as bond enthalpy calculations assume.

Data

Test yourself

13 a) Look back at Figure 2.18 and write out the equation $2H_2(g) + O_2(g) \rightarrow 2H_2O(g)$ showing all the bonds between atoms in the molecules.

b) Look up the mean bond enthalpies for the bonds involved and calculate

- **i)** the energy needed to break one mole of O=O bonds plus two moles of H–H bonds.
- **ii)** the energy given out when four moles of O–H bonds are formed in two moles of water (steam) molecules.
- **iii)** the energy released during the reaction to form two moles of water (steam).

14 Look up the bond enthalpies for the H–H, Cl–Cl and H–Cl bonds.

a) Calculate the overall enthalpy change for the reaction

$$H_2(g) + Cl_2(g) \rightarrow 2HCl(g)$$

b) Draw an energy level diagram for the reaction (similar to Figure 2.18).

15 a) Calculate the average of the successive bond enthalpies for the two O–H bonds in water mentioned in Section 2.7.

b) Compare your answer with the mean bond enthalpy of the O–H bond given in the data section of your CD-ROM.

16 a) Make a table to show the mean bond enthalpies and bond lengths of the C–C, C=C and C≡C bonds.

b) What generalisations can you make based on your table?

17 Use mean bond enthalpies to estimate the enthalpy change when ethene, $H_2C{=}CH_2(g)$, reacts with $H_2(g)$ to form ethane, $CH_3{-}CH_3(g)$.

18 a) Which are likely to give a more accurate answer to a calculation of the enthalpy change for a reaction – mean bond enthalpies or enthalpies of formation?

b) Give a reason for your answer to part **a)**.

19 Look carefully at the mean bond enthalpies for hydrogen and the halogens (chlorine, bromine and iodine).

a) Write an equation for the reaction of hydrogen with chlorine.

b) Explain which bond (H–H or Cl–Cl) you think will break first in the reaction.

c) The mean bond enthalpy for fluorine, E(F–F), is 158 kJ mol^{-1}. How would you expect the reaction of fluorine with hydrogen to compare with the reaction of chlorine with hydrogen?

REVIEW QUESTIONS

1 When Epsom salts ($MgSO_4.7H_2O(s)$) are heated strongly, they decompose forming the anhydrous salt and water.

a) Write a balanced equation with state symbols for the decomposition of Epsom salts. (2)

b) Is the decomposition likely to be exothermic or endothermic? Explain your answer. (2)

c) Why can the enthalpy change for the decomposition of Epsom salts not be measured directly? (1)

d) Using the following values, calculate the standard enthalpy change for the decomposition. (5)

$\Delta H_f^{\ominus}[MgSO_4.7H_2O(s)] = -3389\,kJ\,mol^{-1}$

$\Delta H_f^{\ominus}[MgSO_4(s)] = -1285\,kJ\,mol^{-1}$

$\Delta H_f^{\ominus}[H_2O(l)] = -286\,kJ\,mol^{-1}$

2 Butane (Camping Gaz®), C_4H_{10}, burns readily on a camp cooker. The equation for the reaction is:

$C_4H_{10}(g) + 6\frac{1}{2}O_2(g) \rightarrow 4CO_2(g) + 5H_2O(g)$

a) Rewrite the equation showing all the covalent bonds between atoms in the reactants and products. (4)

b) Make a table showing the bonds broken in the reactants and the bonds formed in the products during the reaction. (2)

c) Use the following mean bond enthalpies to calculate the enthalpy change of the reaction. (4)

$E(C–C) = 347\,kJ\,mol^{-1}$ $E(C–H) = 413\,kJ\,mol^{-1}$

$E(O=O) = 498\,kJ\,mol^{-1}$ $E(C=O) = 740\,kJ\,mol^{-1}$

$E(H–O) = 464\,kJ\,mol^{-1}$

3 An excess of solid sodium hydrogencarbonate was added to 50 cm^3 of 1.0 $mol\,dm^{-3}$ ethanoic acid in an insulated polystyrene container under standard conditions. The temperature fell by 8.0 °C.

a) Complete the following equation for the reaction.

$CH_3COOH(aq) + NaHCO_3(s) \rightarrow$ ____ + ____ + ____ (3)

b) Why do you think the $NaHCO_3$ was added in small portions? (1)

c) Calculate the energy change during the reaction. (Assume that the specific heat capacity of the solution is 4.2 $J\,g^{-1}\,K^{-1}$ and its density is 1.0 $g\,cm^{-3}$. Ignore the mass of sodium hydrogencarbonate.) (2)

d) How many moles of ethanoic acid were used? (1)

e) Calculate the standard enthalpy change of the reaction. Show the correct sign and units. (3)

f) Explain the number of significant figures in your answer. (1)

4 A student suggested that ethane might react with bromine in two different ways in bright sunlight.

Reaction 1: $C_2H_6(g) + Br_2(l) \rightarrow C_2H_5Br(l) + HBr(g)$

Reaction 2: $C_2H_6(g) + Br_2(l) \rightarrow 2CH_3Br(g)$

a) Use the following mean bond enthalpies to calculate the enthalpy changes for the two possible reactions.

$E(C–C) = 347\,kJ\,mol^{-1}$ $E(Br–Br) = 193\,kJ\,mol^{-1}$

$E(C–H = 413\,kJ\,mol^{-1}$ $E(H–Br) = 366\,kJ\,mol^{-1}$

$E(C–Br) = 290\,kJ\,mol^{-1}$ (8)

b) Use your calculation to explain which of the reactions is more likely to occur. (2)

c) Suggest two reasons why your calculated enthalpy changes may not agree with the accurately determined experimental values. (2)

5 Tin is manufactured by heating tinstone, SnO_2, at high temperatures with coke (carbon).

There are two possible reactions for the process.

Reaction 1: $SnO_2(s) + C(s) \rightarrow Sn(s) + CO_2(g)$

Reaction 2: $SnO_2(s) + 2C(s) \rightarrow Sn(s) + 2CO(g)$

a) Calculate the standard enthalpy change for each of the possible reactions using the data below.

$\Delta H_f^{\ominus}[SnO_2(s)] = -581\,kJ\,mol^{-1}$

$\Delta H_f^{\ominus}[CO_2(g)] = -394\,kJ\,mol^{-1}$

$\Delta H_f^{\ominus}[CO(g)] = -110\,kJ\,mol^{-1}$ (6)

b) Use your calculations to explain which of the reactions would be most economic for industry. (2)

6 a) Draw a diagram of the apparatus that you could use to determine the enthalpy change of the reaction between powdered magnesium and excess copper(II) sulfate solution. (3)

b) When excess powdered magnesium was added to 50 cm^3 of 0.04 $mol\,dm^{-3}$ copper(II) sulfate solution, the temperature rose by 5.0 °C.

i) Write a balanced equation with state symbols for the reaction. (1)

ii) Calculate the energy transferred to the copper(II) sulfate solution. (2)

iii) What assumptions have you made in your calculation in part ii)? (2)

iv) Calculate the enthalpy change for the reaction shown in your equation in part i). (3)

3 Atomic and electronic structure

Individual atoms are far too small to be weighed, but their masses can be compared using a mass spectrometer. These measurements from mass spectrometry have not only helped scientists to understand atomic structure and the existence of isotopes, they have also been used in areas as diverse as space research, environmental monitoring and the detection of illegal drug use in sport.

Measurements using modified mass spectrometers can also be used to determine ionisation energies, which have provided evidence for the electronic structures of atoms. The chemistry of every element is dictated by its electronic structure – especially the electrons in its outer shell.

3.1 Atomic number and mass number

Atoms are incredibly small. In spite of this, scientists have been exploring inside atoms since the last few years of the nineteenth century. In the first half of the twentieth century, scientists found that:

- all atoms are made up from three even smaller, sub-atomic particles – protons, neutrons and electrons
- atoms have a central nucleus containing protons and neutrons, surrounded by a much larger space in which electrons are arranged in layers or shells
- the number of protons in an atom is equal to the number of its electrons
- the mass of a proton and the mass of a neutron are almost the same. The mass of an electron is 1840 times smaller than that of a proton.
- protons have a positive charge, neutrons have no charge and electrons have a negative charge. The charge on one proton is opposite in sign, but equal in size to the charge on one electron.

Table 3.1 summarises this information about protons, neutrons and electrons.

Figure 3.1 ▲
Photo of the surface of a gold crystal taken through an electron microscope. Each yellow blob is a separate gold atom – the atoms have been magnified about 40 million times.

Particle	Mass relative to that of a proton	Charge relative to that on a proton	Position in the atom
Proton	1	+1	Nucleus
Neutron	1	0	Nucleus
Electron	$\frac{1}{1840}$	−1	Shells

Table 3.1 ▲
Relative masses, relative charges and positions in atoms of protons, neutrons and electrons

All the atoms of a particular element have the same number of protons, and atoms of different elements have different numbers of protons.

Hydrogen atoms are the simplest of all atoms – they have just one proton and one electron. The next simplest are atoms of helium with two protons and two electrons, then lithium with three protons, and so on. Large atoms have large numbers of protons and electrons. For example, gold atoms have 79 protons and 79 electrons.

The only atoms with one proton are those of hydrogen; the only atoms with two protons are those of helium; the only atoms with three protons are those of lithium, and so on. This means that the number of protons in an atom decides which element it is. Because of this, scientists have a special name for

Definitions

The **atomic number** of an atom is the number of protons in its nucleus.

The term 'proton number' is sometimes used for atomic number.

The **mass number** of an atom is the number of protons plus neutrons in its nucleus.

Protons and neutrons are sometimes called nucleons, so the term 'nucleon number' is an alternative to mass number.

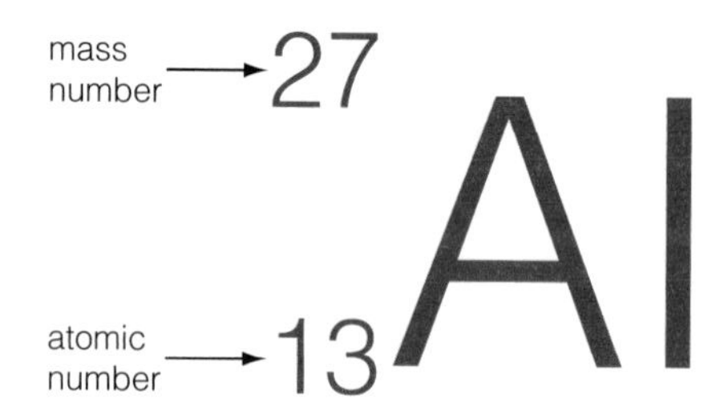

Figure 3.2 ▲
The mass number and atomic number can be shown with the symbol of an atom.

the number of protons in the nucleus of an atom. They call it the atomic number and use the symbol Z to represent it. So, hydrogen has an atomic number of 1 ($Z = 1$), helium has an atomic number of 2 ($Z = 2$), and so on.

Protons do not account for all the mass of an atom – neutrons in the nucleus also contribute. Therefore, the mass of an atom depends on the number of protons plus neutrons. This number is called the mass number of the atom (symbol A).

Hydrogen atoms, with one proton and no neutrons, have a mass number of 1 ($A = 1$). Lithium atoms, with 3 protons and 4 neutrons, have a mass number of 7 ($A = 7$) and aluminium atoms, with 13 protons and 14 neutrons, have a mass number of 27 ($A = 27$).

There is an agreed shorthand for showing the mass number and atomic number of an atom. This is shown for an aluminium atom, $^{27}_{13}Al$, in Figure 3.2. We can also represent ions using this shorthand – for example, the aluminium ion can be written as $^{27}_{13}Al^{3+}$.

Test yourself

1 How many protons, neutrons and electrons are there in the following atoms and ions:
   a) $^{9}_{4}Be$
   b) $^{39}_{19}K$
   c) $^{235}_{92}U$
   d) $^{19}_{9}F^{-}$
   e) $^{40}_{20}Ca^{2+}$?

2 Write symbols showing the mass number and atomic number for these atoms and ions:
   a) an atom of oxygen with 8 protons, 8 neutrons and 8 electrons
   b) an atom of argon with 18 protons, 22 neutrons and 18 electrons
   c) an ion of sodium with a 1+ charge and a nucleus of 11 protons and 12 neutrons
   d) an ion of sulfur with a 2− charge and a nucleus with 16 protons and 16 neutrons.

3.2 Comparing the masses of atoms – mass spectrometry

Individual atoms are far too small to be weighed, but in 1919 F.W. Aston invented the mass spectrometer. This gave chemists an accurate method of comparing the relative masses of atoms and molecules. The relative masses of atoms are called relative atomic masses and the relative masses of molecules are called relative molecular masses (Section 1.5). On the relative atomic mass scale, the masses of atoms and molecules are measured relative to one twelfth the mass of a carbon-12 atom, $^{12}_{6}C$.

A mass spectrometer separates atoms and molecules according to their mass, and also shows the relative numbers of the different atoms and molecules present. Figure 3.3 shows a diagram of a mass spectrometer.

Before atoms can be separated and detected in a mass spectrometer, they must be converted to positive ions in the gaseous or vapour state.

Inside a mass spectrometer there is a high vacuum. This allows ionised atoms and molecules from the chemical being tested to be studied without interference from atoms and molecules in the air.

There are five main stages in a mass spectrometer.

- **Vaporisation** – inject some of the chemical to be tested into the instrument. The high vacuum causes the sample to vaporise if it is not already a gas.

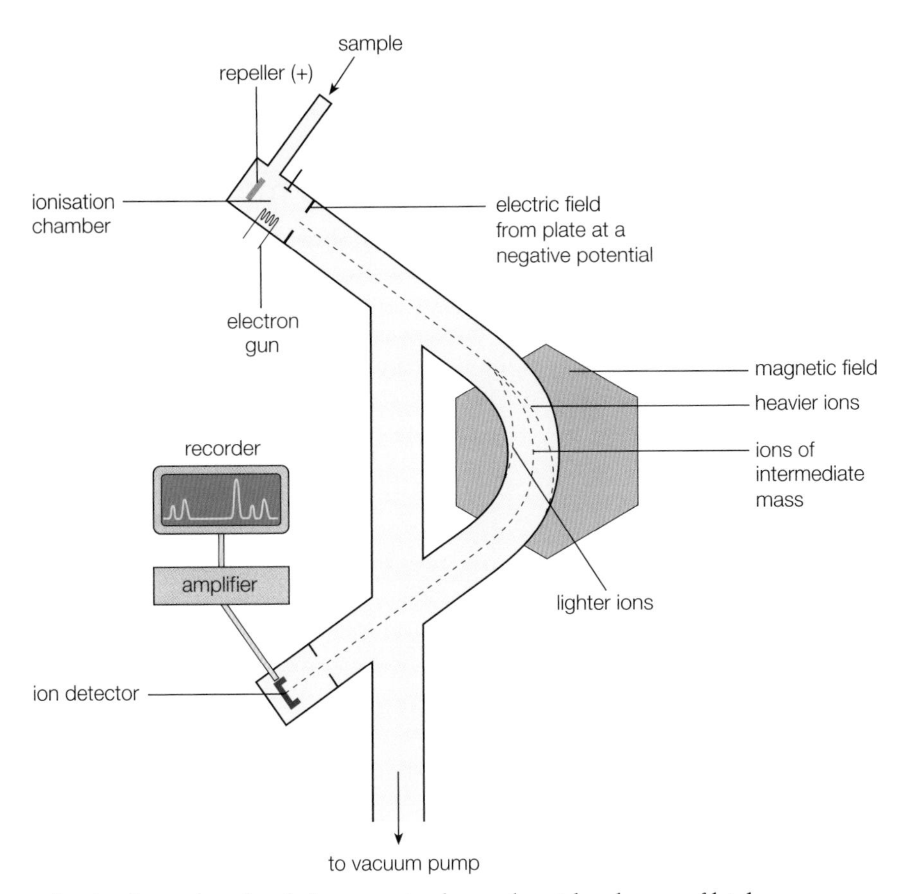

Figure 3.3 ◀
Schematic diagram of a mass spectrometer

- **Ionisation** – bombard the vaporised sample with a beam of high-energy electrons. These knock electrons off atoms or molecules in the sample, forming positive ions:

e^-	+	X	→	X^+	+	e^-	+	e^-
high-energy electron		atom in sample		positive ion		electron knocked out of X		high-energy electron retreating

- **Acceleration** – accelerate the positive ions into the instrument by an electric field.
- **Deflection** – deflect the moving positive ions with a magnetic field. Lighter ions are deflected more than heavier ions with the same charge.
- **Detection** – detect the charged ions as a tiny current and feed the signal to a computer. By increasing the strength of the magnetic field, ions of increasing mass are detected and a mass spectrum similar to that shown in Figure 3.4 can be printed out. By first using a reference compound with a known structure and relative molecular mass, the computer can print a scale on the mass spectrum.

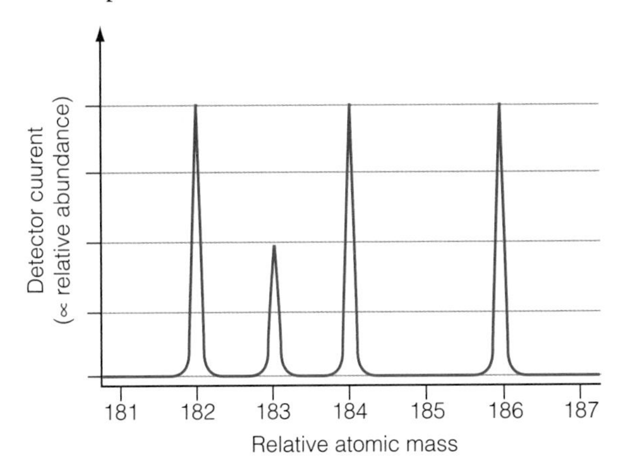

Figure 3.4 ◀
A mass spectrum of the element tungsten

Test yourself

3 Look carefully at Figure 3.4.
   a) How many different ions are detected in the mass spectrum of tungsten?
   b) What are the relative masses of these different ions?
   c) What are the relative proportions of these different ions?

Each of the four peaks on the mass spectrum of tungsten in Figure 3.4 represents a tungsten ion of different mass, and the heights of the peaks give the proportions of the ions present.

3.3 Isotopes and relative isotopic masses

Mass spectrometer traces, like that in Figure 3.4, show that tungsten and most other elements contain atoms that are not exactly alike. When atoms of these elements are ionised in a mass spectrometer, the beam of ions separates into two or more paths producing two or more peaks on the mass spectrum. This shows that the atoms from which the ions formed must have different masses. These atoms of the same element with different masses are called isotopes.

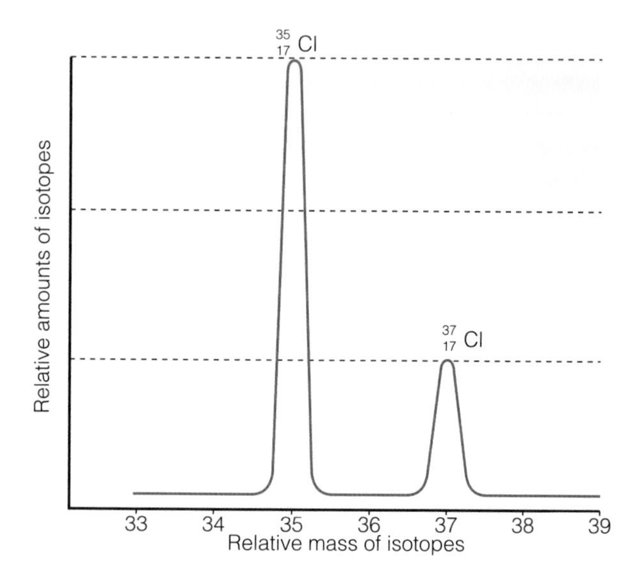

Figure 3.5 ▶
A mass spectrum for chlorine

Look closely at Figure 3.5 – it shows a mass spectrometer print out (mass spectrum) for chlorine. From the two peaks, you can deduce that chlorine consists of two isotopes with relative masses of 35 and 37. These relative masses are best described as relative isotopic masses because they give the relative mass of particular isotopes, relative to atoms of carbon-12. Notice that the relative masses of pure isotopes are called relative *isotopic* masses, whereas the relative masses of the atoms in an element (often containing a mixture of isotopes) are called relative *atomic* masses.

The heights of the two peaks in Figure 3.5 show that the relative proportions of chlorine-35 to chlorine-37 are 3 : 1. So, for every 4 chlorine atoms, 3 are chlorine-35 and 1 is chlorine-37.

The isotope chlorine-35 has a mass number of 35 with 17 protons and 18 neutrons, whereas chlorine-37 has a mass number of 37 with 17 protons and 20 neutrons. Table 3.2 summarises the important similarities and differences in isotopes.

Definitions

Isotopes are atoms of the same element with the same atomic number, but different mass numbers.

Relative isotopic mass is the mass of one atom of an isotope relative to the mass of one atom of carbon-12, for which the relative mass is defined as exactly 12.

Isotopes have the same	Isotopes have different
• number of protons • number of electrons • atomic number • chemical properties	• numbers of neutrons • mass numbers • physical properties

Table 3.2 ▲
Similarities and differences in isotopes

Relative atomic masses and relative molecular masses

The accurate relative atomic masses of most elements in tables of data are not whole numbers. This is because these elements contain a mixture of isotopes.

For example, chlorine contains two isotopes, chlorine-35 and chlorine-37, in the relative proportions of 3 : 1. On average there are 3 atoms of chlorine-35 to every 1 atom of chlorine-37. This is ¾, or 75%, chlorine-35 and ¼, or 25%, chlorine-37.

So, the average mass of a chlorine atom on the ^{12}C scale, which is the relative atomic mass of chlorine, is given by:

$$\frac{(3 \times 35) + (1 \times 37)}{4} = 35.5$$

Mass spectrometers can also be used to study molecules. After injecting a sample into the instrument and vaporising it, bombarding electrons ionise the molecules and also break them into fragments. Because of the high vacuum inside the mass spectrometer, it is possible to study these molecular fragments and ions which do not normally exist.

The peak with the highest mass in the spectrum is produced by ionising molecules without breaking them into smaller pieces. So, the mass of these molecular ions, M^+, is the relative molecular mass of the compound.

e^-	+	M(g)	$\longrightarrow$	M^+	+	$e^- + e^-$
high-energy electron		molecule in sample		molecular ion		electrons

Test yourself

4 Look again at Figure 3.4.
   a) What are the relative proportions of the different isotopes in tungsten?
   b) What is the relative atomic mass of tungsten?

5 Magnesium contains three naturally occurring isotopes, $^{24}_{12}Mg$, $^{25}_{12}Mg$ and $^{26}_{12}Mg$.
   a) How many protons and neutrons are present in the nuclei of each of these isotopes?
   b) What is the relative atomic mass of a sample of magnesium containing 80% $^{24}_{12}Mg$, 10% $^{25}_{12}Mg$ and 10% $^{26}_{12}Mg$?

6 Neon has two isotopes with mass numbers of 20 and 22.
   a) How do you think the boiling temperature of neon-20 will compare with that of neon-22? Explain your answer.
   b) Neon in the air contains 90% neon-20 and 10% neon-22. What is the relative atomic mass of neon in the air?

7 Why do isotopes have the same chemical properties, but different physical properties?

8 Why do samples of naturally occurring uranium obtained from ores in different parts of the world have slightly different relative atomic masses?

9 Look carefully at Figure 3.6.
   a) What is the relative molecular mass of the hydrocarbon?
   b) The fragment of the hydrocarbon with relative mass 15 is a CH_3 group. What do you think the fragments are with relative masses of 29 and 43?
   c) Draw a possible structure for the hydrocarbon.

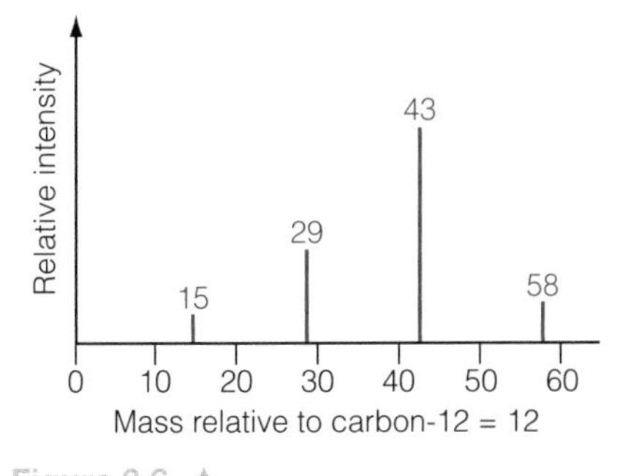

Figure 3.6 ▲ The simplified mass spectrum of a hydrocarbon and its fragments

Notice from your studies in this section that, by carefully interpreting the data from mass spectrometers, chemists can deduce:

- the isotopic composition of elements
- the relative atomic masses of elements
- the relative molecular masses of compounds.

Chemists who separate and synthesise new compounds can also identify the fragments in the mass spectra of these compounds. Then, by piecing the fragments together, they can identify possible structures for the new compounds.

The combination of gas chromatography and mass spectrometry is particularly important in modern chemical analysis. Chromatography is first used to separate the chemicals in an unknown mixture, such as polluted water or similar compounds synthesised for possible use as drugs. Then mass spectrometry is used to detect and identify the separated components.

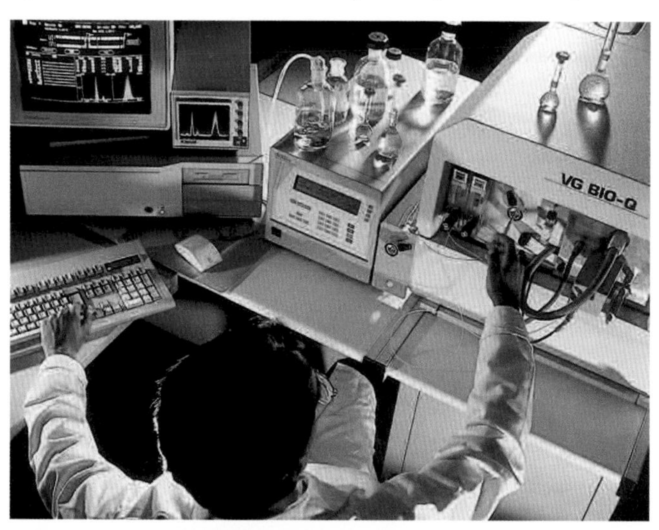

Figure 3.7 ▶ Injecting a sample into a mass spectrometer

Activity

Mass spectrometry in space research and sport

Mass spectrometry provides an incredibly sensitive method of analysis in areas such as space research, medical research, monitoring pollutants in the environment and the detection of illegal drugs in sport.

Space research

In December 1978, space probes landed on Venus for the first time. Because of weight limitations, the low resolution mass spectrometers on board could give measurements of relative masses to only one decimal place. A particle of relative molecular mass 64.1 was identified, but this could not show whether it was SO_2 or S_2.

1 Why could the mass spectrometers not show whether the particle on Venus was SO_2 or S_2? (S = 32.1, O = 16.0)

2 High-resolution mass spectrometers can measure relative masses to three or more decimal places (S = 32.072, O = 15.995). Explain how a high-resolution mass spectrometer could determine whether a particle was SO_2 or S_2.

Detecting the use of anabolic steroids in sport

Since the 1980s, unscrupulous sportsmen and sportswomen have tried to improve their performance by using anabolic steroids. These drugs increase muscle size and strength, which increases the chance of winning. But anabolic steroids also have serious harmful effects on the body. Women develop masculine features and anyone using them may suffer heart disease, liver cancer and depression leading to suicide.

Sporting bodies, such as the International Olympic Committee, have banned the use of anabolic steroids in all sports and have introduced a rigorous testing regime. The testing procedures involve analysis of urine samples using mass spectrometry. Great care is taken during sampling, transport, storage and analysis to ensure that the results of analysis will stand up in court.

Figure 3.9 shows the molecular ion and the largest fragments in the mass spectrum of a banned chemical that is thought to be dihydrocodeine ($C_{18}H_{23}O_3N$).

3 What is the probable relative molecular mass of the banned chemical on the mass spectrum?

4 Is the probable relative molecular mass consistent with that of dihydrocodeine, ($C_{18}H_{23}O_3N$)? Explain your answer.

5 What is the relative mass of the fragment *lost* from one molecule of the banned substance, leaving the fragment of relative mass 284?

6 Dihydrocodeine contains a CH_3O– group and an –OH group. What evidence does the mass spectrum provide for these two groups?

Figure 3.8 ▲
Ben Johnson won the men's 100 m race at the Olympic Games in 1992. Unfortunately, urine tests showed that he had used anabolic steroids – Johnson was stripped of his title and the gold medal.

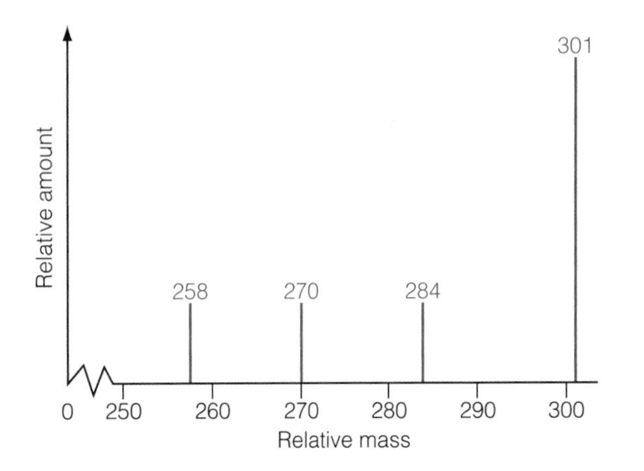

Figure 3.9 ▲
The molecular ion and the largest fragments in the mass spectrum of a banned chemical

3.4 Evidence for the electronic structure of atoms

In a mass spectrometer, a beam of electrons bombards the sample, turning atoms into positive ions. The electrons in the beam must have enough energy to knock electrons off atoms in the sample. By varying the intensity of the beam, it is possible to measure the minimum amount of energy needed to remove electrons from the atoms of an element. From these measurements, scientists can predict the electron structures of atoms.

The energy needed to remove one electron from each atom in a mole of gaseous atoms is known as the first ionisation energy and is given the symbol E_{m1}. The product is one mole of gaseous ions with one positive charge.

Definition

The first ionisation energy of an element is the energy needed to remove one electron from each atom in one mole of gaseous atoms.

So, the first ionisation energy of sodium is the energy required for the process

$$Na(g) \rightarrow Na^+(g) + e^- \quad E_{m1} = +496\,kJ\,mol^{-1}$$

Ionisation energies like this are always endothermic.

Scientists can also determine ionisation energies from the emission spectra of atoms. Using data from spectra, it is possible to measure the energy required to remove electrons from ions with increasing charges. A succession of ionisation energies is obtained, represented by the symbols E_{m1}, E_{m2}, E_{m3} and so on. For example:

$Na(g) \rightarrow Na^+(g) + e^-$ first ionisation energy, $E_{m1} = +\,496\,kJ\,mol^{-1}$

$Na^+(g) \rightarrow Na^{2+}(g) + e^-$ second ionisation energy, $E_{m2} = +\,4563\,kJ\,mol^{-1}$

$Na^{2+}(g) \rightarrow Na^{3+}(g) + e^-$ third ionisation energy, $E_{m3} = +\,6913\,kJ\,mol^{-1}$

There are 11 electrons in a sodium atom so there are 11 successive ionisation energies for this element.

The successive ionisation energies for an element get bigger and bigger. This is not surprising because having removed one electron it is more difficult to remove a second electron from the positive ion formed.

The graph in Figure 3.10 provides evidence to support the theory that the electrons in an atom are arranged in a series of levels or shells around the nucleus.

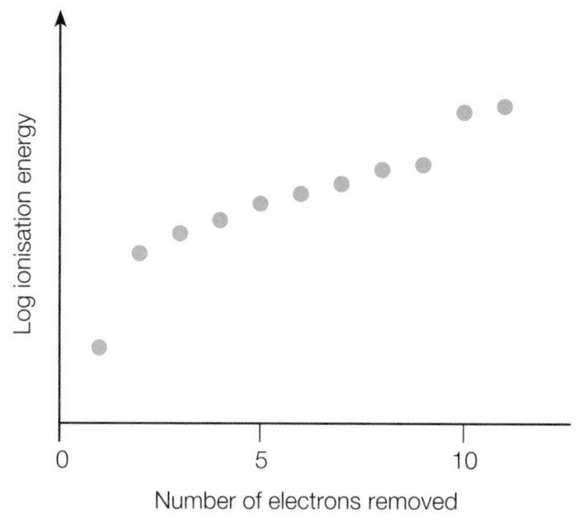

Figure 3.10 ▲
Log ionisation energy against the number of electrons removed for sodium

Note

The shells of electrons at fixed or specific levels are sometimes called *quantum shells*. The word 'quantum' is used to describe something related to a fixed amount or a fixed level.

Notice the big jumps in value between the first and second ionisation energies, and between the ninth and tenth ionisation energies. This suggests that sodium atoms have one electron in an outer shell or energy level furthest from the nucleus. This outer electron is easily removed because it is shielded from the full attraction of the positive nucleus by 10 inner electrons.

Below this outer single electron, sodium atoms appear to have eight electrons in a second shell, all at roughly the same energy level. These eight electrons are closer to the nucleus than the single outer electron.

Finally, sodium atoms have two inner electrons in a shell closest to the nucleus. These two electrons feel the full attraction of the positive nucleus and are hardest to remove with the most endothermic ionisation energies.

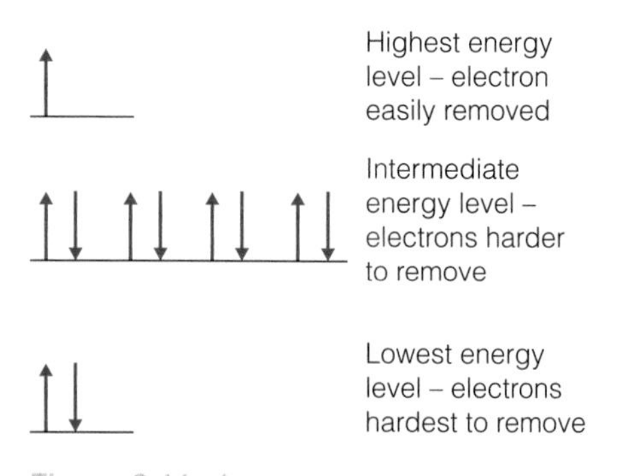

Figure 3.11 ▲
The energy levels of electrons in a sodium atom

This electronic structure for a sodium atom can be represented in an energy level diagram as in Figure 3.11. The electron arrangement in sodium is usually written as 2, 8, 1.

In energy level diagrams such as that in Figure 3.11, the electrons are represented by arrows. When an energy level is filled, the electrons are paired up and in each of these pairs the electrons are spinning in opposite directions. Chemists believe that paired electrons can only be stable when they spin in opposite directions so that the magnetic attraction resulting from their opposite spins can counteract the electrical repulsion from their negative charges.

In energy level diagrams such as Figure 3.11, the opposite spins of the paired electrons are shown by drawing the arrows in opposite directions.

The quantum shells of electrons correspond to the periods of elements in the periodic table. By noting where the first big jump comes in the successive ionisation energies of an element, it is possible to predict the group to which the element belongs. For example, the first big jump in the successive ionisation energies for sodium comes after the first electron is removed. This suggests that sodium has just one electron in its outermost shell and therefore it must be in group 1.

Test yourself

10 The successive ionisation energies of beryllium are 900, 1757, 14 849 and 21 007 kJ mol^{-1}.
  a) What is the atomic number of beryllium?
  b) Why do successive ionisation energies always get more endothermic?
  c) Draw an energy level diagram for the electrons in beryllium, and predict its electron structure.
  d) To which group in the periodic table does beryllium belong?

11 The first five ionisation energies of an element, in kJ mol^{-1}, are 578, 1817, 2745, 11 578, 14 831, 18 378.
  a) How many electrons are there in the outer shell of the atoms of this element?
  b) To which group of the periodic table does this element belong?

12 Sketch a graph of log(ionisation energy) against number of electrons removed when all the electrons are successively removed from a phosphorus atom.

Activity

Evidence for sub-shells of electrons

By studying the first ionisation energies of successive elements in the periodic table, we can compare how easy it is to remove an electron from the highest energy level in different atoms. This provides us with evidence for the arrangement of electrons in sub-shells.

Data

1 Using the data sheet headed 'The first ionisation energies of successive elements in the periodic table' on the Dynamic Learning Student website, plot a graph of the first ionisation energy for the first 20 elements in the periodic table. Put first ionisation energy on the vertical axis and atomic number on the horizontal axis.

2 When you have plotted the points, draw lines from one point to the next to show a pattern of peaks and troughs. Label each point with the symbol of its corresponding element.

3 **a)** Where do the alkali metals in group 1 appear in the pattern?
  b) Where do the noble gases in group 0 appear in the pattern?

4 What similarities do you notice in the pattern for elements in period 2 (lithium to neon) with that for elements in period 3 (sodium to argon)?

5 Identify three sub-groups of points in both period 2 and period 3. How many elements are there in each sub-group?

3.5 Understanding the pattern in ionisation energies

From the study of ionisation energies and spectra, we know that the electrons in atoms are grouped together in energy levels or quantum shells. The numbers 1, 2, 3, etc. are used to label these shells starting nearest to the nucleus.

Each quantum shell can hold only a limited number of electrons. If all the shells of an atom are full, the atom is very stable and it has a highly endothermic first ionisation energy.

The first shell ($n = 1$) nearest the nucleus is full and stable when it contains two electrons. This is the case for helium atoms, which are very stable and unreactive with a higher first ionisation energy than neighbouring elements in the periodic table (Figure 3.12).

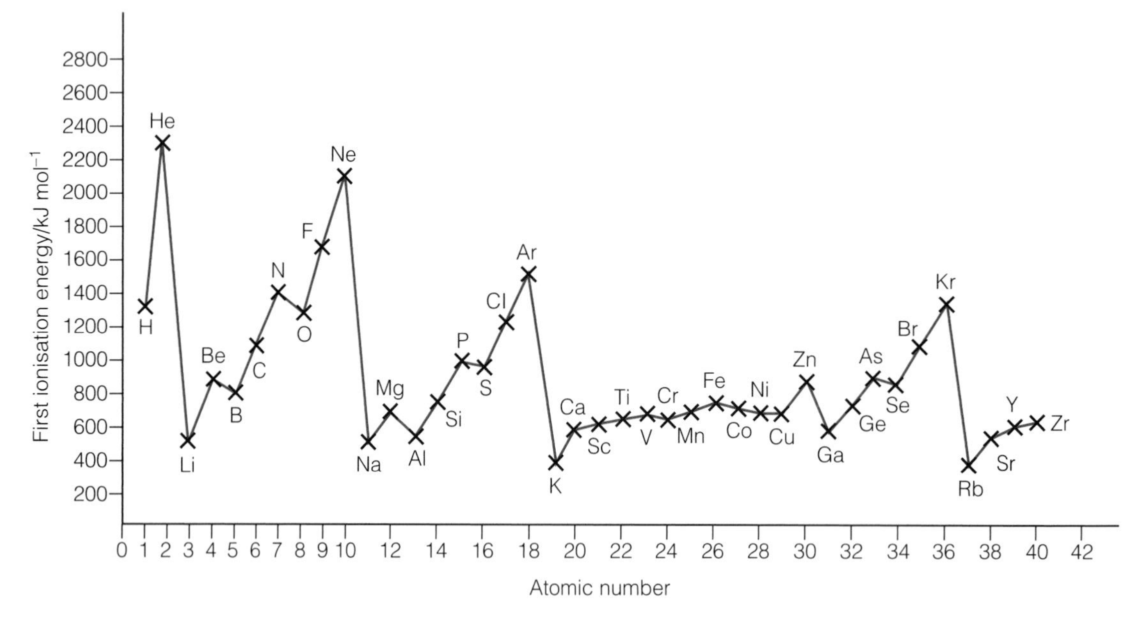

Figure 3.12 ▲
First ionisation energies of the elements plotted against atomic number

The second shell ($n = 2$) is full and stable when it contains eight electrons – this is the case for neon. Neon has a filled first shell with 2 electrons and a filled second shell with 8 electrons. So, its electronic structure is 2, 8 and neon has a higher first ionisation energy than its neighbours in the periodic table, like helium. The high first ionisation energies of these elements show that their electronic structures are very stable and explain why they are so unreactive.

The points between one noble gas and the next in Figure 3.12 can be divided into sub-sections. These sub-sections provide evidence for sub-shells of electrons. After both He and Ne in Figure 3.12, there are deep troughs followed by small intermediate peaks at Be and Mg. These are sub-sections with just two points.

Immediately after Be and Mg there are similar sub-sections of six points (B to Ne and Al to Ar), or two sub-sections of three points. Further along in Figure 3.12, there is a sub-section containing ten points from Sc to Zn.

The points on the graph between one noble gas and the next correspond to one shell of electrons, and the sub-sections of points correspond to sub-shells of electrons.

By studying ionisation energies in this way, chemists have deduced that:

- the $n = 1$ shell can hold 2 electrons in the same sub-shell
- the $n = 2$ shell can hold 8 electrons – 2 in one sub-shell and 6 in a slightly higher sub-shell
- the $n = 3$ shell can hold 18 electrons – 2 in one sub-shell, 6 in a sub-shell slightly higher and 10 electrons in a sub-shell slightly higher still
- the $n = 4$ shell can hold 32 electrons – with sub-shells containing 2, 6, 10 and 14 electrons.

Electronic structures

The sub-shells making up the main shells are given names:

- those that can hold up to 2 electrons are called s sub-shells
- those that can hold up to 6 electrons are called p sub-shells
- those that can hold up to 10 electrons are called d sub-shells
- those that can hold up to 14 electrons are called f sub-shells.

Period 1			H					He
Atomic no.			1					2
Electron shell structure			1					2
Electron sub-shell structure			$1s^1$					$1s^2$
Period 2	Li	Be	B	C	N	O	F	Ne
Atomic no.	3	4	5	6	7	8	9	10
Electron shell structure	2, 1	2, 2	2, 3	2, 4	2, 5	2, 6	2, 7	2, 8
Electron sub-shell structure	$1s^2$ $2s^1$	$1s^2$ $2s^2$	$1s^2$ $2s^22p^1$	$1s^2$ $2s^22p^2$	$1s^2$ $2s^22p^3$	$1s^2$ $2s^22p^4$	$1s^2$ $2s^22p^5$	$1s^2$ $2s^22p^6$
Period 3	Na	Mg	Al	Si	P	S	Cl	Ar
Atomic no.	11	12	13	14	15	16	17	18
Electron shell structure	2, 8, 1	2, 8, 2	2, 8, 3	2, 8, 4	2, 8, 5	2, 8, 6	2, 8, 7	2, 8, 8
Electron sub-shell structure	$1s^2$ $2s^22p^6$ $3s^1$	$1s^2$ $2s^22p^6$ $3s^2$	$1s^2$ $2s^22p^6$ $3s^23p^1$	$1s^2$ $2s^22p^6$ $3s^23p^2$	$1s^2$ $2s^22p^6$ $3s^23p^3$	$1s^2$ $2s^22p^6$ $3s^23p^4$	$1s^2$ $2s^22p^6$ $3s^23p^5$	$1s^2$ $2s^22p^6$ $3s^23p^6$
Period 4	K	Ca						
Atomic no.	19	20						
Electron shell structure	2, 8, 8, 1	2, 8, 8, 2						
Electron sub-shell structure	$1s^2$ $2s^22p^6$ $3s^23p^6$ $4s^1$	$1s^2$ $2s^22p^6$ $3s^23p^6$ $4s^2$						

Figure 3.13 ▲
The electron shell and sub-shell structure of the first 20 elements in the periodic table

The labels s, p, d and f are left over from the early studies of the spectra of different elements. These studies used the words 'sharp', 'principal', 'diffuse' and 'fundamental' to describe different lines in the spectra. The terms have no special significance now.

The electronic structure of an atom can be described in terms of the shells occupied by electrons. In terms of shells, the electron structure of lithium is 2, 1 and that of sodium is 2, 8, 1. However, it is possible to describe the electron structure of an atom more precisely in terms of sub-shells.

When sub-shells (energy sub-levels) are being filled, electrons always occupy the lowest available sub-shell first. Therefore, the order in which the first 6 sub-shells are filled with electrons is:

- 1s in the $n = 1$ shell
- 2s then 2p in the $n = 2$ shell
- 3s then 3p in the $n = 3$ shell
- 4s in the $n = 4$ shell

The electronic structure of hydrogen can therefore be written in sub-shell notation as $1s^1$, representing one electron in the 1s sub-shell. Following on, the electronic structure of helium is $1s^2$, then lithium is $1s^2\ 2s^1$, and so on.

The electron shell and sub-shell structures of the first 20 elements in the periodic table are shown in Figure 3.13.

3.6 Electrons and orbitals

Using complex mathematics, chemists have further developed their ideas about the arrangement of electrons in sub-shells. Knowing the masses and charges on protons and electrons, it is possible to calculate the probability of finding an electron at any point in an atom. These calculations have led chemists to believe that there is a high probability of finding an electron, or a pair of electrons, in certain regions in space around the nucleus of an atom. These regions are called orbitals.

> **Definition**
>
> An **orbital** is a region in space around the nucleus of an atom in which there is a 95% chance of finding an electron, or a pair of electrons with opposite spins.

By pinpointing the likely position of an electron at millions of nanosecond intervals, it is possible to build up a picture showing the electron 'smeared out' over its orbital as a negatively charged cloud. These smeared-out pictures are sometimes described as electron density plots, or electron density maps. The plots are darkest where the electrons are more likely to be and lighter where the electrons are less likely to be.

The overall shapes of orbitals are derived from electron density plots by showing the boundary of the region in which there is a 95% chance of finding an electron or a pair of electrons with opposite spins.

- s sub-shells contain one orbital – best described as a spherical ring, like extra thick peel on an orange (Figure 3.14)
- p sub-shells contain three orbitals, labelled p_x, p_y and p_z – these are dumb-bell shaped and arranged at right angles to each other (Figure 3.14).

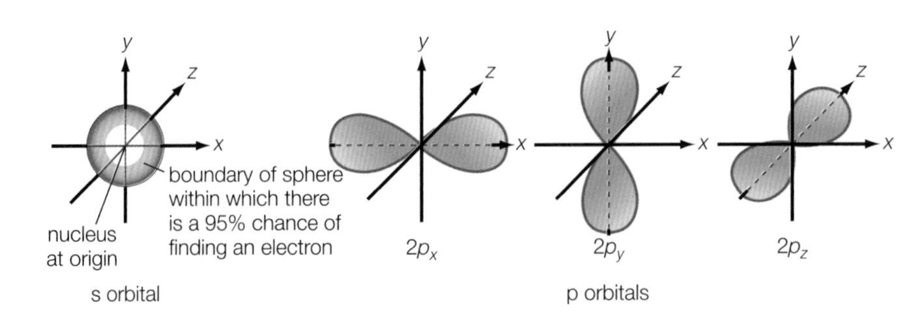

Figure 3.14 ▶ The shapes of s and p orbitals

From these ideas about orbitals, chemists have developed an 'electrons-in-boxes' representation for the electronic structures of atoms.

Using this method, each box represents an orbital and the electrons in an atom fill the orbitals according to a set of three rules:

- electrons go into the orbital with the lowest energy level first
- each orbital can hold just one, or at most two electrons with opposite spins
- when there are two or more orbitals at the same energy level, electrons occupy these orbitals singly before they pair up.

Chemists sometimes use the term 'Aufbau principle' to describe these rules – from the German word 'Aufbau' meaning 'building up'. This is a reminder that electronic structures build up from the bottom, starting with the lowest energy level. The Aufbau principle rationalises information about electronic structures into a pattern that can be used to predict the electron structures of all atoms.

Tutorial

Figure 3.15 shows the electrons-in-boxes representations and the s, p, d, f notations for the electronic structures of beryllium, nitrogen and sodium. These descriptions of the arrangement of electrons in the atoms of elements are called 'electron configurations'.

Definition

Electron configurations describe the number and arrangement of electrons in an atom. A shortened form of electron configuration uses the symbol of the previous noble gas, in square brackets, to stand for the inner shells. So, using this convention, the electron configuration of sodium is [Ne] $3s^1$.

Element	1s	2s	2p			3s	s, p, d, f electron notation
Beryllium	↑↓	↑↓					$1s^2 2s^2$
Nitrogen	↑↓	↑↓	↑	↑	↑		$1s^2 2s^2 2p^3$
Sodium	↑↓	↑↓	↑↓	↑↓	↑↓	↑	$1s^2 2s^2 2p^6 3s^1$

(Column heading: Electrons-in-boxes notation of electronic structure)

Figure 3.15 ▲
Electrons-in-boxes representations and s, p, d, f notations for the electronic structure of beryllium, nitrogen and sodium

The development of knowledge and understanding about electronic structures illustrates how chemists use the results of their experiments, such as the measurements of ionisation energies, to explain the structure and properties of materials. It also illustrates the important distinction between evidence and experimental data on the one hand, and ideas, theories and explanations on the other.

Ionisation energies and spectra have provided chemists with evidence and information that has caused them to develop and modify their models and theories about electron structure. Early ideas about electrons arranged in shells have been developed to take in the evidence for sub-shells, and then modified to include ideas about orbitals.

Test yourself

13 Write out the electron structure in terms of shells (for sodium this would be 2, 8, 1) for the following elements:
a) lithium
b) oxygen
c) neon
d) silicon.

14 Write the electronic sub-shell structure for the elements in question 13 – for sodium this would be $1s^2 2s^2 2p^6 3s^1$.

15 Draw the electrons-in-boxes representations for the following elements:
a) boron
b) fluorine
c) phosphorus
d) potassium.

16 Identify the elements with the following electron structures in their outermost shells:
a) $1s^2$
b) $2s^2 2p^2$
c) $3s^2$
d) $3s^2 3p^4$.

3.7 Electron structures and the periodic table

The periodic table helps chemists to bring order and patterns to the vast amount of information they have discovered about all the elements and their compounds.

In the modern periodic table, elements are arranged in order of atomic number. The horizontal rows in the table are called 'periods' – each period ends with a noble gas. The vertical columns in the table are called 'groups' which can be divided into four blocks – the s block, p block, d block and f block – based on the electron structures of the elements (Figure 3.16).

So, the modern arrangement of elements in the periodic table reflects the underlying electronic structures of the atoms, while the more sophisticated model of electron structure in terms of orbitals allows chemists to explain the

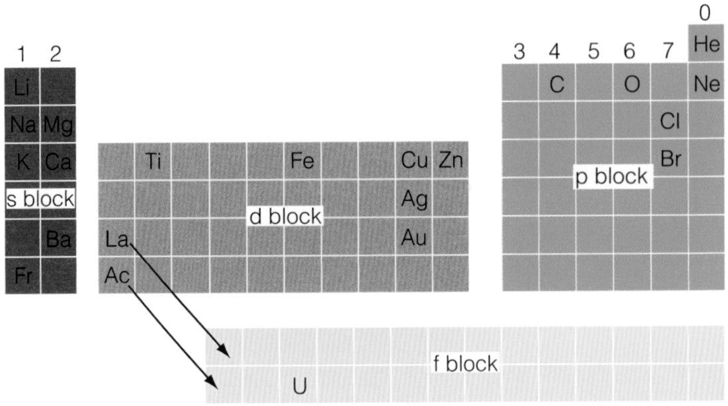

Figure 3.16 ▶
The s, p, d and f blocks in the periodic table

properties of elements more effectively. The four blocks in the periodic table are shown in different colours in Figure 3.16.

- The *s block* comprises the reactive metals in group 1 and group 2 – such as potassium, sodium, calcium and magnesium. In these metals, the last electron added goes into an s orbital in the outer shell.
- The *p block* comprises the elements in groups 3, 4, 5, 6, 7 and 0 on the right of the periodic table. These elements include relatively unreactive metals such as tin and lead plus all the non-metals. In these elements, the last electron added goes into a p orbital in the outer shell.
- The *d block* elements occupy a rectangle across periods 4, 5, 6 and 7 between group 2 and group 3. The d block elements are all metals – including titanium, iron, copper and silver – in which the last electron added goes into a d orbital. These metals are much less reactive than the s block metals in groups 1 and 2. Within the d block there are marked similarities across the periods, as well as the usual vertical similarities. The d block elements are sometimes loosely called 'transition metals'.
- The *f block* elements occupy a low rectangle across periods 6 and 7 within the d block, although they are usually placed below the main table to prevent it becoming too wide to fit the page. Like the d block elements, those in the f block are all metals. Here the last added electron is in an f orbital. The f block elements are often called the lanthanides and actinides because they are the 14 elements immediately following lanthanum, La, and actinium, Ac, in the periodic table. Another name used for the f block elements is the 'inner transition elements'.

In the periodic table, the elements in each group have similar properties because they have similar electron structures. This important point is well illustrated by the alkali metals in group 1.

Group 1
The alkali metals
Lithium Li 2, 1 ($1s^2 2s^1$)
Sodium Na 2, 8, 1 ($1s^2 2s^2 2p^6 3s^1$)
Potassium K 2, 8, 8, 1 ($1s^2 2s^2 2p^6 3s^2 3p^6 4s^1$)

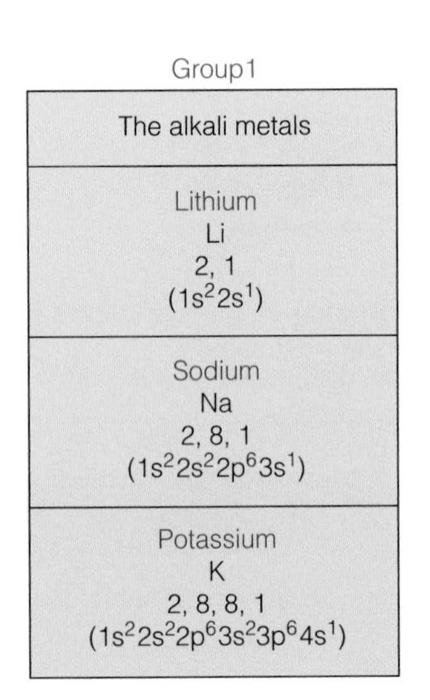

Figure 3.17 ▲
Electron structures of the first three alkali metals

Group 1: the alkali metals

Look at Figure 3.17 – notice that each alkali metal has one s electron in its outer shell. This similarity in their electron structures explains why they have similar properties.

Alkali metals:

- are very reactive because they lose their single outer electron so easily
- form ions with a charge of 1+ (Li^+, Na^+, K^+, etc.) so the formulae of their compounds are similar
- form very stable ions with an electron structure like that of a noble gas.

As the atomic numbers of the alkali metals increase, the outer electron is further from the positive nucleus. This means that the outer electron is held less strongly by the nucleus. So, the electron is lost more readily, and this explains why the alkali metals become more reactive as their atomic number increases.

The chemical properties of all other elements are also determined by their electronic structures. Chemistry is largely about the electrons in the outer shells of atoms. The reactivity of an element depends on the number of electrons in the outer shell and how strongly they are held by the nuclear charge. This is a fundamental feature of chemistry and one that we will return to again and again. It is also an essential principle which governs the way in which chemists think and work.

3.8 Periodic properties

Modern versions of the periodic table are all based on the one suggested by the Russian chemist Dmitri Mendeléev in 1869. When Mendeléev arranged the elements in order of atomic mass, he saw a repeating pattern in their properties. A repeating pattern is a periodic pattern – hence the terms 'periodic properties' and 'periodicity'.

Perhaps the most obvious repeating pattern in the periodic table is from metals on the left, through elements with intermediate properties (called metalloids), to non-metals on the right. Graphs of the physical properties of the elements – such as melting temperatures, electrical conductivities and first ionisation energies – against atomic number also show repeating patterns. Using the models of bonding between atoms and molecules, we can explain the properties of elements and the repeating patterns in the periodic table.

Test yourself

17 Why are sodium and potassium so alike?

18 Why are the noble gases so unreactive?

19 a) Write down the electron shell structures and sub-shell structures of fluorine and chlorine in group 7.

b) Why do you think fluorine and chlorine are so reactive with metals?

c) Why do the compounds of fluorine and chlorine with metals have similar formulae?

Note

The International Union of Pure and Applied Chemistry (IUPAC) now recommends that the groups in the periodic table should be numbered from 1 to 18. Groups 1 and 2 are the same as before. Groups 3 to 12 are the vertical families of d block elements. The groups traditionally numbered 3 to 7 and 0 then become groups 13 to 18.

Melting temperatures of the elements

Figure 3.18 shows the periodic pattern revealed by plotting the melting temperatures of elements against atomic number.

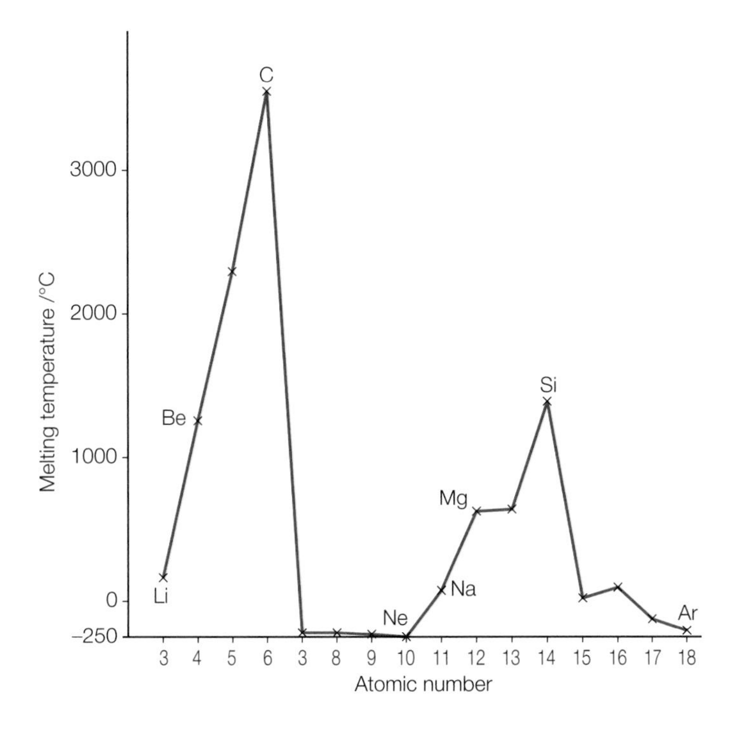

Figure 3.18 Periodicity in the melting temperatures of the elements

The melting temperature of an element depends on both its structure and the type of bonding between its atoms. In metals, the bonding between atoms is strong (Section 4.3), so their melting temperatures are usually high. The more electrons each atom contributes from its outermost shell to the shared delocalised electrons, the stronger the bonding and the higher the melting temperature.

Therefore, melting temperatures rise from group 1 to group 2 to group 3. In group 4, the elements carbon and silicon have giant covalent structures. The bonds in these structures are highly directional, so most of the bonds must break before the solid melts. This means that the melting temperatures of group 4 elements are very high and at the peaks of the graph in Figure 3.18.

The non-metal elements in groups 5, 6, 7 and 0 form simple molecules. The intermolecular forces between these simple molecules are weak, so these elements have low melting temperatures.

First ionisation energies of the elements

Figure 3.19 shows the clear periodic trend in the first ionisation energies of the elements. The general trend is that first ionisation energies increase from left to right across a period.

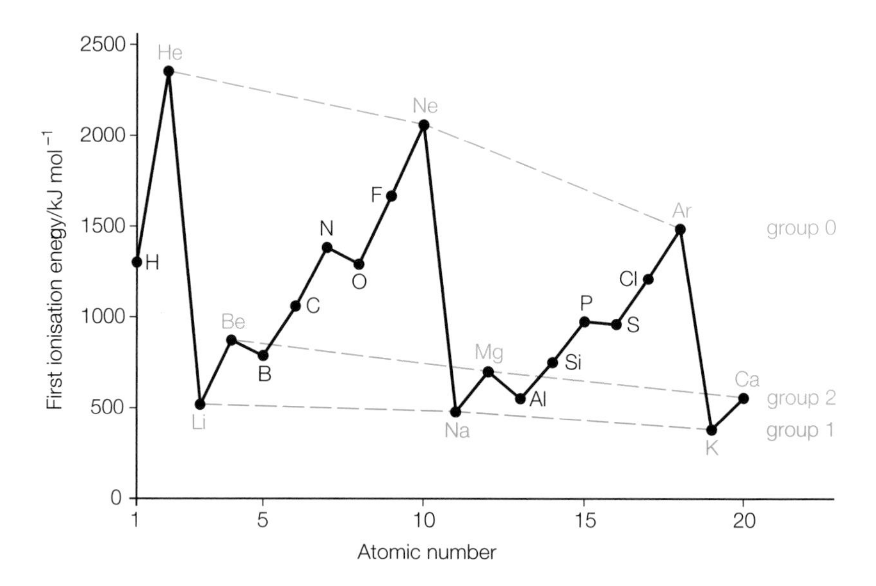

Figure 3.19 ▶ Periodicity in the first ionisation energies of the elements

The ionisation energy of an atom is determined by three atomic properties.

- *The distance of the outermost electron from the nucleus.* As this distance increases, the attraction of the positive nucleus for the negative electron decreases and this tends to reduce the ionisation energy.
- *The size of the positive nuclear charge.* As the positive nuclear charge increases, its attraction for outermost electrons increases and this tends to increase the ionisation energy.
- *The shielding effect of electrons.* Electrons in inner shells exert a repelling effect on electrons in the outer shells of an atom. This reduces the pull of the nucleus on the electrons in the outer shell. Thanks to shielding, the 'effective nuclear charge' attracting electrons in the outer shell is much less than the full positive charge of the nucleus. As expected, the shielding effect increases as the number of inner shells increases.

Moving from left to right across any period, the nuclear charge increases as electrons are added to the same outer shell. The increasing nuclear charge tends to pull the outer electrons closer to the nucleus, while the shielding effect of full inner shells is constant. However, the increased nuclear charge and the reduced distance of the outer electrons from the nucleus makes the outer electrons more difficult to remove and the first ionisation energy increases.

But notice in Figure 3.19 that the rising trend in ionisation energies across a period is not smooth. There is a 2-3-3 pattern, which reflects the way in which electrons feed into s and p orbitals (Section 3.6). Earlier we noted that filled electron shells are associated with extra stability. There is also some extra stability associated with filled sub-shells and with half-filled p sub-shells in which charge is distributed evenly into three orbitals.

So, beryllium's $1s^2\,2s^2$ electron structure in which all sub-shells are filled is more stable than that of boron which is $1s^2\,2s^2\,2p^1$.

A similar situation arises with nitrogen which has a higher first ionisation energy than oxygen. The electron structure of nitrogen is $1s^2\,2s^2\,2p^3$ and that of oxygen is $1s^2\,2s^2\,2p^4$. The half-filled 2p sub-shell in nitrogen, with one electron in each of the three 2p orbitals and its evenly distributed charge, is more stable than the 2p sub-shell in oxygen which contains four electrons. This results in a higher first ionisation energy for nitrogen than oxygen.

Test yourself

20 The table below shows the groups and formulae of chlorides for the elements in periods 2 and 3.

1	2	3	4	5	6	7
LiCl	$BeCl_2$	BCl_3	CCl_4	NCl_3	OCl_2	FCl
NaCl	$MgCl_2$	$AlCl_3$	$SiCl_4$	PCl_3	SCl_2	ClCl

a) Why have the entries for group 0 been omitted?
b) What periodic pattern is shown by the formulae of these chlorides?
c) The formulae OCl_2 and FCl are normally written as Cl_2O and ClF, respectively. Why is this?

21 a) Copy the style of the table in question 20 to show the formulae of the oxides of the elements in periods 2 and 3.
b) Is there a periodic pattern in the formulae of these oxides?

22 Use data on the Dynamic Learning Student website to explore whether there is any pattern in the boiling temperatures of the chlorides of the elements in periods 2 and 3.

23 Why do you think the first ionisation energies of elements decrease with atomic number in every group of the periodic table?

24 a) Write down the electron sub-shell structures for magnesium and aluminium.
b) Suggest a reason why the first ionisation energy of aluminium is less than that of magnesium.

REVIEW QUESTIONS

1 Antimony has two main isotopes – antimony-121 and antimony-123. A forensic scientist was asked to help a crime investigation by analysing the antimony in a bullet. This was found to contain 57.3% of ^{121}Sb and 42.7% of ^{123}Sb.

a) Explain what you understand by the term 'relative atomic mass'. (3)

b) Calculate the relative atomic mass of the sample of antimony from the bullet. Write your answer in an appropriate number of significant figures. (3)

c) State one similarity and one difference, in terms of sub-atomic particles, between the isotopes. (2)

2 a) During mass spectrometry the following processes occur: *acceleration, deflection, detection, ionisation* and *vaporisation.*

Write these processes in the order in which they occur in a mass spectrometer. (3)

www
Extension questions

b) The isotopes of magnesium, ^{24}Mg, ^{25}Mg and ^{26}Mg, can be separated by mass spectrometry.

i) Explain what you understand by the term 'isotope'. (2)

ii) Copy and complete the table below to show the composition of the ^{24}Mg and ^{26}Mg isotopes. (2)

	Protons	Neutrons	Electrons
^{24}Mg			
^{26}Mg			

iii) Copy and complete the electronic configuration of an atom of ^{24}Mg.

$1s^2$ __________ (1)

3 The table below shows the electron structures and first ionisation energies of the first five elements in period 2 of the periodic table.

Element	Electron structure	First ionisation energy/kJ mol^{-1}
Li	$1s^2 2s^1$	520
Be	$1s^2 2s^2$	900
B	$1s^2 2s^2 2p^1$	801
C	$1s^2 2s^2 2p^2$	1086
N	$1s^2 2s^2 2p^3$	1402

a) Describe the *general* trend in first ionisation energies from Li to N. (1)

b) Explain this general trend. (3)

c) Explain why boron, B, has a lower first ionisation energy than beryllium, Be. (3)

4 The table below shows the first and second ionisation energies of sodium and potassium in group 1 of the periodic table.

	First ionisation energy/kJ mol^{-1}	Second ionisation energy/kJ mol^{-1}
Sodium	496	4563
Potassium	419	3051

a) Write an equation, with state symbols, for the second ionisation energy of sodium. (2)

b) Why are the second ionisation energies of sodium and potassium larger than their first ionisation energies? (3)

c) Why are the first and second ionisation energies of potassium smaller than those of sodium? (4)

d) The first five successive ionisation energies, in kJ mol^{-1}, of an element, X, in period 3 of the periodic table are 578, 1817, 2745, 11 578, 14 831.

i) Identify element X. (1)

ii) Explain how you obtained your answer. (2)

5 The table below shows the melting temperatures of the elements in period 3 of the periodic table.

Element	Melting temperature/°C
Na	98
Mg	649
Al	660
Si	1410
P	44
S	119
Cl	−101
Ar	−189

The trend in the melting temperatures across period 3 and other periods is described as a periodic property.

a) What is the general pattern in melting temperatures across periods in the periodic table? (2)

b) How is this general trend related to the different types of elements? (1)

c) What do you understand by the term 'periodic property'? (2)

d) State two other properties which can be described as periodic in relation to the periodic table. (2)

6 a) This question concerns the following five species:

$^{16}_{8}O^{2-}$ $^{19}_{9}F^{-}$ $^{20}_{10}Ne$ $^{23}_{11}Na$ $^{25}_{12}Mg^{2+}$

i) Which two species have the same number of neutrons? (2)

ii) Which two species have the same ratio of neutrons to protons? (2)

iii) Which species does not have 10 electrons? (1)

b) The first ionisation energies of three consecutive elements in the periodic table are 1251, 1521 and 419 kJ mol^{-1}.

Draw a sketch graph of these ionisation energies against atomic number. Continue the sketch to show the pattern of the first ionisation energies of the next three elements in the periodic table, assuming that no transition metals are involved. (2)

c) The first five ionisation energies of an element are 738, 1451, 7733, 10 541, 13 629 kJ mol^{-1}. Explain why the element cannot have an atomic number less than 12. (4)

4 Bonding and structure

One of the central aims of chemistry is to explain the properties of elements and compounds in terms of their bonding and structure. In this topic, we will be studying the arrangement and structure of atoms, molecules and ions in different materials, and the theories of bonding which account for the forces which hold atoms or ions together.

4.1 Investigating structure and bonding

The word 'structure' has different levels of meaning in science. On a grand scale, engineers design the structures of buildings and bridges; on the smallest scale, chemists and physicists explore the inner structure of atoms. Not surprisingly, scientists use different models and different theories to explain the structure and properties of materials at these different levels. In the same way, tourists to London must use different types of maps to negotiate its road and rail systems.

Scientists have developed increasingly sophisticated models to account for the structure, bonding and properties of materials as their knowledge has increased. No single model can be used to explain the properties of elements and compounds at all levels. Similarly, no single map can be used by all visitors to London. We must live with these limitations and appreciate which map and which model are the most appropriate in different contexts.

In this topic, we will use Dalton's model of atoms and ions as discreet, tiny spheres to explain crystal structures, and then apply the model of electron shells to explain metallic, ionic and covalent bonding.

The regular shapes of crystals suggest an underlying arrangement of the atoms, ions or molecules in their structure. Until the early part of the twentieth century, scientists could only guess at the arrangement of invisible atoms in crystals. Then, Sir Lawrence Bragg (1890–1971) realised that X-rays could be used to investigate crystal structures because their wavelengths are about the same as the distances between atoms in a crystal.

A narrow beam of X-rays is directed at a crystal of the substance being studied (Figure 4.1).

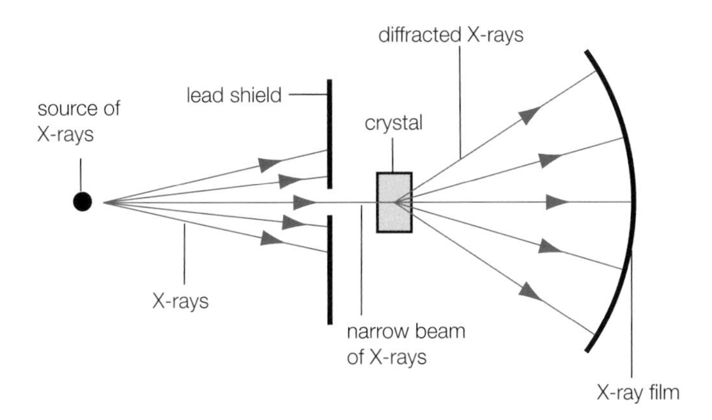

Figure 4.1 ◀
Using X-rays to study the structure of atoms or ions in a crystal

The atoms or ions in the crystal scatter the X-rays producing a pattern of diffracted rays. The diffracted X-rays were originally photographed using X-ray film but can now be recorded electronically. From the diffraction pattern, it is possible to deduce the three-dimensional crystal structure by studying the pattern of dots.

If we know how the atoms or ions in a substance are arranged (the structure) and how they are held together (the bonding) then we can explain its properties.

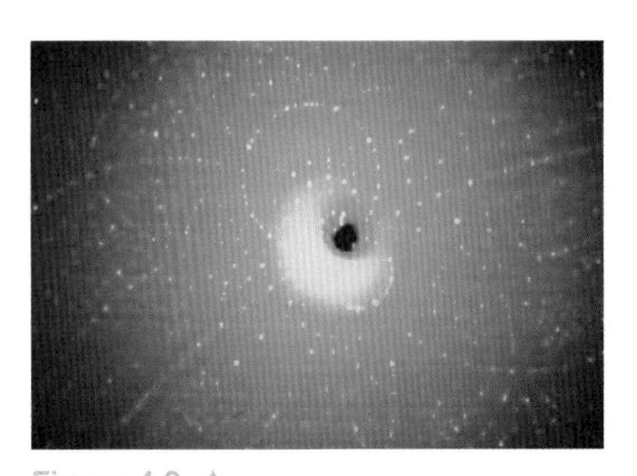

Figure 4.2 ▲
An X-ray diffraction pattern of lysozyme, a protein found in egg white

Figure 4.3 ▲
Crystals of rock salt (sodium chloride, NaCl)

For example, copper is composed of closely packed atoms with freely moving outer electrons. These electrons move through the structure when copper is connected to a battery, so it is a good conductor of electricity. Atoms in the closely packed structure can slide over each other and because of this copper can be drawn into wires. These properties of copper lead to its use in electrical wires and cables.

Notice how:

- the structure and bonding of copper determine its properties
- the properties of copper determine its uses.

The links between structure, bonding and properties help us to explain the uses of different substances and materials. They explain why metals are used as conductors, why graphite is used in pencils and why clay is used to make pots (Figure 4.4).

Figure 4.4 ▶
Wet clay is easily moulded by the potter because water molecules can get between its flat two-dimensional particles. When the clay is fired, all the water is driven out and atoms in one layer bond to those in the layers above and below. This gives the clay a three-dimensional structure making it hard and rigid for use as pots and ceramics.

4.2 Two types of structure

Broadly speaking there are two types of structure – giant structures and simple molecular structures.

Materials with giant structures form crystals in which all the atoms or ions are linked by a network of strong bonds extending throughout the crystal. These strong bonds result in giant structures with high melting and boiling temperatures.

Substances with simple molecular structures consist of small groups of atoms. The bonds linking the atoms in the molecules (intramolecular forces) are relatively strong, but the forces between molecules (intermolecular forces) are weak. These weak intermolecular forces allow the molecules to be separated easily. So molecular substances have low melting and boiling temperatures.

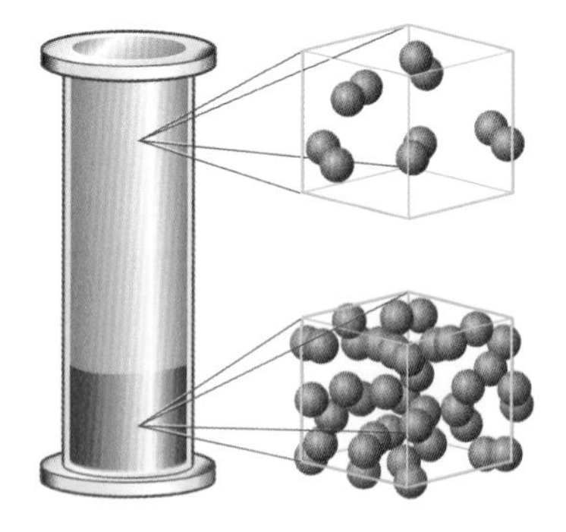

Figure 4.5 ▲
Molecules in bromine liquid and vapour. Many molecular elements and compounds are liquids or gases at room temperature because little energy is needed to overcome the weak forces between their molecules.

Test yourself

1 Look up the melting and boiling temperatures of the following elements and decide whether they have giant or simple molecular structures: boron, fluorine, silicon, sulfur, manganese, iodine.

2 Look at the crystals of rock salt in Figure 4.3.
   a) What shape are most of the crystals of rock salt?
   b) How do you think the ions are arranged in rock salt?

www Data

Figure 4.6 ◀
Metal cables in the electricity grid supported by steel pylons – a reminder that metals are strong, bendable and good conductors of electricity. Ceramic insulators between the conducting cables and the pylons prevent the electric current leaking away to earth.

The main types of giant structures are metals, ionic solids, polymers, ceramics (pottery) and glasses. All of these materials are solids that depend for their properties on three types of strong bonding – metallic bonding, ionic bonding and covalent bonding.

These three types of strong bonding will be the main focus of our interest in the following sections of this topic. For each type of bonding, the strength of the bond depends on electrostatic attractions between positive and negative charges.

4.3 Metallic bonding and structures

Metals are very important and useful materials. Just look around you and notice the uses of different metals – in vehicles, bridges, pipes, taps, radiators, cutlery, pans, jewellery and ornaments. X-ray studies show that the atoms in most metals are packed together as close as possible. This arrangement is called 'close packing'.

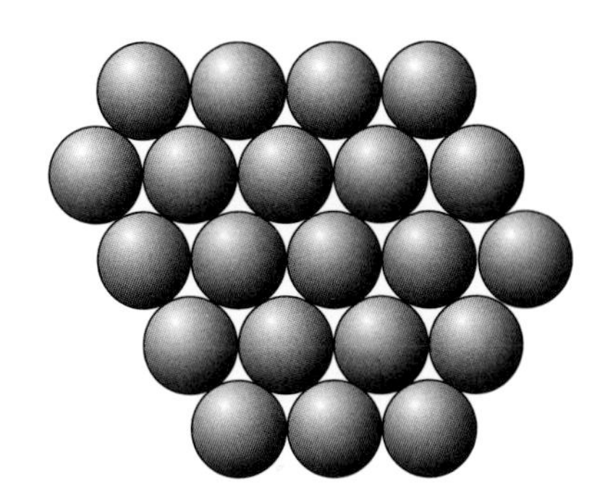

Figure 4.7 ▲
Close packing of atoms in one layer of a metal

Figure 4.7 shows a model of a few atoms in one layer of a metal crystal. Notice that each atom in the middle of the layer 'touches' six other atoms in the same layer.

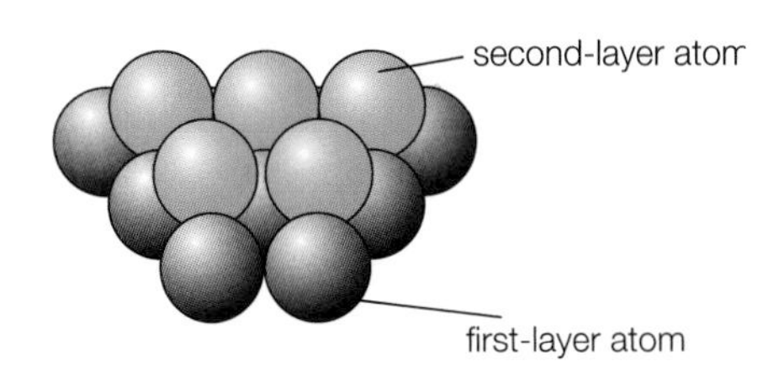

Figure 4.8 ▲
Atoms in two layers of a metal crystal

When a second layer is placed on top of the first, atoms in the second layer sink into the dips between atoms in the first layer (Figure 4.8).

This close packing allows atoms in one layer to get as close as possible to those in the next layer, forming a giant lattice of closely packed atoms in a regular pattern. In this giant lattice, electrons in the outer shell of each metal atom are free to drift through the whole structure. These electrons do not have fixed positions – they are described as 'delocalised'.

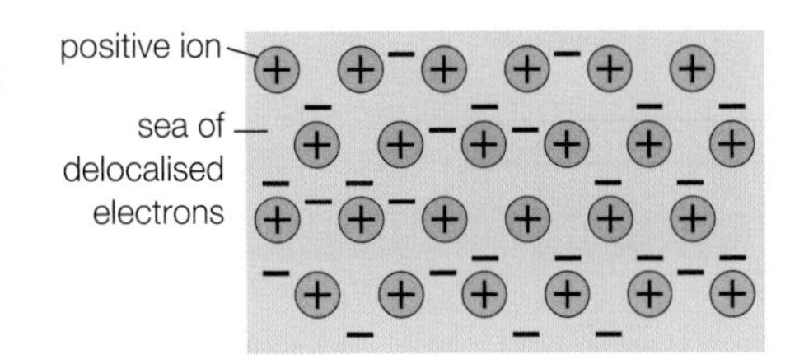

Figure 4.9 ▲
Metallic bonding results from the strong attractions between metal ions and the sea of delocalised electrons.

So, metals consist of giant lattices of positive ions with electrons moving around and between them in a 'sea' of delocalised negative charge (Figure 4.9). The strong electrostatic attractions between the positive metal ions and the 'sea' of delocalised electrons result in strong forces between the metal atoms.

Definitions

Chemists use the word **lattice** to describe a regular arrangement of atoms or ions in a crystal.

Delocalised electrons are bonding electrons which are not fixed between two atoms in a bond. They are free to move and shared by several or many more atoms.

Metallic bonding is the strong attraction between a lattice of positive metal ions and a 'sea' of delocalised electrons.

The properties of metals

In general, metals:

- have high melting and boiling temperatures
- have high densities
- are good conductors of heat and electricity
- are malleable – can be bent or hammered into different shapes.

All the properties of metals can be explained and interpreted in terms of their close-packed structure and delocalised electrons.

- *High melting and boiling temperatures* – metal atoms are closely packed with strong forces of attraction between the positive ions and delocalised electrons. So, it takes a lot of energy to move the positive ions away from their positions in the giant lattice and move around each other, allowing the metal to melt. It takes even more energy to separate individual atoms in the metal at the boiling temperature.
- *Good conductors of heat* – when a metal is heated, energy is transferred to the electrons. The delocalised electrons move around faster and conduct the heat (energy) rapidly to other parts of the metal.
- *Malleable* – the bonds between metal atoms are strong, but they are not directional because the delocalised electrons can drift throughout the lattice and attract any of the positive ions. When a force is applied to a metal, lines or layers of atoms can slide over each other. This is known as 'slip'. After slipping, the atoms settle into close-packed positions again. Figure 4.10 shows the positions of atoms before and after slip. This is what happens when a metal is bent or hammered into different shapes.

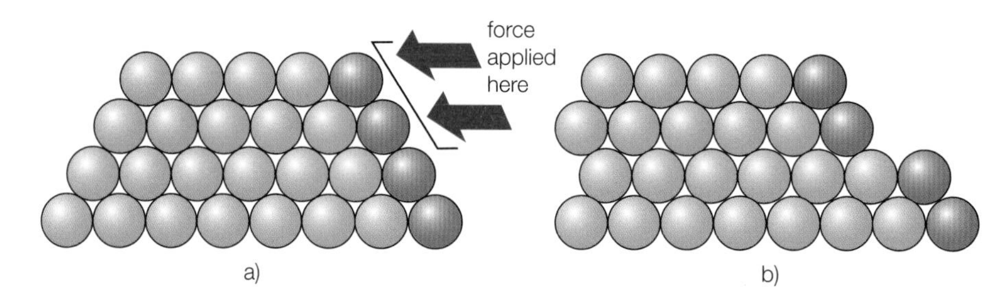

Figure 4.10 ▶
Positions of atoms in a metal a) before and b) after 'slip' has occurred.

Figure 4.11 ▲
Blacksmiths rely on the malleability of metals to hammer and bend them into useful shapes.

Test yourself

3 Why are the electrons in the outermost shell of metal atoms described as 'delocalised'?

4 Explain why most metals:
   a) have high densities
   b) are good conductors of electricity.

5 Name a metal or alloy and its particular use to illustrate each of the following typical properties of metals:
   a) shiny
   b) conduct electricity
   c) bend without breaking
   d) high tensile strength.

6 Consider the patterns of metal properties in the periodic table.
   a) Which metals have:
      i) relatively high densities
      ii) relatively low densities?
   b) Which metals have:
      i) relatively high melting temperatures
      ii) relatively low melting temperatures?

Activity

Choosing metals for different uses

Various properties of six metals are shown in Table 4.1.

Metal	Density /g cm^{-3}	Tensile strength/10^7 N m^{-2}	Melting temperature/°C	Electrical resistivity /10^{-8} ohm m	Thermal conductivity /J s^{-1} cm^{-1} K^{-1}	Cost per tonne/£
Aluminium	2.7	8	660	2.5	2.4	960
Copper	8.9	33	1083	1.6	3.9	1200
Iron	7.9	21	1535	8.9	0.8	130
Silver	10.5	25	962	1.5	4.2	250 000
Titanium	4.5	23	1660	43.0	0.2	27 000
Zinc	7.1	14	420	5.5	1.1	750

Table 4.1 ▲

1 Use the information in Table 4.1 to explain the following statements.

a) Copper is used in most electrical wires and cables, but high tension cables in the National Grid are made of aluminium.

b) Bridges are built from steel which is mainly iron even though the tensile strength of iron is lower than that of some other metals.

c) Metal gates and dustbins are made from steel coated with zinc (galvanised).

d) Silver is no longer used to make our coins.

e) Aircraft are now constructed from an aluminium/titanium alloy rather than pure aluminium.

f) The base of high quality saucepans is copper rather than steel (iron).

2 If the atoms in a metal pack closer then the density should be higher, the bonds between atoms should be stronger and so the melting temperature should be higher. This suggests there should be a relationship between the density and melting temperature of a metal.

Use the data in the table to check if there is a relationship between density and melting temperature. State 'yes' or 'no' and explain your answer.

3 The explanation of both electrical and thermal conductivity in metals uses the concept of delocalised electrons. This suggests that there should be a relationship between the electrical and thermal conductivities of metals.

Use the data in the table to check if there is a relationship. (Hint: electrical resistivity is the reciprocal of electrical conductivity.) State 'yes' or 'no' and explain your answer.

Figure 4.12 ▲
Hot sodium reacting with chlorine gas

4.4 Ionic bonding and structures

Atoms into ions

Compounds of metals with non-metals, such as sodium chloride and calcium oxide, are composed of ions. When compounds form between metals and non-metals, the metal atoms lose electrons and become positive ions (cations). At the same time, the non-metal atoms gain electrons and become negative ions (anions). For example, when sodium reacts with chlorine (Figure 4.12) each sodium atom loses its one outer electron to form a sodium ion, Na^+, which has the same electron structure as the noble gas neon. Chlorine atoms gain these electrons to form chloride ions, Cl^-, with the same electron configuration as the noble gas argon (Figure 4.13).

In many cases when atoms react to form ions, they gain or lose electrons in such a way that the ions formed have the same electron configuration as a noble gas. This transfer of electrons involves redox (Topic 10).

Chemists describe diagrams like that in Figure 4.13 as dot-and-cross diagrams, in which the electrons belonging to one reactant are shown as dots and those belonging to the other reactant are shown as crosses. But remember, all electrons are the same – dots and crosses are simply used to show which electrons come from the metal and which come from the non-metal. Dot-and-cross diagrams are useful because they provide a balance sheet for keeping track of the electrons when ionic compounds form.

Figure 4.14 shows simplified dot-and-cross diagrams for the formation of sodium chloride and calcium fluoride in which only the outer shell electrons are drawn.

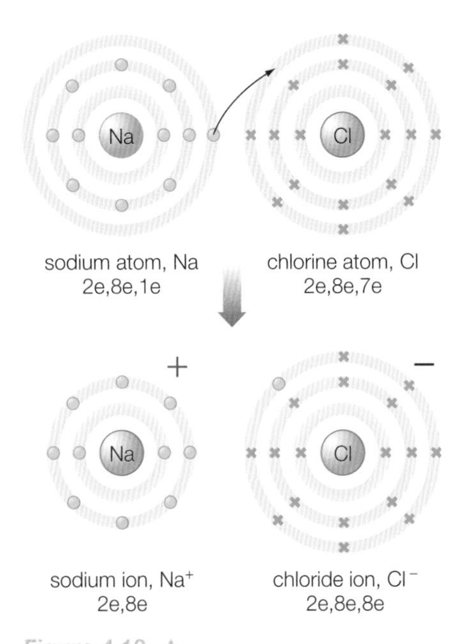

Figure 4.13 ▲
Formation of ions when sodium reacts with chlorine

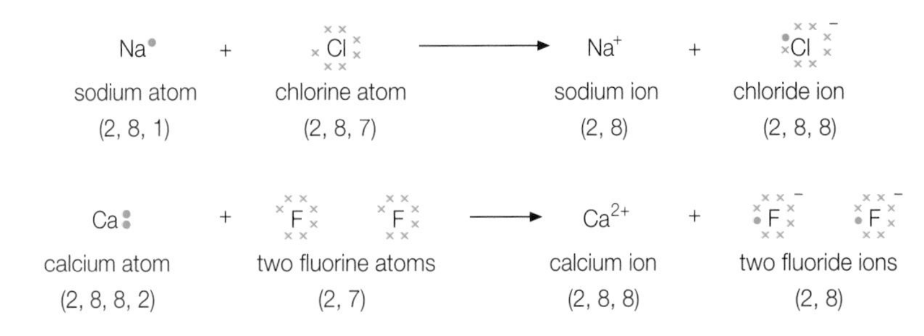

Figure 4.14 ▲
Dot-and-cross diagrams for the formation of sodium chloride and calcium fluoride showing only the electrons in the outer shells of the reacting atoms

Test yourself

7 Draw dot-and-cross diagrams for:
  a) lithium fluoride
  b) magnesium chloride
  c) lithium oxide
  d) calcium oxide.

8 With the help of a periodic table, predict the charges on ions of each of the following elements: caesium, strontium, gallium, selenium and astatine.

9 Why do metals form positive ions, while non-metals form negative ions?

Ionic bonding

When metals react with non-metals, the ions produced form ionic crystals. These ionic crystals are giant lattices containing billions of positive and negative ions packed together in a regular pattern.

Figure 4.15 shows how the ions are arranged in one layer of sodium chloride (NaCl) and Figure 4.16 shows a three-dimensional model of its structure.

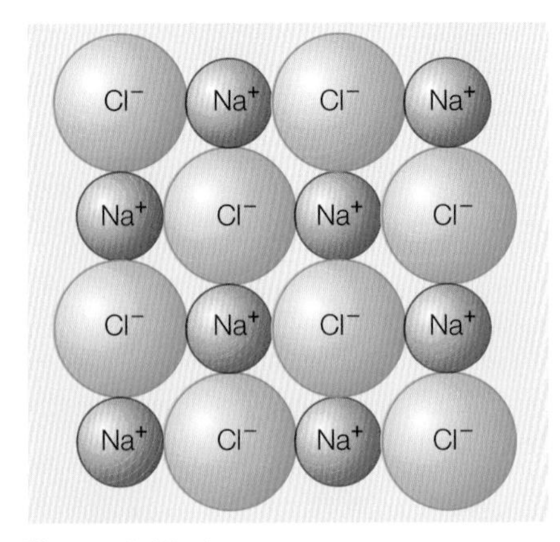

Figure 4.15 ▲
The arrangement of ions in one layer of a sodium chloride crystal

Notice that each Na^+ ion is surrounded by Cl^- ions, and that each Cl^- ion is surrounded by Na^+ ions. This means that overall there are strong net electrostatic attractions between ions in all directions throughout the lattice. These electrostatic attractions between oppositely charged ions are described as ionic bonding.

Many other compounds have the same structure as sodium chloride including the chlorides, bromides and iodides of Li, Na, K and the oxides and sulfides of Mg, Ca, Sr and Ba.

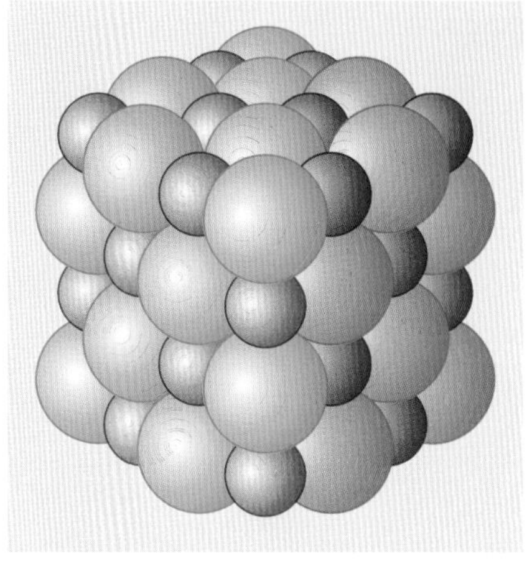

Figure 4.16 ▲
A three-dimensional model of the structure of sodium chloride. The smaller red balls represent Na^+ ions and the larger green balls represent Cl^- ions.

Properties of ionic compounds

Strong ionic bonds hold the ions firmly together in ionic compounds. This explains why ionic compounds:

- are hard crystalline substances
- have high melting and boiling temperatures
- are often soluble in water and other polar solvents, but insoluble in non-polar solvents (Sections 8.3 and 9.4)
- do not conduct electricity when solid because their ions cannot move away from fixed positions in the giant lattice
- conduct electricity when they are melted or dissolved in water because the charged ions are then free to move.

For example, when molten sodium chloride conducts electricity, positive sodium ions move towards the negative terminal (cathode) while negative chloride ions move towards the positive terminal (anode). When the sodium ions reach the cathode, they gain electrons and become sodium atoms:

Cathode (−): $2Na^+(l) + 2e^- \rightarrow 2Na(l)$

When chloride ions reach the anode, they lose electrons and become chlorine molecules:

Anode (+): $2Cl^-(l) \rightarrow 2e^- + Cl_2(g)$

This process is described as electrolysis. It reverses the changes that happen when sodium chloride forms from its elements (Figure 4.12).

> **Definition**
>
> **Electrolysis** is the decomposition of a compound by electricity – the compound is usually decomposed into its constituent elements. The compound which is decomposed is called an **electrolyte** and we say that it has been electrolysed.

Ionic radii

X-ray diffraction methods are used to study ionic compounds and to measure the spacing between ions in crystals. From the diffraction patterns, it is possible to calculate the radii of individual ions. The radius of the positive ion of an element is smaller than its atomic radius because it loses electrons from its outer shell when turning into an ion. The radius of the negative ion of an element is larger than its atomic radius because electrons are added to the outer shell (Figure 4.17).

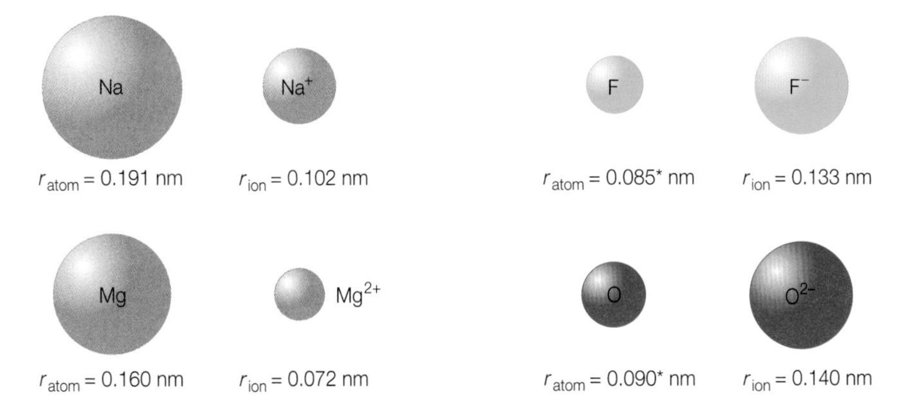

Figure 4.17 ◀
Comparing the radii of atoms and ions (*estimated values)

Test yourself

10 Look carefully at Figures 4.15 and 4.16.
  a) How many Cl^- ions surround one Na^+ ion in a layer of the NaCl crystal?
  b) How many Cl^- ions surround one Na^+ ion in the three-dimensional crystal?
  c) How many Na^+ ions surround one Cl^- ion in the three-dimensional crystal?
  d) The structure of crystalline sodium chloride is described as 6 : 6 co-ordination. Why is this?

11 The melting temperature of sodium fluoride is 993 °C, but that of magnesium oxide is 2852 °C.
  a) Write the formulae of these two compounds, showing charges on the ions.
  b) Suggest why the melting temperature of magnesium oxide is so much higher than that of sodium fluoride.

12 Write equations for the reactions at the cathode and anode during electrolysis of the following compounds:
  a) molten potassium bromide.
  b) molten magnesium chloride.

Activity

Identifying and explaining the trends in ionic radii

Table 4.2 shows the radii of ions of the elements in group 1, and those of the consecutive elements from nitrogen to aluminium in the periodic table.

Ions of group 1	Li^+	Na^+	K^+	Rb^+	Cs^+	
Radius/nm	0.074	0.102	0.138	0.149	0.170	
Ions of N to Al	N^{3-}	O^{2-}	F^-	Na^+	Mg^{2+}	Al^{3+}
Radius/nm	0.171	0.140	0.133	0.102	0.072	0.053

Table 4.2 ▲
Ionic radii

Look carefully at the data in Table 4.2.

1 a) Describe the trend in ionic radii in group 1 of the periodic table.
  b) Explain the trend in ionic radii in group 1.

2 Do you think this trend will be repeated in other groups of the periodic table? State 'yes' or 'no' and explain your answer.

3 Use the s, p, d, f notation to describe the electron configuration of:
  a) a nitride ion, N^{3-}
  b) a fluoride ion, F^-
  c) a sodium ion, Na^+
  d) an aluminium ion, Al^{3+}.

4 a) All the ions of consecutive elements in the periodic table from N^{3-} to Al^{3+} are described as 'isoelectronic'. What do you think this means?
  b) Describe the trend in ionic radii for the isoelectronic ions from N^{3-} to Al^{3+}.
  c) Explain the trend in ionic radii for the isoelectronic ions from N^{3-} to Al^{3+}.

4.5 Energy changes and ionic bonding

Figure 4.13 shows the formation of the ions of sodium chloride from its elements, but it is simplified in many ways. It ignores the fact that when sodium reacts with chlorine:

- sodium starts as a giant structure of atoms
- energy is required to separate the sodium atoms (the enthalpy change of atomisation)
- energy is required to remove electrons (the first ionisation energy) to form sodium ions
- chlorine consists of molecules
- energy is required to separate the chlorine atoms (the enthalpy change of atomisation)
- energy is involved (the electron affinity) in adding one electron to each chlorine atom to form a chloride ion.

So, where does the energy come from to turn sodium metal and chlorine molecules into sodium chloride? It turns out that ionic crystals are stable because of the large release of energy when oppositely charged ions come together forming a crystal lattice.

This is called the lattice energy, $\Delta H^{\ominus}_{lattice}$, of the ionic compound. For sodium chloride, it is summarised by the equation

$$Na^+(g) + Cl^-(g) \rightarrow NaCl(s) \quad \Delta H^{\ominus}_{lattice}[NaCl(s)] = -787\ kJ\ mol^{-1}$$

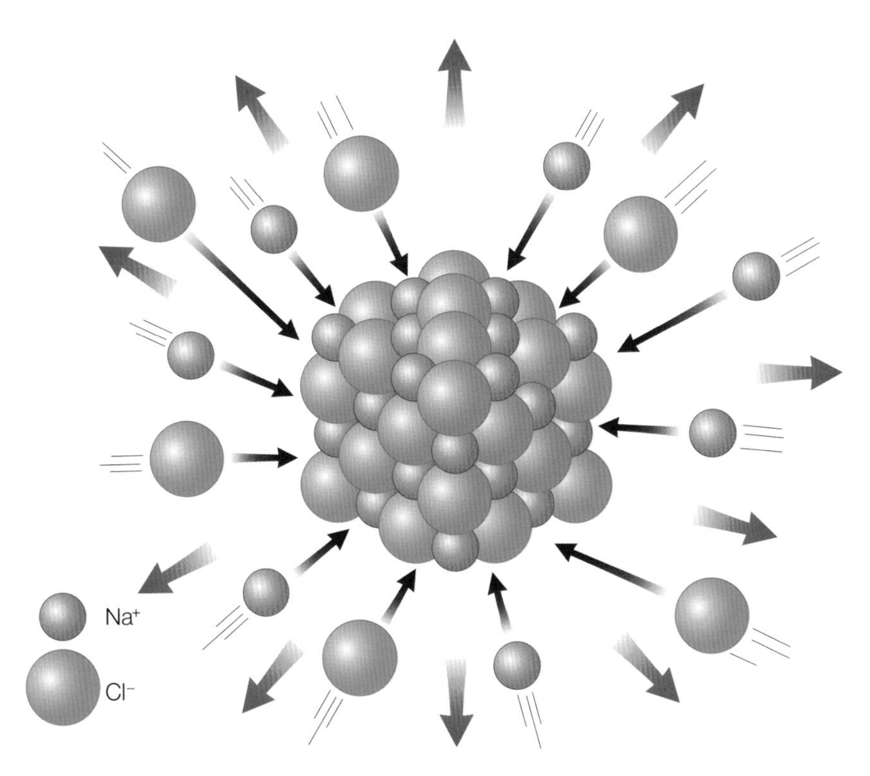

Figure 4.18 Lattice energy is the energy which would be given out to the surroundings (red arrows) if 1 mole of a compound could form directly from free gaseous ions coming together and arranging themselves into a crystal lattice (black arrows).

During the early part of the twentieth century, the German scientists Max Born (1882–1970) and Fritz Haber (1868–1934) analysed the energy changes in the formation of ionic compounds. Their work resulted in Born–Haber cycles, which are thermochemical cycles for investigating the stability and bonding in ionic compounds.

The strength of ionic bonds, measured as lattice energies in kJ mol^{-1}, arises from the energy given out as millions upon millions of positive and negative ions come together to form a crystal lattice.

Definitions

The **first electron affinity** of an element is the energy change when each atom in one mole of gaseous atoms gains one electron to form one mole of gaseous ions with a single negative charge.

For example, $O(g) + e^- \rightarrow O^-(g)$ $E^{\ominus}_{aff.1} = -141\ kJ\ mol^{-1}$

The gain of the first electron is exothermic, but adding the second electron to a negatively charged ion is endothermic.

For example, $O^-(g) + e^- \rightarrow O^{2-}(g)$ $E^{\ominus}_{aff.2} = +798\ kJ\ mol^{-1}$

The **lattice energy** of an ionic solid is the standard enthalpy change when one mole of the compound forms from free gaseous ions.

The energy released is greater and the force of attraction between the ions is stronger if:

- the charges on the ions are large
- the ions are small allowing them to get closer to each other.

It is important to distinguish between the lattice energy of a compound and its standard enthalpy change of formation (Section 2.4), which involves the formation of a compound from its elements. For sodium chloride, this is summarised by the equation

$$Na(s) + \frac{1}{2}Cl_2(g) \rightarrow NaCl(s) \quad \Delta H_f^\ominus[NaCl(s)] = -411\,kJ\,mol^{-1}$$

Tutorial

4.6 Born–Haber cycles

Born–Haber cycles are an application of Hess's Law (Section 2.5). They enable chemists to calculate lattice energies which cannot be measured directly.

A Born–Haber cycle identifies all the enthalpy changes which contribute to the standard enthalpy change of formation of a compound.

These changes, shown in Figure 4.19, involve:

- the energy required to create free gaseous ions by atomising and then ionising the elements
- the energy given out (the lattice energy) when the ions come together to form a crystal.

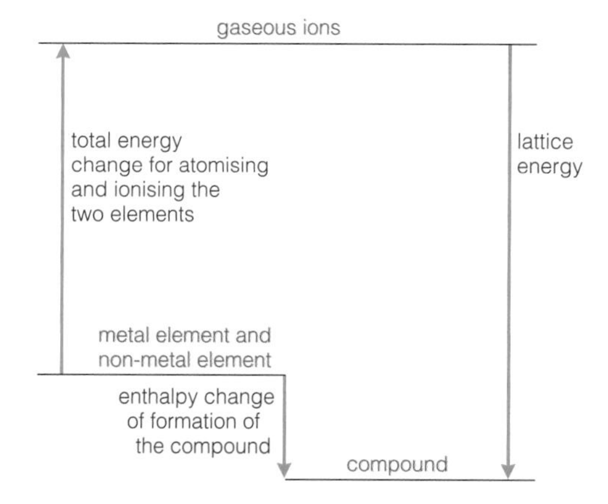

Figure 4.19 ▶ The main structure of a Born–Haber cycle

A Born–Haber cycle is usually set out as an enthalpy level diagram. All the terms in the cycle can be measured experimentally except the lattice energy. So, by using Hess's Law, it is possible to calculate the lattice energy. Figure 4.20 shows a Born–Haber cycle for sodium oxide (Na_2O).

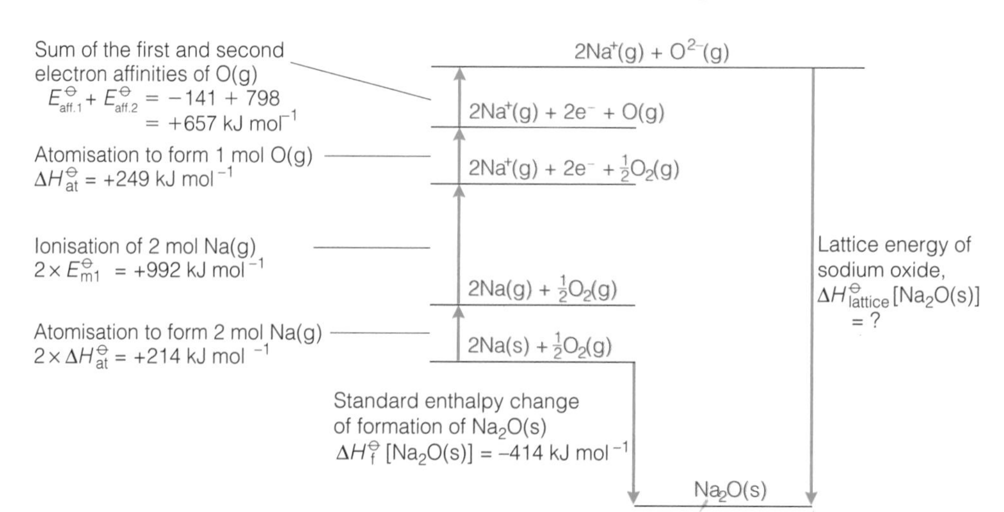

Figure 4.20 ▶ The Born–Haber cycle for sodium oxide

Starting with the elements sodium and oxygen, the measured value of the standard enthalpy change of formation of sodium oxide has been written downwards on the cycle showing that it is exothermic. Above that, the terms and values for the atomisation and then ionisation of sodium are written upwards as endothermic processes. Notice also that the amount of sodium required is 2 mol because there are 2 moles of sodium in 1 mole of sodium oxide.

These terms for sodium are followed by those required for the conversion of half a mole of oxygen molecules [$\frac{1}{2}O_2(g)$] to one mole of oxide ions [$O^{2-}(g)$]. This involves the atomisation of oxygen followed by its first and second electron affinities.

From the experimentally determined values, which we now have in the cycle, it is possible to calculate the lattice energy.

Worked example

Applying Hess's Law to the cycle in Figure 4.20 and remembering that an exothermic change in one direction becomes an endothermic change, with the opposite sign, in the reverse direction:

$$\Delta H^{\ominus}_{\text{lattice}}[Na_2O(s)] = (-657 - 249 - 992 - 214 - 414)\,\text{kJ mol}^{-1}$$
$$= -2526\,\text{kJ mol}^{-1}$$

Note

Lattice energies, as usually defined, are negative. This means that the descriptions 'larger' and 'smaller' can be ambiguous in comparing lattice energies. For this reason, it is better to describe one lattice energy as more or less exothermic than another.

Test yourself

13 Look carefully at Figure 4.21, which is a Born–Haber cycle for magnesium chloride.

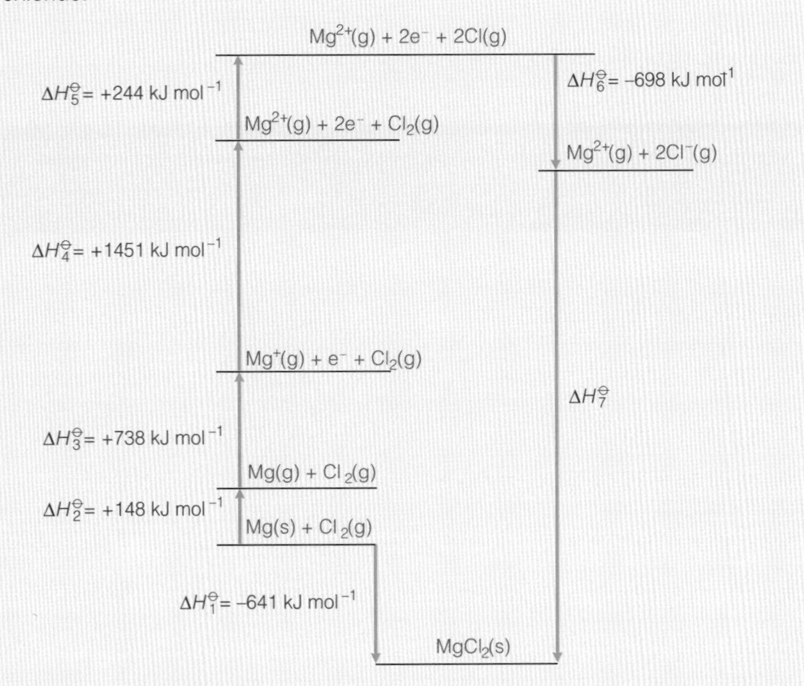

Figure 4.21

a) Identify the enthalpy changes $\Delta H^{\ominus}_1$, $\Delta H^{\ominus}_2$, $\Delta H^{\ominus}_3$, $\Delta H^{\ominus}_4$, $\Delta H^{\ominus}_5$, $\Delta H^{\ominus}_6$ and $\Delta H^{\ominus}_7$.

b) Calculate the lattice energy of magnesium chloride.

14 Explain why a Born–Haber cycle is an application of Hess's Law.

15 Why are lattice energies always exothermic?

Tutorial

4.7 Testing the ionic model – ionic or covalent?

One way in which chemists can test their theories and models is by comparing the predictions from their theoretical models with the values obtained by experiment.

Born–Haber cycles are very helpful in this respect because they enable chemists to test the ionic model and check whether the bonding in a compound is truly ionic. The experimental lattice energy calculated from a Born–Haber cycle can be compared with the theoretical value calculated using the laws of electrostatics and assuming that the only bonding in the crystal is ionic.

Using the laws of electrostatics, it is possible to calculate a theoretical value for the lattice energy of an ionic compound by summing the effects of all the attractions and repulsions between the ions in the crystal lattice (Figure 4.22).

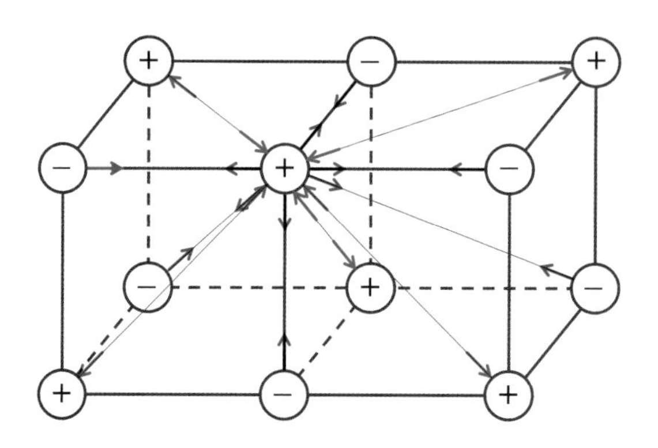

Figure 4.22 ▶ Some of the many attractions (red) and repulsions (blue) between ions which must be taken into account in calculating a theoretical value for the lattice energy of an ionic crystal

Note

Electrostatic forces operate between the ions in a crystal. Oppositely charged ions attract each other and ions with like charges repel each other. The size of the electrostatic force, F, between two charges is given by:

$$F \propto \frac{Q_1 \times Q_2}{d^2}$$

- the bigger the charges, Q_1 and Q_2, the stronger the force
- the greater the distance, d, between the two charges, the smaller the force. This has a big effect because it is the square of the distance that matters.

Table 4.3 shows the experimentally determined lattice energies and the theoretical lattice energies for a number of compounds.

Compound	Experimental lattice energy from a Born–Haber cycle /kJ mol^{-1}	Theoretical lattice energy calculated assuming that the only bonding is ionic/kJ mol^{-1}
NaCl	−780	−770
NaBr	−742	−735
NaI	−705	−687
KCl	−711	−702
KBr	−679	−674
KI	−651	−636
AgCl	−905	−833
MgI_2	−2327	−1944

Table 4.3 ▲

Pure ionic bonding arises solely from the electrostatic forces between the ions in a crystal. Notice in Table 4.3 that there is close agreement between the theoretical and experimental values of the lattice energies of the sodium and potassium halides – the difference is less than 3% in all these compounds,. This shows that ionic bonding can account almost entirely for the bonding in sodium and potassium halides.

But look at the experimental and theoretical lattice energies of silver chloride and magnesium iodide in Table 4.3. In these two compounds, the experimental values are much larger than the theoretical values which assume the only bonding is ionic. The bonding is significantly stronger than that predicted by a pure ionic model. This suggests that there is some covalent bonding as well as ionic bonding in these substances.

Polarisation of ions

In ionic compounds, positive metal ions will attract the outermost electrons of negative ions, pulling electrons into the space between the ions. This distortion of the electron clouds around anions by positively charged cations is an example of polarisation. Polarisation of negative anions by positive metal ions gives rise to some electron sharing – that is, to a degree of covalent bonding. And the values in Table 4.3 show that both silver chloride and magnesium iodide, although mainly ionic, have some covalent bonding.

Figure 4.23 shows three examples of ionic bonding with increasing degrees of electron sharing as a positive cation polarises the neighbouring negative ion. In general, results show that:

- the polarising power of a cation depends on its charge and its radius
- the polarisability of an anion depends on its size.

Definition

Polarisation is the distortion of the electron cloud in a molecule or ion by a nearby charge.

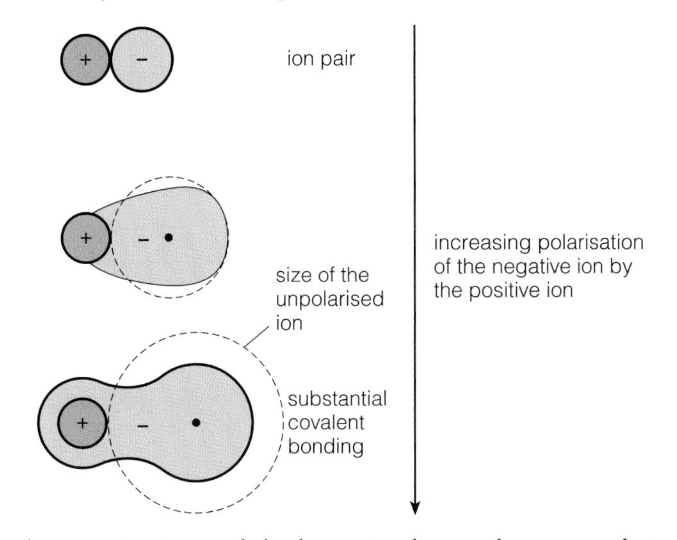

Figure 4.23 ◀ Ionic bonding with increasing degrees of electron sharing as a positive cation polarises the neighbouring negative anion. (Dotted circles show the size of the unpolarised ions.)

The larger the negative ion and the larger its charge, the more polarisable it becomes. So, iodide ions are more polarisable than fluoride ions and fluorine, with a small, singly charged fluoride ion, forms compounds which are more ionic than any other non-metal.

Test yourself

16 Use the data in Table 4.2 to explain the following.
  a) The polarising power of Mg^{2+} is greater than that of Li^+.
  b) The polarising power of Li^+ is greater than that of K^+.
  c) The polarising power of Al^{3+} is greater than that of Na^+.
  d) The polarisability of N^{3-} is greater than that of F^-.

17 The lattice energy of LiF is $-1031\,kJ\,mol^{-1}$ and that of LiI is $-759\,kJ\,mol^{-1}$.
  a) Why is the lattice energy of LiI less exothermic than the lattice energy of LiF?
  b) Which compound would you expect to have the closer agreement between the Born–Haber experimental value of its lattice energy and its theoretical value based on the ionic model?
  c) Explain your answer to part **b)**.

18 Here are four values for lattice energy, in $kJ\,mol^{-1}$: −3791, −3299, −3054, −2725.

The four ionic compounds with these four values are BaO, MgO, BaS and MgS. Match the formulae with the values and explain your choices.

The stability of ionic compounds

Almost all the compounds of metals with non-metals are ionic, and these compounds have standard enthalpy changes of formation which are exothermic. This means that the compounds are at a lower energy level and therefore more stable than their elements.

If you look at the Born–Haber cycles in Figures 4.20 and 4.21, you will see that an ionic compound will have an exothermic standard enthalpy of formation if its negative lattice energy can outweigh the total energy needed to produce gaseous ions from the elements.

Using a Born–Haber cycle with theoretical values for lattice energies, it is possible to calculate the standard enthalpy change of formation for compounds which do not normally exist. For example, consider the Born–Haber cycle for the hypothetical compound MgCl in Figure 4.24.

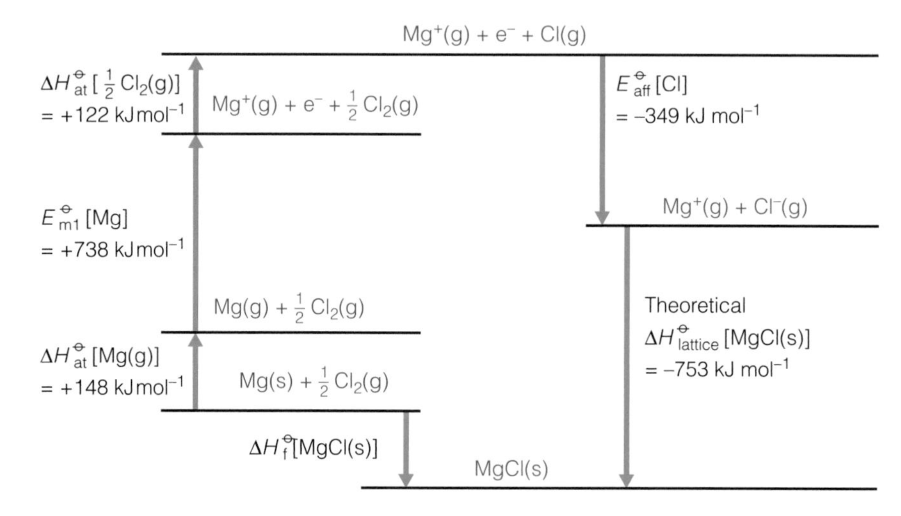

Figure 4.24 ▶ A Born–Haber cycle for the hypothetical compound MgCl

Applying Hess's Law to Figure 4.24, we can calculate a value for the standard enthalpy of formation of MgCl(s).

$$\Delta H_f^{\ominus}[\text{MgCl(s)}] = (+148 + 738 + 122 - 349 - 753)\,\text{kJ mol}^{-1}$$
$$= -94\,\text{kJ mol}^{-1}$$

The negative (exothermic) value of $\Delta H_f^{\ominus}$[MgCl(s)] suggests that MgCl(s) would be more stable than its elements Mg(s) and Cl_2(g). However, we should also check its stability with respect to the usual compound that magnesium forms with chlorine, $MgCl_2$(s).

We can do this by calculating the standard enthalpy change for the reaction

$$2\text{MgCl(s)} \rightarrow \text{MgCl}_2\text{(s)} + \text{Mg(s)}$$

knowing that $\Delta H_f^{\ominus}[\text{MgCl}_2\text{(s)}] = -641\,\text{kJ mol}^{-1}$.

Figure 4.25 shows a Hess cycle for the reaction.

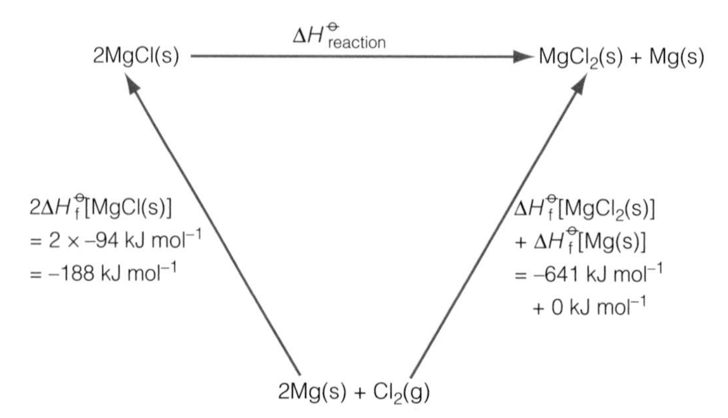

Figure 4.25 ▶ A Hess cycle for the reaction $2\text{MgCl(s)} \rightarrow \text{MgCl}_2\text{(s)} + \text{Mg(s)}$

Applying Hess's Law to the cycle in Figure 4.25:

$$\Delta H^{\ominus}_{reaction} = -(-188\,kJ\,mol^{-1}) + (-641 + 0)\,kJ\,mol^{-1}$$
$$= +188 - 641$$
$$= -453\,kJ\,mol^{-1}$$

This result shows that $MgCl_2(s)$ and Mg(s) are much more stable than MgCl(s). So, $MgCl_2(s)$ will form in preference to MgCl in any reaction between magnesium and chlorine.

Test yourself

19 A Born–Haber cycle for the hypothetical compound $MgCl_3$ suggests that $\Delta H^{\ominus}_f[MgCl_3(s)] = +3950\,kJ\,mol^{-1}$ and its estimated lattice energy is $-5440\,kJ\,mol^{-1}$.

a) What does the value of $\Delta H^{\ominus}_f[MgCl_3(s)]$ tell you about the stability of $MgCl_3(s)$?

b) Why is the value of $\Delta H^{\ominus}_f$ so endothermic?

c) Write an equation to summarise the lattice energy of $MgCl_3(s)$.

d) Why is the lattice energy of $MgCl_3(s)$ more negative than that of $MgCl_2(s)$?

4.8 Covalent bonding and structures

Living things are composed mainly of water and a huge variety of molecular compounds of carbon with other non-metals such as hydrogen, oxygen and nitrogen. These compounds differ greatly from minerals in the rocks in the Earth's crust. Most of these minerals consist of giant structures of the non-metals silicon and oxygen with other elements. So chemists must understand the structure and bonding of non-metals and their compounds if they are to explain the properties of both the organic and inorganic worlds.

Simple molecular structures

In most non-metal elements, atoms are joined together in small molecules such as hydrogen, (H_2), nitrogen (N_2), phosphorus (P_4), sulfur (S_8) and chlorine (Cl_2).

Most of the compounds of non-metals with other non-metals also have simple molecular structures. This is true of simple compounds such as water, carbon dioxide, ammonia, methane and hydrogen chloride. It is also true of the many thousands of carbon compounds (see Topic 5).

Figure 4.26 ◀ The clouds and atmosphere contain simple molecular substances such as oxygen, nitrogen and water vapour. The oceans and hydrosphere contain ionic substances such as sodium chloride dissolved in water. Rocks in the Earth (the lithosphere) contain giant structures of atoms and ions.

The covalent bonds holding atoms together within these simple molecular structures are strong, so the molecules do not break up into atoms easily. However, the forces between the individual molecules (intermolecular forces) are weak, so it is quite easy to separate them. This means that molecular substances are often liquids or gases at room temperature and that molecular solids are usually easy to melt and evaporate (Figure 4.27).

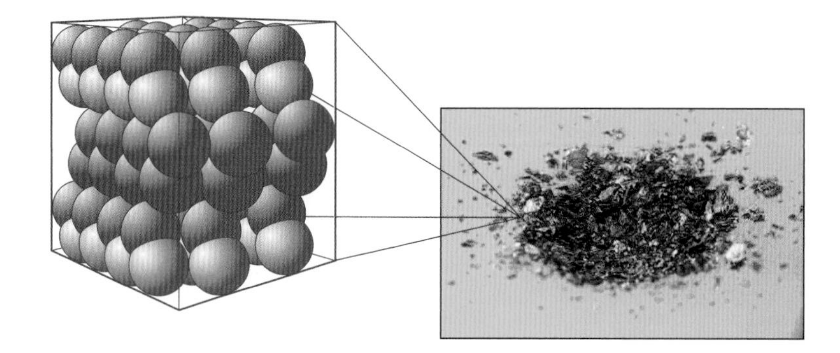

Figure 4.27 ▶
The structure of iodine showing the arrangement of I_2 molecules. The forces between I_2 molecules are so weak that iodine sublimes, changing directly from solid to vapour at low temperatures.

Properties of simple molecular substances
In general, simple molecular elements and compounds:

- are usually gases, liquids or soft solids at room temperature
- have relatively low melting and boiling temperatures
- do not conduct electricity as solids, liquids or gases because they contain neither ions nor free electrons to carry the electric charge
- are usually more soluble in non-polar solvents, such as hexane, than in water – and the solutions do not conduct electricity.

Giant covalent structures

A few non-metal elements – including diamond, graphite and silicon – consist of giant structures of atoms held together by covalent bonding.

The covalent bonds in diamond are strong and point in a definite direction, so diamonds are very hard and have very high melting temperatures. Diamond does not conduct electricity because the electrons in its covalent bonds are fixed (localised) between pairs of atoms (Figure 4.28).

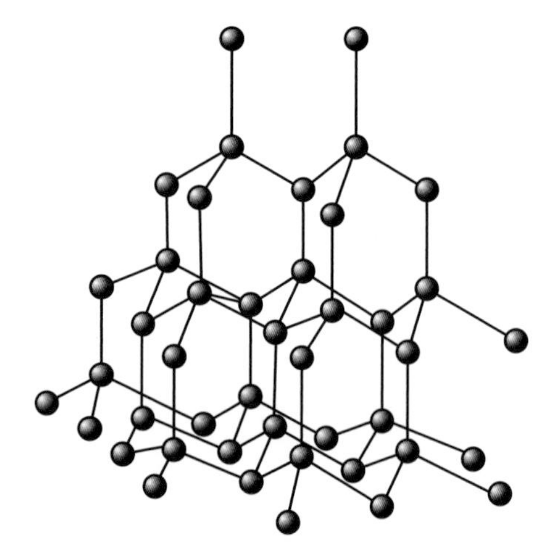

Figure 4.28 ▲
Part of the giant covalent structure in diamond – each carbon atom is linked to four other atoms in a network extending throughout the giant structure.

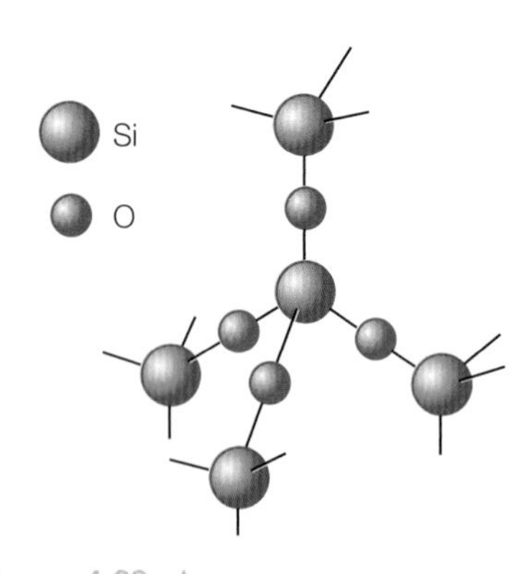

Figure 4.29 ▲
Part of the giant structure of silicon dioxide in the mineral quartz. Silicon atoms are arranged in the same way as carbon atoms in diamond, but with an oxygen atom between each pair of silicon atoms. Sandstone and sand consist mainly of silicon dioxide.

Some compounds of non-metals, such as silicon dioxide and boron nitride, also have giant structures with covalent bonding. These compounds are also hard, non-conductors.

Test yourself

20 Compare the melting and boiling temperatures of the non-metals boron, fluorine, white phosphorus, germanium and bromine. Which of these elements are simple molecular and which have giant covalent structures?

21 a) Compare the formulae and physical properties of carbon dioxide and silicon dioxide.
b) Explain the differences in terms of the structures of the two compounds.

Figure 4.30 ▲
Amethyst is crystalline quartz (silicon dioxide, SiO_2) coloured purple due to the presence of iron(III) ions. Amethyst has both ionic and covalent bonds.

4.9 Covalent bonding

Strong covalent bonding holds the atoms of non-metals together in molecules and giant structures.

Covalent bonds form when atoms share electrons – a single covalent bond consists of a shared pair of electrons. The atoms are held together by the electrostatic attraction between the positive charges on their nuclei and the negative charge on the shared electrons.

The electron configuration of fluorine is $1s^2 2s^2 2p^5$ or more simply 2, 7, with seven electrons in the outer shell. When two fluorine atoms combine to form a molecule, they share two electrons. The electron configuration of each atom is then like that of neon, the nearest noble gas (Figure 4.31).

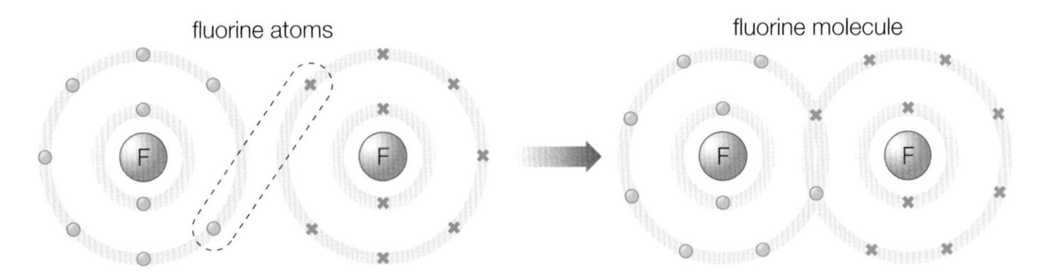

Figure 4.31 ▲
Covalent bonding in a fluorine molecule

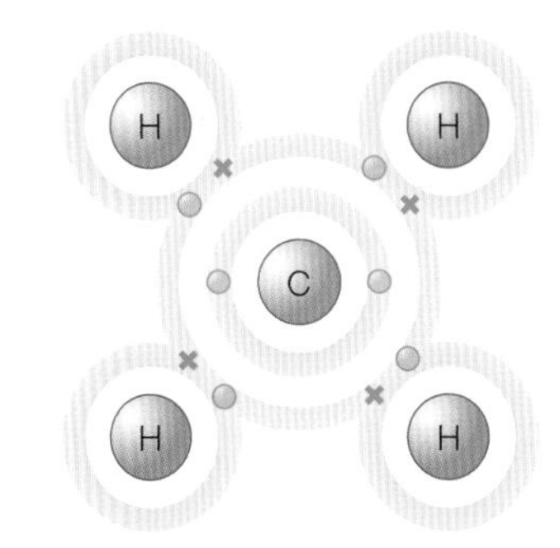

Figure 4.32 ▲
Covalent bonding in methane

Chemists draw a line between symbols to represent a covalent bond (Section 1.2). So they write a fluorine molecule as F–F. This is the structural formula, showing the atoms and bonding. The molecular formula of fluorine is F_2.

Covalent bonds also link the atoms in non-metal compounds. Figure 4.32 shows the covalent bonding in methane.

Dot-and-cross diagrams, showing only the electrons in outer shells, provide a simple way of representing covalent bonding. Three of these dot-and-cross diagrams are shown in Figure 4.33. Remember that each non-metal usually forms the same number of covalent bonds (Section 1.2), which makes it easier to work out the structures of molecules (Table 1.2).

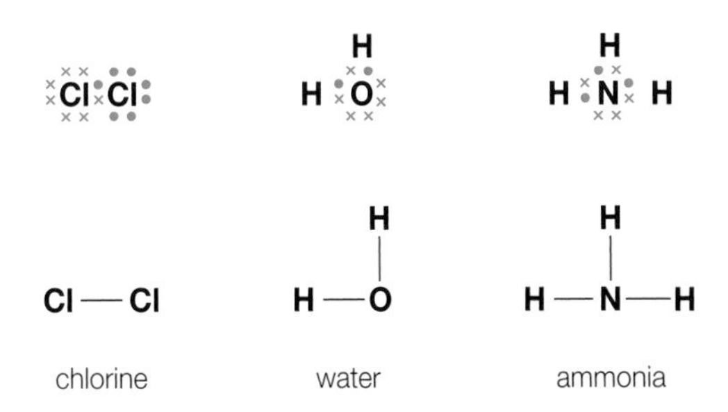

Figure 4.33 ▲
Dot-and-cross diagrams showing the single covalent bonds in molecules. A simpler way of showing the bonding in molecules is also included. This shows each covalent bond as a line between two symbols.

Multiple bonds

www Tutorials

One shared pair of electrons makes a single bond. Double bonds and triple bonds are also possible with two or three shared pairs, respectively.

There are two covalent bonds between the two oxygen atoms in an oxygen molecule, and two covalent bonds between both the oxygen atoms and the carbon atom in carbon dioxide (Figure 4.34). With two electron pairs involved in the bonding, there is a region of high electron density between the two atoms joined by a double bond.

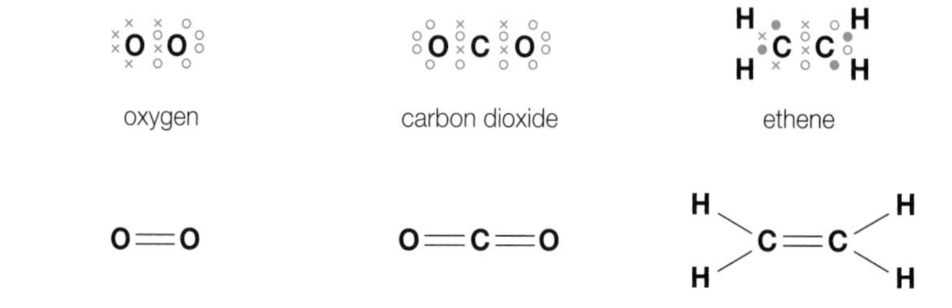

Figure 4.34 ▶
Three molecules with double covalent bonds

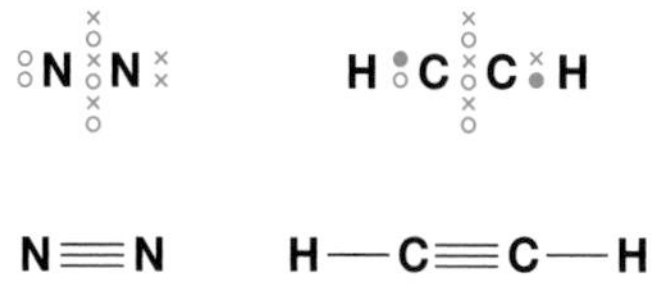

N≡N   H—C≡C—H

Figure 4.35 ▲
Two molecules with triple covalent bonds

Lone pairs of electrons

In many molecules, there are atoms with pairs of electrons in their outer shells which are not involved in the bonding between atoms in the molecule. Chemists call these 'lone pairs' of electrons.

Lone pairs of electrons:

- affect the shapes of molecules (Topic 7)
- form dative covalent bonds
- are important in the chemical reactions of some compounds including water and ammonia.

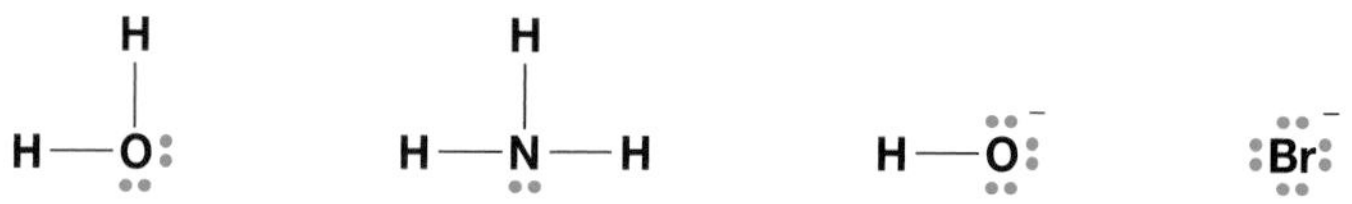

Figure 4.36 ▶
Molecules and ions with lone pairs of electrons

Definition

A **dative covalent bond** is a bond in which two atoms share a pair of electrons, both the electrons being donated by one atom.

Dative covalent bonds

In a covalent bond, two atoms share a pair of electrons. Usually each atom supplies one electron to make up the pair. Sometimes, however, one atom provides both the electrons and chemists call this a dative covalent bond. The word 'dative' means 'giving' and one atom gives both the electrons to make the covalent bond. An alternative name for a dative covalent bond is a co-ordinate bond. Once formed there is no difference between a dative covalent bond and any other covalent bond.

Ammonia forms a dative covalent bond when it reacts with a hydrogen ion to make an ammonium ion, NH_4^+ (Figure 4.37). A dative bond is represented by an arrow in displayed formulae like that of NH_4^+ in Figure 4.37. The arrow points from the atom donating the electron pair to the atom receiving them.

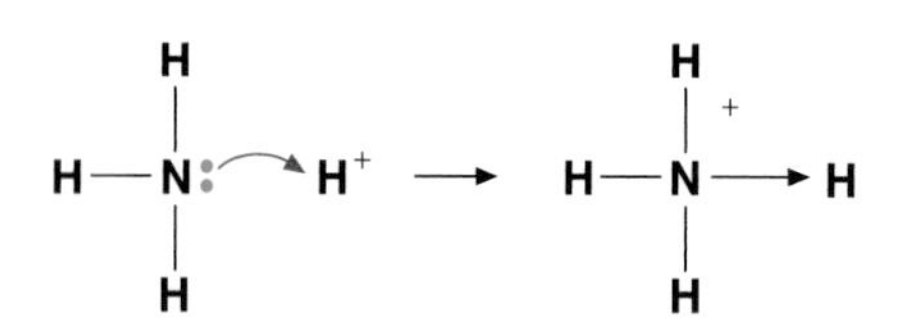

Figure 4.37 ▲
Formation of an ammonium ion, NH_4^+

Dative covalent (co-ordinate) bonding also accounts for the structures of the oxonium ion, H_3O^+ (in which water molecules combine with H^+ ions), nitric acid and carbon monoxide (Figure 4.38).

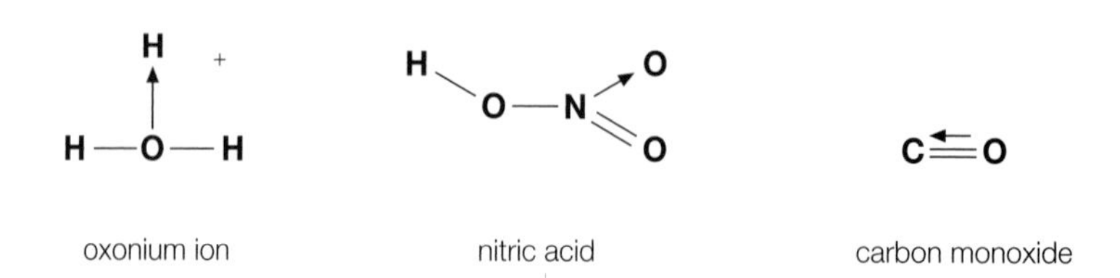

Figure 4.38 ▶
Dative covalent (co-ordinate) bonds in the oxonium ion, nitric acid and carbon monoxide

Test yourself

22 Look at the electron density map of hydrogen, H_2, in Figure 4.39.
   a) What kind of bond exists between the hydrogen atoms in a hydrogen molecule?
   b) What evidence does the electron density map provide for the existence of a bond between the hydrogen atoms in a hydrogen molecule?
   c) How do the physical properties of giant atomic (giant covalent) structures, such as diamond and silicon dioxide, provide evidence for strong covalent bonds between non-metal atoms?

23 Draw diagrams showing all the electrons in the shells and in covalent bonds in:
   a) hydrogen, H_2
   b) hydrogen chloride, HCl
   c) ammonia, NH_3.

24 Draw dot-and-cross diagrams to show the covalent bonding in:
   a) hydrogen sulfide, H_2S
   b) ethane, C_2H_6
   c) carbon disulfide, CS_2
   d) nitrogen trifluoride, NF_3
   e) phosphine, PH_3.

25 Identify the atoms with lone pairs of electrons in these molecules and state the number of lone pairs:
   a) ammonia
   b) water
   c) hydrogen fluoride
   d) carbon dioxide.

26 a) In aqueous solution, acids donate H^+ ions to water molecules forming H_3O^+ ions. Draw a dot-and-cross diagram to show the formation of an H_3O^+ ion.
   b) Boron fluoride forms molecules with the formula BF_3. Draw a dot-and-cross diagram for BF_3 and then explain why BF_3 molecules readily react with ammonia molecules, NH_3, to form the compound NH_3BF_3.

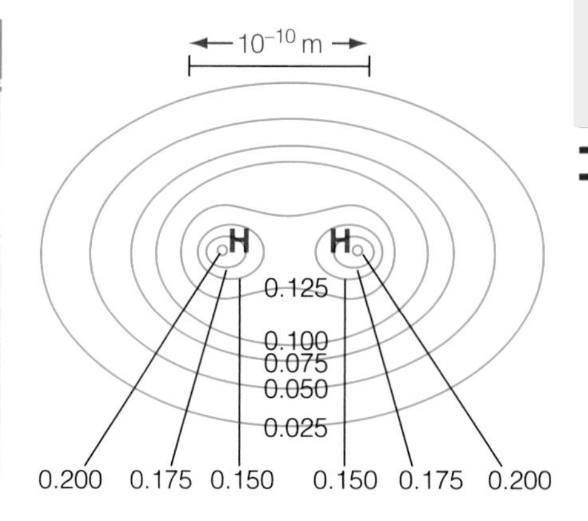

Figure 4.39 ▲
An electron density map for hydrogen, H_2. The units for the contours are electrons per 10^{-30} m³.

Tutorials

REVIEW QUESTIONS

Extension questions

1 This question is about calcium and calcium oxide.
   a) i) Describe the bonding in calcium. (3)
      ii) Explain why calcium is a good conductor of electricity. (2)
   b) Draw dot-and-cross diagrams for the ions in calcium oxide showing *all* the electrons and the ionic charges. (4)
   c) Under what conditions does calcium oxide conduct electricity? Explain your answer. (6)

2 The table in the next column shows the melting temperatures of the elements in period 3 of the periodic table.
   a) Explain why the melting temperature of sodium is much lower than that of magnesium. (3)
   b) Phosphorus and sulfur exist as molecules of P_4 and S_8, respectively. Explain their difference in melting temperatures. (2)

Element	Melting temperature/K
Na	371
Mg	922
Al	933
Si	1683
P	317
S	386
Cl	−101
Ar	84

   c) State the type of structure and the nature of the bonding in each of the following elements:
      i) aluminium
      ii) silicon
      iii) chlorine. (6)

3 a) Define the term 'first electron affinity'. (3)

b) The equation below represents the change occurring during the second electron affinity of nitrogen

$$N^-(g) + e^- \rightarrow N^{2-}(g)$$

Explain why the second electron affinity for all elements is endothermic. (2)

c) Write equations for the first and third electron affinities of nitrogen. (2)

4 The following data can be used in a Born–Haber cycle for copper(II) iodide, CuI_2.

Enthalpy change of atomisation of iodine,
$\Delta H^{\ominus}_{at}[\frac{1}{2}I_2(s)] = +107\,kJ\,mol^{-1}$

Enthalpy change of atomisation of copper,
$\Delta H^{\ominus}_{at}[Cu(s)] = +338\,kJ\,mol^{-1}$

First ionisation energy of copper,
$E^{\ominus}_{m1}[Cu(g)] = +746\,kJ\,mol^{-1}$

Second ionisation energy of copper,
$E^{\ominus}_{m2}[Cu(g)] = +1958\,kJ\,mol^{-1}$

Electron affinity of iodine,
$E^{\ominus}_{aff}[I(g)] = -295\,kJ\,mol^{-1}$

Lattice energy of copper(II) iodide, $CuI_2(s)$,
$\Delta H^{\ominus}_{lattice}\,[CuI_2(s)] = -963\ kJ\,mol^{-1}$

a) Use the outline of the Born–Haber cycle for copper(II) iodide in Figure 4.40 to:

i) write the formulae and state symbols of the species that should appear in boxes A, B and C

ii) name the enthalpy changes D, E and F. (6)

b) Define the term lattice energy. (2)

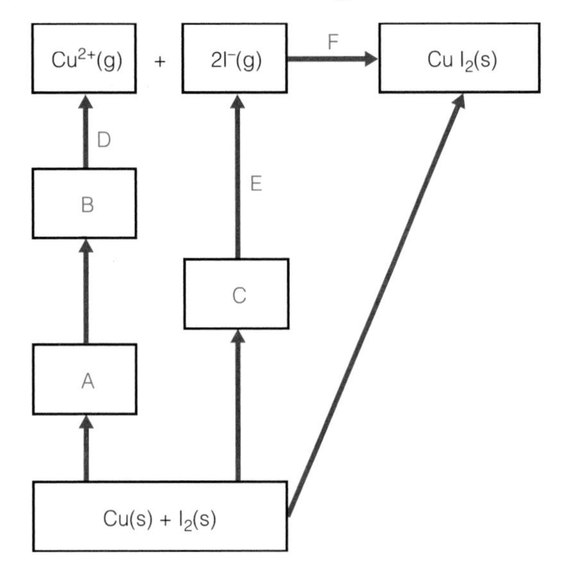

Figure 4.40 ▲

c) Use Figure 4.40 and the data supplied to calculate the enthalpy change of formation of copper(II) iodide. Give a sign and units in your answer. (3)

d) When the lattice energy of copper(II) iodide is calculated from ionic radii and charges, the result is about 14% smaller than the one obtained from the Born–Haber cycle.

i) What does this suggest about the nature of the bonding in copper(II) iodide? (1)

ii) Draw a diagram to show how the smaller copper ion influences the larger iodide ion. (1)

5 a) Using sodium chloride, hydrogen chloride and copper, explain what is meant by covalent, ionic and metallic bonding. (9)

b) Compare and explain the conduction of electricity by sodium chloride and copper in terms of structure and bonding. (6)

6 a) Draw dot-and-cross diagrams, with outer shell electrons only, to show the bonding in ammonia and water. (2)

b) On mixing with water, ammonia reacts to form an alkaline solution containing ammonium ions and hydroxide ions.

i) Write an equation for this reaction including state symbols. (2)

ii) The ammonium ion has dative covalent bonding. Explain the term 'dative covalent bonding'. (2)

iii) Draw a dot-and-cross diagram of the ammonium ion and label the dative covalent bond. (2)

5 Introduction to organic chemistry

Carbon is an amazing element. The number of compounds containing carbon and hydrogen is well over ten million. This is far more than the number of compounds of all the other elements put together.

Most compounds containing carbon also contain hydrogen. The main sources of these compounds are organic – living or once-living materials in animals and plants. Because of this, the term 'organic chemistry' is used to describe the branch of chemistry concerned with the study of compounds containing C–H bonds. This covers most of the compounds of carbon. Simple carbon compounds which don't contain C–H bonds, such as carbon dioxide and carbonates, are usually included in the study of inorganic chemistry.

Figure 5.1 ◀
Children playing with their toys. From the cells in their bodies to the fibres in their clothes and the plastics in their toys, almost everything in this photo consists of organic chemicals – molecules containing carbon–hydrogen bonds.

5.1 Carbon – a special element

There are two main reasons why carbon can form so many compounds.

The first reason is that carbon atoms have an exceptional ability to form chains, branched chains and rings of varying size. No other element can form long chains of its atoms in the same way as carbon.

The second reason why carbon can form so many compounds is the relative inertness and unreactive nature of the C–C and C–H bonds (Section 6.3).

When carbon atoms form a chain or a ring linked by single covalent bonds, no more than two of the bonds on each atom are used. This leaves at least two other bonds on each carbon atom which can bond with other atoms. Carbon often forms bonds with hydrogen, oxygen, nitrogen and halogen atoms (Figure 5.3).

A knowledge of organic chemicals enables chemists to extract, synthesise and manufacture a wide range of important products including fuels, plastics, medicines, anaesthetics and antibiotics.

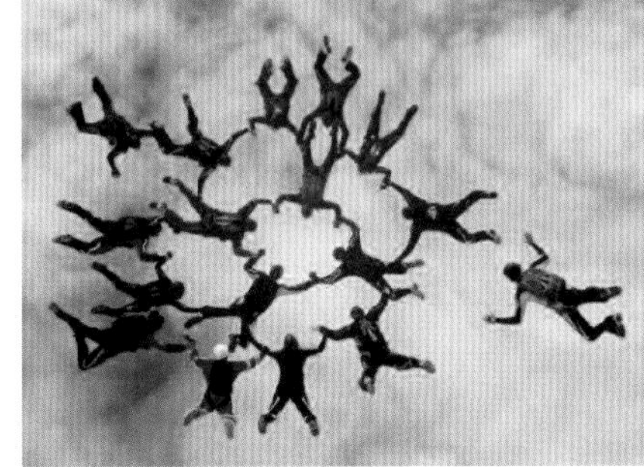

Figure 5.2 ▲
Sky divers can use their arms and legs to form four links to one another. Like carbon atoms, they can form chains and rings.

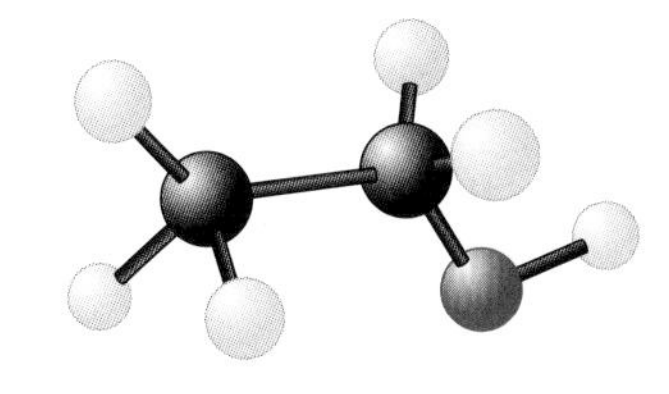

Figure 5.3 ▲
The structure of ethanol, CH_3CH_2OH. Ethanol is commonly called 'alcohol'. Its structure shows two carbon atoms linked to each other and to hydrogen and oxygen atoms by covalent bonds.

Big molecules

Most molecules in living things are molecules of carbon compounds, so biochemistry and molecular biology are important applications of organic chemistry. The compounds found in living cells include carbohydrates, fats, proteins and nucleic acids. The molecules of these compounds are large –

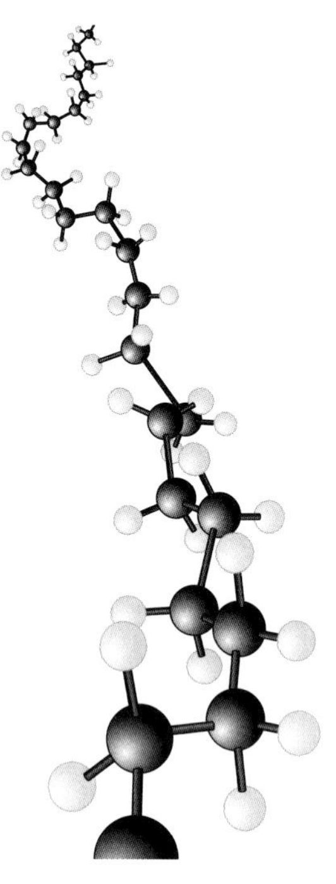

Figure 5.4 ▲
A short section of a polythene molecule

some of them are very large. For example, cellulose, the carbohydrate in cotton, is a natural polymer made of very long chains of glucose units linked together. Its relative molecular mass is about one million.

Organic chemists can synthesise other long-chain molecules by linking together thousands of small molecules to make polymers. These synthetic polymers include polythene, pvc (polyvinylchloride), polystyrene and nylon (Figure 5.4).

With so many organic compounds to study and understand, we need a way of simplifying and organising our knowledge.

Chemists have found a method of classifying organic compounds into families or series, each of which has a distinctive group of atoms called a functional group (Section 5.2). Examples of these families include the alkanes (Sections 6.2 and 6.3), the alkenes (Sections 6.6 to 6.10), alcohols (Sections 15.1 to 15.4) and halogenoalkanes (Sections 15.4 to 15.7). Each of these families of similar compounds with the same functional group is sometimes called a homologous series.

www Data

Test yourself

1. What is organic chemistry?
2. State two reasons why carbon can form so many compounds.
3. Use the data section on the Dynamic Learning Student website to look up the mean bond enthalpies (bond energies) of various single bonds between similar elements.
   a) How does the strength of the single C–C bond compare with other single bonds between two atoms of the same non-metal?
   b) How do you think the relative strength of the C–C bond will affect the number of carbon compounds?

Definition

A **functional group** is the group of atoms which gives an organic compound its characteristic properties and reactions.

5.2 Functional groups

Ethane (CH_3CH_3) and ethanol (CH_3CH_2OH) have very different properties despite their similar structures (Figure 5.5). Ethane is a gas, ethanol is a liquid; ethane does not react with sodium, but ethanol reacts vigorously forming hydrogen. Clearly, the –OH group in ethanol has a big effect on its properties.

The –OH group in ethanol is an example of a functional group – the group of atoms which gives an organic compound its characteristic properties. The functional group in a molecule is responsible for most of its reactions, while the hydrocarbon chain which makes up the rest of the organic compound is relatively unreactive (Figure 5.6).

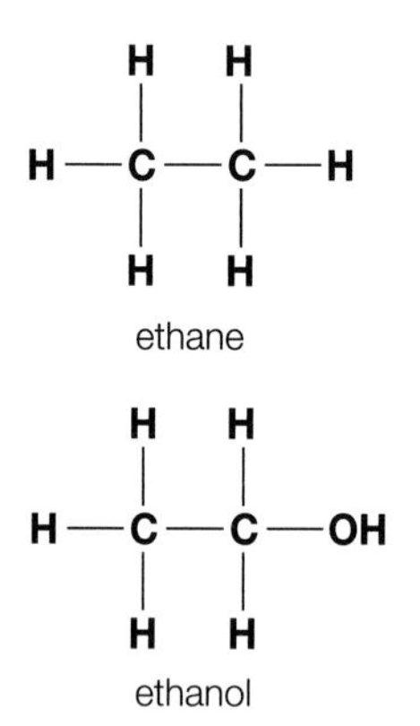

Figure 5.5 ▲
Structures of ethane and ethanol

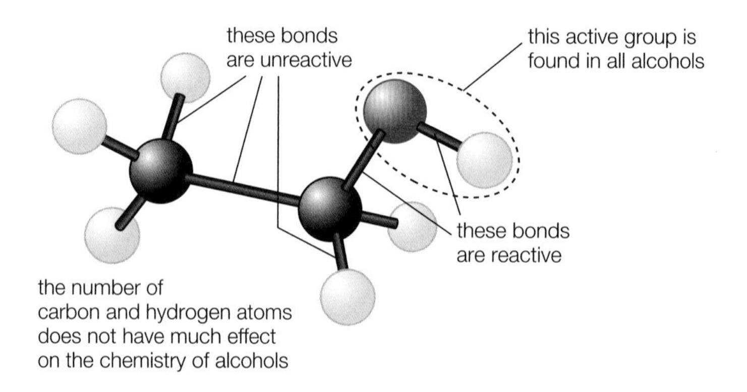

Figure 5.6 ▲
The structure of ethanol showing the reactive functional group and the unreactive hydrocarbon skeleton

Functional groups, such as –OH, have more or less the same effect whatever the size of the hydrocarbon skeleton to which they are attached.

This makes the study of organic compounds much simpler because all molecules containing the same functional group will have similar properties. In this respect, molecules with the same functional group can be regarded as a chemical family and compared to a group of elements in the periodic table.

Ethanol is a member of the series of compounds called alcohols, all of which contain the –OH functional group. Ethene, $CH_2{=}CH_2$, is a member of the series of compounds called alkenes which contain the $>C{=}C<$ functional group. The common functional groups and series of organic compounds which we will meet in this AS course are shown in Table 5.1.

Functional group	Name of the series of compounds	Example
–OH	Alcohols	CH_3CH_2OH ethanol
$>C{=}C<$	Alkenes	$CH_2{=}CH_2$ ethene
–Hal	Halogenoalkanes	CH_3CH_2Cl chloroethane
C–O–C	Ethers	CH_3OCH_3 methoxymethane
–C(=O)H (–CHO)	Aldehydes	CH_3CHO ethanal
$>C{=}O$	Ketones	CH_3COCH_3 propanone
–C(=O)OH (–COOH)	Carboxylic acids	CH_3COOH ethanoic acid
$-NH_2$	Primary amines	$CH_3CH_2NH_2$ ethylamine

Table 5.1 Common functional groups and their series of compounds

Some organic molecules have two or more functional groups. Lactic acid in sour milk, for example, has both an –OH group and a –COOH group (Figure 5.7). In its reactions, lactic acid sometimes acts like an alcohol, sometimes like an acid and sometimes it shows the properties of both types of compound.

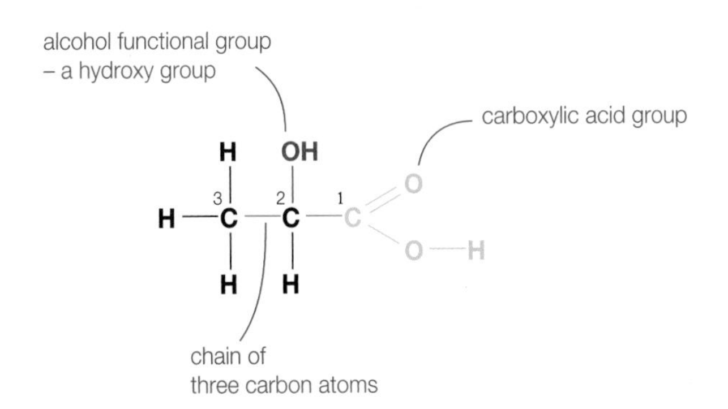

Figure 5.7 The structure of lactic acid (2-hydroxypropanoic acid)

Test yourself

4 a) Why do all alcohols have similar properties?
   b) Why is there a gradual change of the same property in alcohols from ethanol (C_2H_5OH) to hexanol ($C_6H_{13}OH$)?

5 Identify the functional groups in the following compounds and the series to which they belong:
   a) CH_3CH_2CHO
   b) CH_3CH_2I
   c) $CH_2{=}CHCH_2Cl$

5.3 Empirical, molecular and structural formulae

Empirical formulae

The terms 'empirical formula' and 'molecular formula' were introduced in Section 1.1. The empirical formula of a compound is the formula found by experiment. In Section 1.6, we used the combined masses of elements in a compound to calculate its empirical formula. The formulae we obtain by this method show the simplest whole number ratio of the atoms of different elements in a compound.

Worked example

0.15 g of a liquid was analysed and found to contain 0.06 g of carbon, 0.01 g of hydrogen and 0.08 g of oxygen. What is the empirical formula of the liquid?

	C	H	O
Masses of elements combined	0.06 g	0.01 g	0.08 g
Molar masses			
$12\,g\,mol^{-1}$	$1\,g\,mol^{-1}$	$16\,g\,mol^{-1}$ Amounts of elements	combined
$\frac{0.06\,g}{12\,g\,mol^{-1}}$	= 0.005 mol	$\frac{0.01\,g}{1\,g\,mol^{-1}}$	= 0.01 mol
$\frac{0.08\,g}{16\,g\,mol^{-1}}$	= 0.005 mol	Ratio of moles of elements	1

So, the empirical formula of the compound is CH_2O.

Figure 5.8 ▶ Chemists involved in medical research studying a computer model of a protein. The purple ribbon shows the general structure of the molecule. The green pattern shows the atoms at the active site of the molecule which helps cells respond to hormones.

Molecular formulae

The molecular formula of a compound shows the actual number of atoms of each element in one molecule. The molecular formula of ammonia is NH_3 and that of ethanol is C_2H_6O. The term 'molecular formula' only applies to substances that consist of molecules.

For molecular compounds, the relative molecular mass shows whether or not the molecular formula is the same as the empirical formula. A molecular formula is always a simple multiple of the empirical formula.

For example, analysis shows that the empirical formula of octane in petrol is C_4H_9, but the mass spectrum of octane shows that its relative molecular mass is 114.

The relative mass of the empirical formula is given by:

$$M_r(C_4H_9) = (4 \times 12) + (9 \times 1)$$
$$= 57$$

The relative molecular mass is twice this value – so, the molecular formula is twice the empirical formula.

$\therefore$ The molecular formula of octane $= C_8H_{18}$

Although the empirical and molecular formulae of organic compounds can be determined by analysis along the lines just described, the modern way to find molecular formulae is by mass spectrometry (Section 17.1).

Test yourself

6 A sample of a hydrocarbon was burned completely in oxygen. All the carbon in the sample was converted to 1.69 g of carbon dioxide, and all the hydrogen was converted to 0.346 g of water.
   a) What is the percentage of carbon in carbon dioxide?
   b) What is the mass of carbon in 1.69 g of carbon dioxide?
   c) What is the percentage of hydrogen in water?
   d) What is the mass of hydrogen in 0.346 g of water?
   e) Use the masses of carbon and hydrogen from parts b) and d) to calculate the empirical formula of the hydrocarbon.

7 A compound containing only carbon, hydrogen and oxygen was analysed. It consisted of 38.7% carbon and 9.68% hydrogen.
   a) What percentage of oxygen does it contain?
   b) What is its empirical formula?
   c) The relative molecular mass of the compound is 62. What is its molecular formula?

8 A hydrocarbon which consists of 82.8% carbon has an approximate relative molecular mass of 55.
   a) What is its empirical formula?
   b) What is its molecular formula?

Structural formulae

The molecular formula of a compound gives the numbers of atoms of the different elements in one molecule, but it does not show how the atoms are arranged. To understand the properties of a compound, we need to know its structural formula – this shows which atoms, or groups of atoms, are attached to each other. For example, the molecular formula of ethanol is C_2H_6O – but this does not show how the two carbon atoms, six hydrogen atoms and one oxygen atom are arranged. Its structural formula, however, is written as CH_3CH_2OH. This shows that ethanol has a CH_3 group attached to a CH_2 group which, in turn, is attached to an OH group.

Often, it is clearer to draw the full structural formula showing all the atoms and all the bonds (Figure 5.9). This type of formula is called a displayed formula.

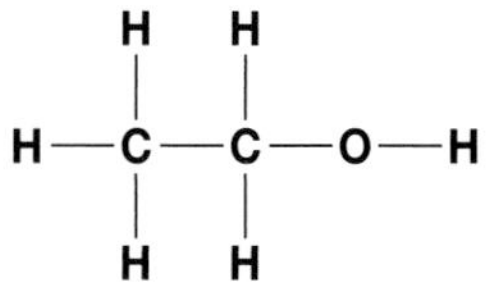

```
     H   H
     |   |
H — C — C — O — H
     |   |
     H   H
```

Figure 5.9 ▲
The displayed formula of ethanol

Definitions

A **structural formula** shows in minimal detail which atoms, or groups of atoms, are attached to each other in one molecule of a compound.

A **displayed formula** shows all the atoms and all the bonds between them in one molecule of a compound.

A **skeletal formula** shows the functional groups fully, but the hydrocarbon part of a molecule simply as lines between carbon atoms, omitting the symbols for carbon and hydrogen atoms.

Sometimes, skeletal formulae are used – these show only the carbon–carbon bonds and functional groups in a compound (Figure 5.10).

Molecular formula	Structural formula	Displayed formula	Skeletal formula
C_3H_8	$CH_3CH_2CH_3$	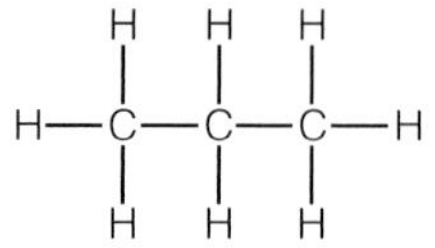 H H H H—C—C—C—H H H H	 ⌃
C_2H_6O	CH_3CH_2OH	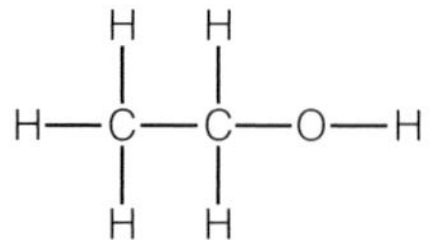 H H H—C—C—O—H H H	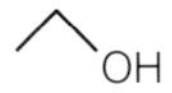 ⌃OH

Figure 5.10 ▲
Alternative formulae for propane and ethanol

Skeletal formulae are outline formulae only – they provide a useful shorthand for large and complex molecules. However, skeletal formulae need careful study because they show the hydrocarbon part of a molecule as nothing more than lines for the bonds between carbon atoms and for the bonds from carbon atoms to functional groups. The symbols for carbon and hydrogen atoms in the carbon skeleton are omitted. In contrast, functional groups are shown in full.

Activity

The alkanes – an important series of organic compounds

Methane (CH_4), ethane (C_2H_6), propane (C_3H_8) and butane (C_4H_{10}) are the first four members of the series of alkanes. Most of their empirical, molecular, structural and displayed formulae are shown in Table 5.2.

Name	Methane	Ethane	Propane	Butane
Empirical formula	CH_4		C_3H_8	C_2H_5
Molecular formula	CH_4	C_2H_6		C_4H_{10}
Structural formula		CH_3CH_3	$CH_3CH_2CH_3$	
Displayed formula	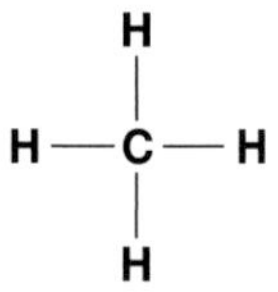 H H—C—H H		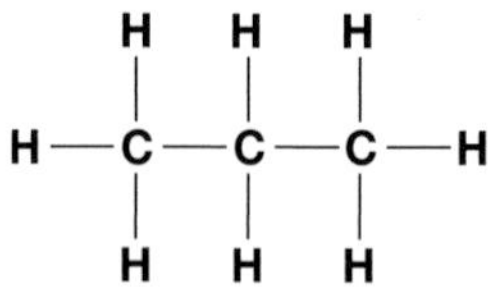 H H H H—C—C—C—H H H H	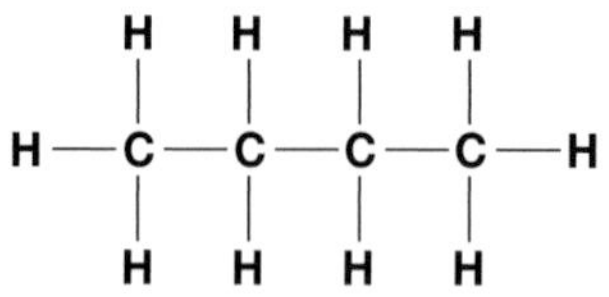 H H H H H—C—C—C—C—H H H H H

Table 5.2 ▲
Empirical, molecular, structural and displayed formulae of the simplest alkanes

1 Copy and complete the table by adding the missing formulae.

2 The names of all alkanes end in -ane. The names of the first four alkanes in Table 5.2 do not follow a logical system. All other straight-chain alkanes are named using a Greek numerical prefix for the number of carbon atoms in one molecule with the ending -ane. So, C_5H_{12} is pentane and C_7H_{16} is heptane. The prefixes are the same as those used for geometrical figures (pentagon, etc.).

What is the name for:

a) $CH_3CH_2CH_2CH_2CH_2CH_3$

b) $CH_3CH_2CH_2CH_2CH_2CH_2CH_2CH_3$?

3 Which of the following molecular formulae are alkanes?

C_2H_2 C_3H_8 C_4H_8 C_8H_{18} $C_{10}H_{20}$

4 Look at the formulae of methane, ethane, propane and butane in Table 5.2.

What is the difference in numbers of carbon and hydrogen atoms between:

a) one molecule of methane and one molecule of ethane

b) one molecule of ethane and one molecule of propane

c) one molecule of propane and one molecule of butane?

5 It is possible to write a general formula for alkanes in the form of C_xH_y.

a) Suppose x equals n. If an alkane has n carbon atoms, how many hydrogen atoms will it have? (Hint: In long-chain alkanes, every carbon atom has 2 hydrogen atoms, except the end carbon atoms, each of which has 1 extra hydrogen atom.)

b) What is the value of y in terms of n?

c) Write the general formula for alkanes in terms of C, H and n?

6 a) Draw the skeletal formula of butane.

b) Why is it not possible to draw a skeletal formula of methane?

5.4 Naming simple organic compounds

The International Union of Pure and Applied Chemistry (IUPAC) is the recognised authority for naming chemical compounds. IUPAC has developed systematic names based on a set of rules. These IUPAC rules make it possible to work out the structure of a compound from its name and to work out its name from its structure. The names of organic compounds are based on the longest straight chain or main ring of carbon atoms in the carbon skeleton.

CH_3 CH_2 CH_2 CH_2 CH_2 CH_3

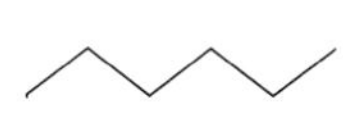

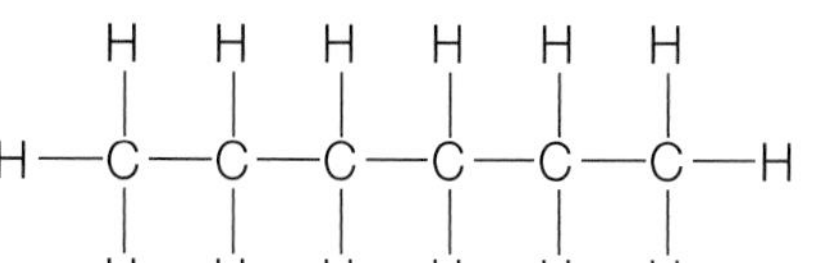

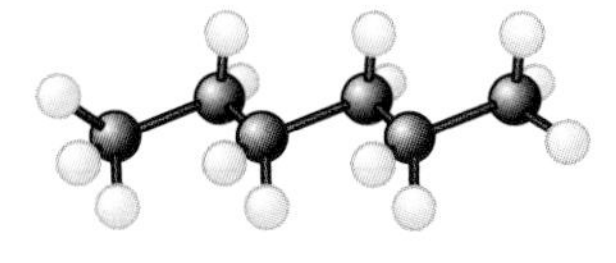

Figure 5.11 Four ways of representing the structure of the alkane with an unbranched chain of six carbon atoms. This is hexane.

Naming alkanes

In naming an alkane, it is important to follow the IUPAC rules.

1 Look for the longest unbranched chain of carbon atoms in the carbon skeleton of the molecule and name that part of the compound. So:

$\mathbf{CH_3CH_2CH_2CH_2CH_3}$ is pentane

$$\begin{array}{ccc} CH_3CH_2CH_2 & CH & CH_3 \\ & | & \\ & CH_3 & \end{array}$$ is pentane with a CH_3 group attached and

$$\begin{array}{ccccccccccccc} CH_3 & - & CH_2 & - & CH_2 & - & CH & - & CH & - & CH_2 & - & CH_3 \\ &&&&&& | && | &&&& \\ &&&&&& CH_2 && CH_3 &&&& \\ &&&&&& | &&&&&& \\ &&&&&& CH_3 &&&&&& \end{array}$$ is heptane with one CH_3 and one CH_3CH_2 group attached.

2 Identify the alkyl groups attached to the longest unbranched chain. The simplest alkyl group is the methyl group, CH_3, which is methane with one hydrogen atom removed. Alkyl groups are alkane molecules minus one hydrogen atom (Table 5.3).

Alkyl group	Formula
Methyl	CH_3-
Ethyl	CH_3CH_2-
Propyl	$CH_3CH_2CH_2-$
Butyl	$CH_3CH_2CH_2CH_2-$

Table 5.3 ▲
The structures of alkyl groups

So $$\begin{array}{ccc} CH_3CH_2CH_2 & CH & CH_3 \\ & | & \\ & CH_3 & \end{array}$$ has a methyl side group

and $$\begin{array}{ccccccccccccc} CH_3 & - & CH_2 & - & CH_2 & - & CH & - & CH & - & CH_2 & - & CH_3 \\ &&&&&& | && | &&&& \\ &&&&&& CH_2 && CH_3 &&&& \\ &&&&&& | &&&&&& \\ &&&&&& CH_3 &&&&&& \end{array}$$ has an ethyl side group and a methyl side group.

3 Number the carbon atoms in the main chain to identify which carbon atoms the side groups are attached to.

4 Name the compound using the name of the longest unbranched chain, prefixed by the names of the side groups and the numbers of the carbon atoms to which they are attached. The numbering of the carbon atoms can be from either the left or the right to give the name with the lowest numbers.

So $$\begin{array}{ccc} CH_3CH_2CH_2 & CH & CH_3 \\ & | & \\ & CH_3 & \end{array}$$ is 2-methylpentane – *not* 4-methylpentane.

and $$\begin{array}{ccccccccccccc} CH_3 & - & CH_2 & - & CH_2 & - & CH & - & CH & - & CH_2 & - & CH_3 \\ &&&&&& | && | &&&& \\ &&&&&& CH_2 && CH_3 &&&& \\ &&&&&& | &&&&&& \\ &&&&&& CH_3 &&&&&& \end{array}$$ is 4-ethyl-3-methylheptane – *not* 4-ethyl-5-methylheptane.

5 When there is more than one type of side group, they should be arranged alphabetically.

So $$\begin{array}{ccccccccccccc} CH_3 & - & CH_2 & - & CH_2 & - & CH & - & CH & - & CH_2 & - & CH_3 \\ &&&&&& | && | &&&& \\ &&&&&& CH_2 && CH_3 &&&& \\ &&&&&& | &&&&&& \\ &&&&&& CH_3 &&&&&& \end{array}$$ is 4-ethyl-3-methylheptane – *not* 3-methyl-4-ethylheptane.

6 When there are two or more of the same side group, add the prefix 'di', 'tri', 'tetra' and so on.

So

```
CH3–CH2–CH–CH–CH3
        |   |
       CH3 CH3
```

is 2,3-dimethylpentane – *not* 3,4-dimethylpentane and *not* 2-methyl-3-methylpentane.

Test yourself

9 Name the following alkanes.

a) $CH_3(CH_2)_6CH_3$

b)

```
CH3–CH–CH–CH–CH2–CH3
    |   |   |
   CH3 CH3 CH3
```

c)

```
CH3CH2CH2CH CH3
          |
          CH2
          |
          CH3
```

d)

```
        CH3
        |
CH3CH2C CH3
        |
        CH3
```

10 Draw the displayed formulae of the following alkanes:

a) 3,3,4-trimethylheptane

b) 2-methylbutane

c) 3-ethyl-2-methyl-5,5-dipropyldecane.

Tutorial

Naming alkenes

Ethene (CH_2=CH_2) and propene (CH_3CH=CH_2) are the first two members of the homologous series of alkenes with the functional group >C=C<.

Alkenes are named using the same general rules as alkanes, with the suffix -ene instead of -ane, sometimes prefixed by a number to indicate the position of the double bond in the chain.

With ethene and propene there is no need to number the carbon atoms because the double bond must be between carbon atoms 1 and 2. But with a chain of four or more carbon atoms, the double bond may be in more than one position.

Thus, the molecule CH_2=$CHCH_2CH_3$ is named but-1-ene. Although the double bond links carbon atoms 1 and 2, the number 1 is used because it is the lower.

Using the same rule, CH_3CH=$CHCH_3$ is named but-2-ene.

The methods of naming organic compounds with other functional groups will be explained as they arise.

Test yourself

11 Name the following alkenes.

a)

```
CH3—C=CH2
    |
    CH3
```

b) $CH_3CH_2CH_2CH$=$CHCH_3$

c)

```
CH3–CH2–C=C–CH3
        |  |
       CH3 CH3
```

12 Draw the displayed formulae of the following alkenes:

a) 2-methylbut-2-ene

b) 3,4-dimethylpent-1-ene.

5.5 Isomerism

Another reason why carbon forms so many compounds is that it is sometimes possible to join the same atoms together in different ways. Consider, for example, the molecular formula C_4H_{10}. You probably realise already that this could be butane – but there is another compound, 2-methylpropane, which also has the molecular formula C_4H_{10} (Figure 5.12). Compounds like butane and 2-methylpropane, which have the same molecular formula but different structural formulae, are called structural isomers.

Definition

Structural isomers are compounds with the same molecular formula but different structural formulae.

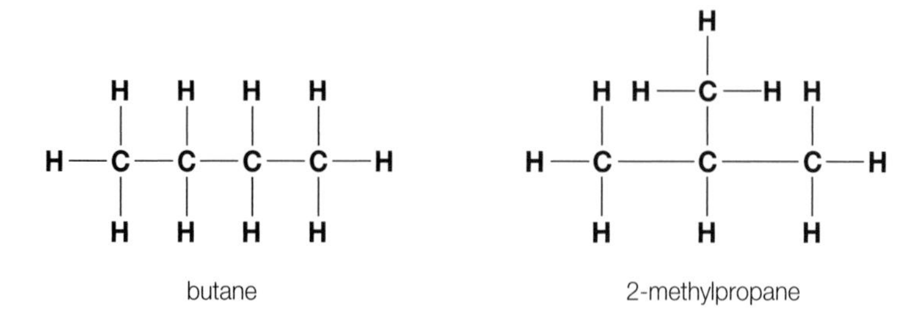

Figure 5.12 ▶
Isomers of C_4H_{10}

It is useful to divide structural isomers into three different types – chain isomers, position isomers and functional group isomers.

- *Chain isomers* have different chains of carbon atoms (Figure 5.12).
- *Position isomers* have different positions of the same functional group (Figure 5.13).

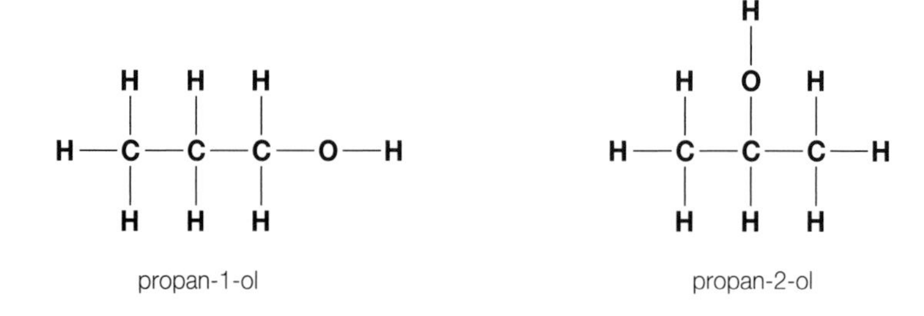

Figure 5.13 ▶
Propan-1-ol and propan-2-ol are both alcohols like ethanol, CH_3CH_2OH. All alcohols contain the –OH group. In propan-1-ol and propan-2-ol the –OH group is in different positions on the carbon chain.

- *Functional group isomers* have different functional groups (Figure 5.14).

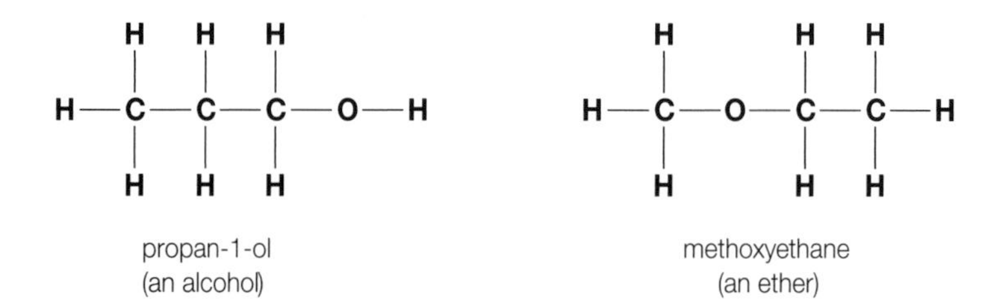

Figure 5.14 ▶
Propan-1-ol is an alcohol with the –OH functional group. Methoxyethane is an ether with the C–O–C functional group. Both these compounds have the same molecular formula, C_3H_8O.

Notice that the word describing the type of structural isomers (chain, position and functional group) tells you how they differ from each other.

Test yourself

13 a) Draw the carbon chains of the three chain isomers with the molecular formula C_5H_{12}. (Hint: One obvious chain is C–C–C–C–C.)
b) Use a molecular model kit to construct models of the three isomers, and look at the rotatable models of their structures on the student website.
c) Draw the displayed formulae of the three isomers and name them.

Figure 5.15 ▲
Most vehicles rely on petrol (gasoline) for fuel. Petrol is a mixture of more than 100 different alkanes with between 5 and 10 carbon atoms. There are three isomers of C_5H_{12}, five of C_6H_{14}, 18 of C_8H_{18} and 75 of $C_{10}H_{22}$.

5.6 Introduction to the mechanisms of organic reactions

The mechanism of a reaction shows, step by step, the bonds which break and the new bonds which form as reactants turn into products. Chemists have used great ingenuity to work out mechanisms. They began with simple reactions but are now using their knowledge to explain what happens during industrial processes and in living cells, where enzymes control biochemical processes.

There are two ways in which a bond can break – homolytically or heterolytically.

Homolytic bond breaking

A covalent bond involves a shared pair of electrons. If the bond breaks homolytically then each atom keeps one electron – this is 'equal splitting' (homolytic fission').

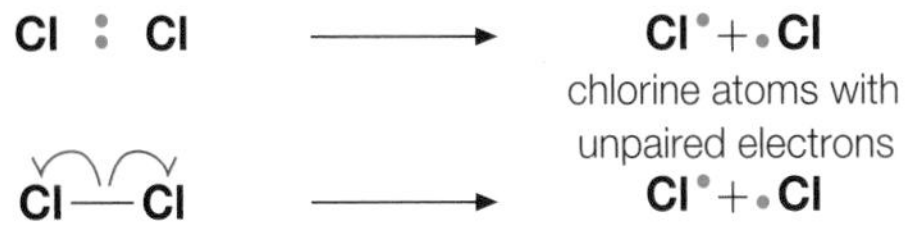

Cl : Cl ⟶ Cl• + •Cl
chlorine atoms with unpaired electrons

Cl—Cl ⟶ Cl• + •Cl

Figure 5.16 ◀
Homolytic bond breaking. Note the use of curly half-arrows to show what happens to the electrons as the bond breaks. The covalent bond breaks and the atoms separate, each taking one of the shared pair of electrons.

This type of splitting produces fragments with unpaired electrons called free radicals. Free radicals exist for a short time during a reaction, but quickly react to form new products. So, free radicals are usually very short-lived. They are intermediates which form during a reaction but then disappear as the reaction is completed.

Chemists show the formation of free radicals using curly half-arrows and use a single dot to represent the unpaired electron. Other paired electrons in the outer shells are not usually shown.

Free radicals often occur in reactions taking place in the gas phase or in a non-polar solvent. Ultraviolet light can speed up free-radical reactions.

Examples of free-radical processes include the thermal cracking of hydrocarbons (Section 6.4), the burning of petrol and other alkanes (Section 6.3) and the substitution reactions of alkanes with halogens (Section 6.3). Free-radical reactions are important high in the atmosphere where gases are exposed to intense ultraviolet radiation from the Sun. The reactions which form and destroy the ozone layer are free-radical reactions (Section 19.5).

Note

The prefix 'homo' means 'the same' or 'similar'. Chemical terms which include this prefix include homolytic fission, homogeneous catalyst and homologous series.

The prefix 'hetero' means 'different'. Chemical terms which include this prefix include heterolytic fission and heterogeneous catalyst.

Heterolytic bond breaking

When a covalent bond breaks heterolytically, one atom takes both of the electrons from the bond, leaving the other atom with none.

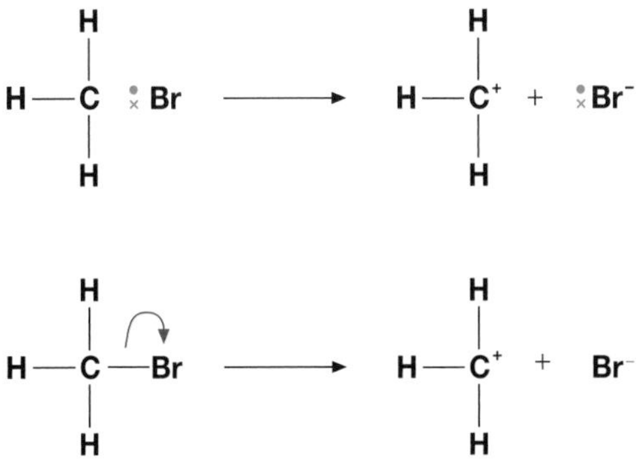

H_3C : Br ⟶ H_3C^+ + $:Br^-$

H_3C—Br ⟶ H_3C^+ + Br^-

Figure 5.17 ◀
Heterolytic bond breaking. Note the use of a curly full arrow to show what happens to the electrons as the bond breaks. The covalent bond breaks and the atoms separate, with one atom taking both electrons in the shared pair. The head of the arrow points to where the electron pair will be after the change.

Heterolytic bond breaking (heterolytic fission) produces ionic intermediates, such as CH_3^+ and Br^- in reactions like that shown in Figure 5.17. This type of bond breaking is favoured when reactions take place in polar solvents such as water. Often, the bond which breaks is already polar (Section 8.3) with a δ+ end and a δ− end like the C−Br bond in Figure 5.18.

Figure 5.18 ▶
An ionic reagent attacking a polar bond leading to heterolytic bond breaking

Some of the reagents which initiate reactions seek out the δ+ end of polar bonds – these are called nucleophiles.

Other reagents seek out the δ− end of polar bonds or the electron dense regions in molecules – these are called electrophiles.

Nucleophiles

Nucleophiles are molecules or ions with a lone pair of electrons which can form a new covalent bond (Figure 5.19). They are electron-pair donors. Nucleophiles are reagents which attack molecules that have a partial positive charge, δ+, so they seek out positive charges – they are 'nucleus loving'.

The substitution reactions of halogenoalkanes involve nucleophiles (Section 15.6).

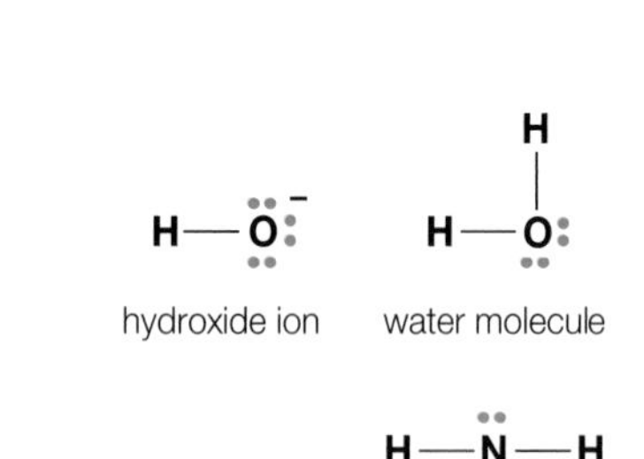

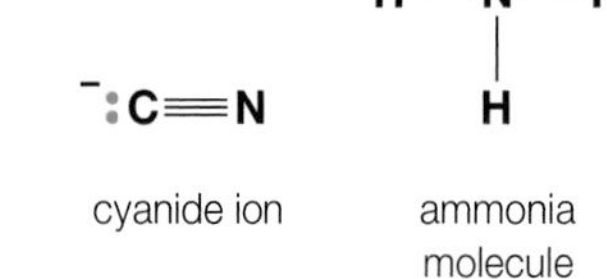

ammonia molecule

Figure 5.19 ▲
Examples of nucleophiles

Electrophiles

Electrophiles are molecules or ions that attack parts of molecules which are rich in electrons – negative ions or negative centres, δ−, in molecules. They are 'electron-loving' reagents. Electrophiles form a new bond by accepting a pair of electrons from the molecule attacked during a reaction.

An example of an electrophile is the H atom at the δ+ end of the H–Br bond in hydrogen bromide. See, for example, the electrophilic addition reactions of alkenes (Section 6.9).

Definitions

Free radicals are highly reactive single atoms, or groups of atoms, with unpaired electrons.

Nucleophiles are ions or molecules with a lone pair of electrons that attack positive ions or positive centres in molecules.

Electrophiles are reactive ions and molecules which attack negative ions or negative centres in molecules.

Test yourself

14 Write equations showing how:
  a) a bromine molecule breaks homolytically
  b) a hydrogen bromide molecule breaks heterolytically
  c) a C–H bond in a methane molecule breaks homolytically.

15 In each of the following examples decide whether the reagent attacking the carbon compound is a free radical, a nucleophile or an electrophile:
  a) $CH_3CH_2I + H_2O \rightarrow CH_3CH_2OH + HI$
  b) $CH_2{=}CH_2 + HBr \rightarrow CH_3CH_2Br$
  c) $Cl\bullet + CH_4 \rightarrow CH_3Cl + H\bullet$
  d) $CH_3CH_2Br + CN^- \rightarrow CH_3CH_2CN + Br^-$

REVIEW QUESTIONS

Extension questions

1 a) Carbon is able to form a vast number of chemical compounds. Suggest two reasons for this. (2)

b) Petrol is a mixture of hydrocarbons containing between 6 and 10 carbon atoms. Some of these hydrocarbons are structural isomers.

i) Explain the term 'structural isomers'. (2)

ii) Some of the hydrocarbons in petrol are alkanes. Write the molecular formula of an alkane that could be present in petrol. (1)

c) Petrol also contains cycloalkanes. Draw the structure of cyclohexane and write the general formula of cycloalkanes. (2)

2 The compound X below, containing two functional groups, can be extracted from oil of violets.

CH_3CH_2 and H on one carbon, CH_2OH and H on the other: $C{=}C$

X

Figure 5.20 ▲

a) State the empirical and molecular formula of X and draw its skeletal formula. Explain the term 'functional group' and name the functional groups present in X. (7)

b) X reacts with hydrogen and a nickel catalyst in the gas phase to produce compound Y with the formula $CH_3(CH_2)_3CH_2OH$.

i) What is the systematic name of Y? (1)

ii) Write an equation, including state symbols, for the reaction of X with hydrogen to form Y. (2)

iii) Y has two structural isomers which are also position isomers. Name these two position isomers of Y. (2)

iv) Y also has structural isomers with a different functional group. Write the structural formula of one of these isomers. (1)

3 a) Crude oil is a mixture of many hydrocarbons. Using fractional distillation it can be separated into fractions which can be refined to produce hydrocarbons such as dodecane.

i) What is meant by the term 'hydrocarbon'? (1)

ii) One molecule of dodecane contains twelve carbon atoms. What is the molecular formula of dodecane? (1)

iii) What is the empirical formula of dodecane? (1)

b) Decane, $C_{10}H_{22}$, is a straight-chain alkane. It reacts with chlorine in a free-radical reaction to form the compound $C_{10}H_{21}Cl$.

i) Explain the term 'free radical'. (2)

ii) Write an equation, including curly half-arrows, for the formation of chlorine free radicals from Cl_2. (2)

iii) What type of bond fission is involved in the formation of chlorine free radicals? (1)

iv) How many different structural isomers can be produced when decane reacts with chlorine to form $C_{10}H_{21}Cl$? (1)

v) Draw the structural formula of one of these structural isomers. What is its name? (2)

6 Hydrocarbons: alkanes and alkenes

Hydrocarbons are important because they make up the majority of crude oil – the source of most fuels and the main raw material for the chemical industry. Hydrocarbons are compounds containing only hydrogen and carbon. The most common and most important hydrocarbons are alkanes and alkenes.

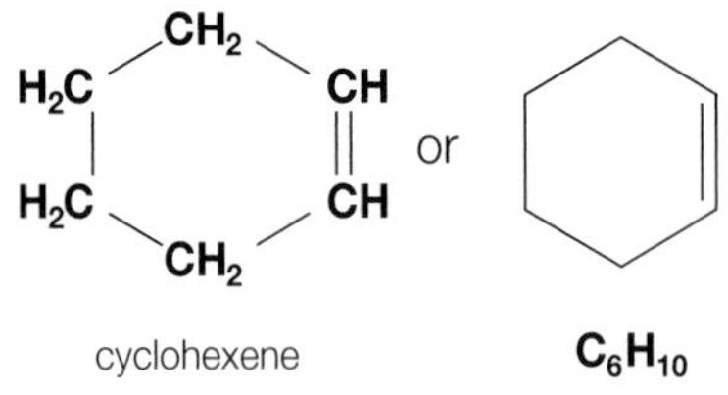

6.1 Types of hydrocarbons

There are three important types of hydrocarbons.

- *Aliphatic hydrocarbons* are those with no rings of carbon atoms – their chains of carbon atoms may be branched or unbranched. Alkanes and alkenes are examples of aliphatic compounds.
- *Alicyclic hydrocarbons* are those with rings of carbon atoms, such as cycloalkanes and cycloalkenes (Figure 6.1). Alicyclic hydrocarbons are often simply called cyclic hydrocarbons.
- *Arenes* are hydrocarbons, such as benzene, methylbenzene and naphthalene. They are ring compounds in which there are delocalised electrons (Figure 6.2). Arenes are sometimes called 'aromatic hydrocarbons' because of their smells (aromas).

cyclohexene or C_6H_{10}

Figure 6.1 ▲
Structures of the cyclic hydrocarbon cyclohexene, which is a cycloalkene.

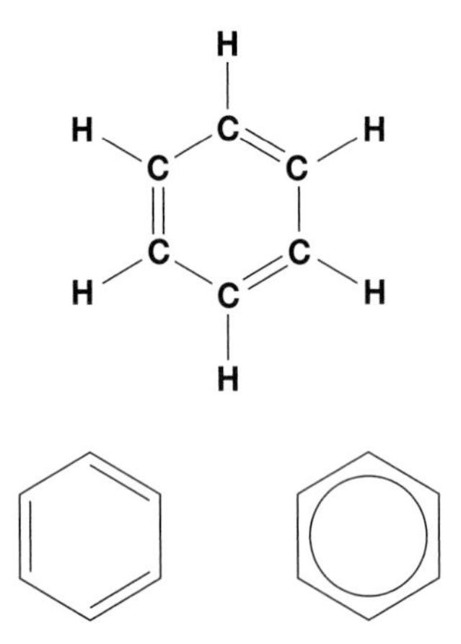

Figure 6.2 ▲
Representations of the structure of benzene. At one time, chemists thought that the ring structure in benzene had three double and three single bonds. X-ray studies have shown that all six bonds in the ring are identical and that each carbon atom contributes one electron to a cloud of delocalised electrons. This has led to the third structure with a ring inside a hexagon.

Definitions

Aliphatic hydrocarbons are straight-chain and branched hydrocarbons with no rings of carbon atoms.

Alicyclic hydrocarbons are hydrocarbons with at least one ring of carbon atoms.

Arenes are hydrocarbons with a ring or rings of carbon atoms in which there are delocalised electrons.

6.2 Alkanes

Alkanes are the hydrocarbons which make up most of crude oil and natural gas. Alkanes form a series of organic compounds with the general formula C_nH_{2n+2}.

Alkanes are saturated compounds – they have only single bonds between the atoms in their molecules. The term 'saturated' is also used for compounds with saturated hydrocarbon chains, such as saturated fats and fatty acids in food.

Figure 6.3 ▲
The saturated fats in foods such as fish and chips, beefburgers and doughnuts contain alkyl groups with long chains of carbon atoms. These substances, if eaten in excess, lead to high levels of cholesterol in the blood which causes furring and blocking of the arteries (arteriosclerosis).

Physical properties

Alkanes are composed of simple molecules which are held together by only weak intermolecular forces (Section 9.1). As the molecules get larger, the intermolecular forces increase, and therefore the melting and boiling temperatures rise as the number of carbon atoms per molecule increases.

At room temperature and pressure, alkanes in the range C_1 to C_4 are gases, those in the range C_5 to C_{17} are liquids, while those from C_{18} upwards are solids.

Test yourself

1 a) Write the molecular formulae and names of the first four members of the alkanes.
b) What is the difference in molecular formula from one alkane to the next?

2 Why is the general formula of alkanes C_nH_{2n+2}?

6.3 Chemical reactions of the alkanes

The bond enthalpies of C–C and C–H bonds are relatively high, so the bonds in alkanes are difficult to break. In addition, these bonds are non-polar (Section 9.1). This means that alkanes are very unreactive with ionic reagents in water – such as acids, alkalis, oxidising agents and reducing agents. There are, however, three important reactions of alkanes involving homolytic bond breaking and free radicals (Section 5.6). These three important reactions are burning (combustion), halogenation and cracking.

Burning (combustion)

Many common fuels consist mainly of alkanes. In a plentiful supply of air or oxygen, the alkanes are completely oxidised to carbon dioxide and water. The reaction is highly exothermic.

$$C_4H_{10}(g) + 6\tfrac{1}{2}O_2(g) \rightarrow 4CO_2(g) + 5H_2O(l) \quad \Delta H_c^\ominus = -2876\,\text{kJ mol}^{-1}$$

butane in Gaz®

Figure 6.4 ▲
Red Calor gas cylinders contain propane for use as a fuel.

If the air is in short supply, the products include soot (carbon) and highly toxic carbon monoxide as well as carbon dioxide. Alkanes are kinetically stable in the air (oxygen), but they are energetically (thermodynamically) unstable with respect to the products of oxidation.

The combustion of alkanes involves a free-radical mechanism, which occurs rapidly in the gas phase. This means that liquid and solid alkanes must vaporise before they burn and it explains why less volatile alkanes burn less easily.

The burning of alkanes is immensely important in any advanced, technological society. It is used to generate energy of one kind or another in power stations, furnaces, domestic heaters, cookers, candles and vehicles. Unfortunately this mass burning of alkanes is now accepted to be the major cause of global warming and the increased greenhouse effect (Section 19.3).

Reactions with chlorine and bromine (halogenation)

Alkanes react with chlorine and bromine, either on heating or on exposure to ultraviolet light. During the reactions, hydrogen atoms in the alkane molecules are replaced (substituted) by halogen atoms. These are described as substitution reactions.

Any of the hydrogen atoms in an alkane may be replaced, and the reaction can continue until all the hydrogen atoms have been substituted by halogen atoms. Consequently, the product is a mixture of compounds.

In strong sunlight, methane and chlorine react explosively. The initial products are chloromethane, CH_3Cl, and hydrogen chloride.

Definition

A **substitution reaction** is one in which an atom, or group of atoms, is replaced (substituted) by another atom, or group of atoms.

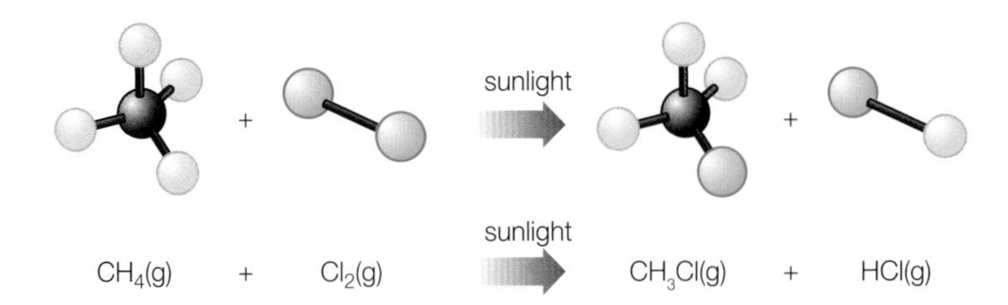

Figure 6.5 ▶
The equation and models representing the initial substitution reaction of methane with chlorine

The reaction involves breaking some bonds – for which energy must be supplied – and making new bonds – when energy is released. Possible reaction mechanisms can be tested using bond enthalpies (energies) and this leads to a probable reaction mechanism.

The reaction between methane and chlorine will not occur in the dark because the molecules do not have enough energy for bonds to break when they collide. But in ultraviolet light, the energy provided by absorbed photons is 400 kJ mol^{-1}. This is enough to cause homolytic fission of chlorine molecules into free radicals:

$$Cl_2 \rightarrow Cl\bullet + Cl\bullet \quad \Delta H = +242\,kJ\,mol^{-1}$$

But this is not enough for the homolytic fission of methane, which requires 435 kJ mol^{-1}:

$$CH_4 \rightarrow CH_3\bullet + H\bullet \quad \Delta H = +435\,kJ\,mol^{-1}$$

and definitely not enough for the heterolytic fission of either chlorine or methane:

$$Cl_2 \rightarrow Cl^+ + Cl^- \quad \Delta H = +1130\,kJ\,mol^{-1}$$

$$CH_4 \rightarrow H^+ + CH_3^- \quad \Delta H = +1700\,kJ\,mol^{-1}$$

These figures suggest that ultraviolet light starts the reaction by splitting chlorine molecules into chlorine atoms (free radicals). This stage is called initiation.

The chlorine atoms, each with an unpaired electron, are highly reactive. They remove hydrogen atoms from methane molecules to form hydrogen chloride and a new free radical, $CH_3\bullet$:

$$Cl\bullet + CH_4 \rightarrow HCl + CH_3\bullet \quad \Delta H = +4\,kJ\,mol^{-1}$$

The $CH_3\bullet$ free radical now reacts with more chlorine to form chloromethane, CH_3Cl, and generate another chlorine free radical:

$$CH_3\bullet + Cl_2 \rightarrow CH_3Cl + Cl\bullet \quad \Delta H = -97\,kJ\,mol^{-1}$$

The new Cl• free radical can react with another CH_4 molecule and the last two reactions can be repeated again and again until either all the Cl_2 or all the CH_4 is used up. These two repeated reactions create a chain reaction and are described as propagation stages.

Propagation ends when two free radicals combine. This is the termination stage of the reaction, which is very exothermic:

$$Cl\bullet + Cl\bullet \rightarrow Cl_2 \quad \Delta H = -242\,kJ\,mol^{-1}$$

$$CH_3\bullet + Cl\bullet \rightarrow CH_3Cl \quad \Delta H = -339\,kJ\,mol^{-1}$$

$$CH_3\bullet + CH_3\bullet \rightarrow CH_3CH_3 \quad \Delta H = -346\,kJ\,mol^{-1}$$

The three stages in the free-radical substitution of methane with chlorine (initiation, propagation and termination) are summarised in Figure 6.6.

Definitions

Free-radical chain reactions involve three stages:

initiation – the step which produces free radicals

propagation – steps which form products and more free radicals

termination – steps which remove free radicals by turning them into molecules.

Figure 6.6 ▶
Stages in the free-radical substitution of methane with chlorine

Stage 1 Initiation $Cl : Cl \xrightarrow{light} Cl\bullet + Cl\bullet$

Stage 2 Propagation
$Cl\bullet + CH_4 \rightarrow HCl + CH_3\bullet$
$CH_3\bullet + Cl_2 \rightarrow CH_3Cl + Cl\bullet$
(chain reaction)

Stage 3 Termination
$Cl\bullet + Cl\bullet \rightarrow Cl_2$
$CH_3\bullet + Cl\bullet \rightarrow CH_3Cl$
$CH_3\bullet + CH_3\bullet \rightarrow CH_3CH_3$

0 s

10 s

16 s

Figure 6.7 ▲
The effect of light on a mixture of bromine and hexane after 0,10 and 16 seconds

Test yourself

3 Why is a series of organic compounds, such as the alkanes, comparable to a group of elements in the periodic table?

4 a) Write an equation for the complete combustion of propane in Calor gas.
b) What are the products formed when propane burns in a poor supply of oxygen?

5 Why is it dangerous to allow a car engine to run in a garage with the door closed?

Data

6 a) Use the data sheet on the Dynamic Learning Student website to find the boiling temperatures of propane and butane.
b) Why is it wise for campers to use Calor gas (propane) rather than Gaz (butane) for cooking during the winter?

7 Write equations for all four possible substitution reactions when chlorine reacts with methane and name the products.

8 a) Why does a mixture of bromine in hexane remain orange in the dark, but fade and become colourless in sunlight?
b) Write an equation for the reaction in part a).
c) Why can acidic fumes be detected above the solution once the colour has faded?

6.4 Fuels from crude oil

Crude oil, also known as petroleum, is arguably the most important naturally occurring raw material. It provides a very large proportion of our energy needs, and is the source of most of our organic chemicals – including plastics, fibres, drugs and pesticides.

Crude oil is a complex mixture of hydrocarbons, most of which are alkanes. Crude oil has no uses in its raw form. The challenge for refineries is to produce the various oil products in the proportions required by industrial and domestic users.

Generally, crude oil contains too much of the high boiling fractions with larger molecules, and not enough of the low boiling fractions with the smaller molecules needed for fuels such as petrol. In order to satisfy the demand for very different products, crude oil undergoes three main processes – fractional distillation, cracking and reforming.

Fractional distillation

Fractional distillation is the first stage in refining crude oil. This produces fuels and lubricants as well as feedstocks for the petrochemical industry. The continuous process operates on a large scale, separating crude oil into different fractions.

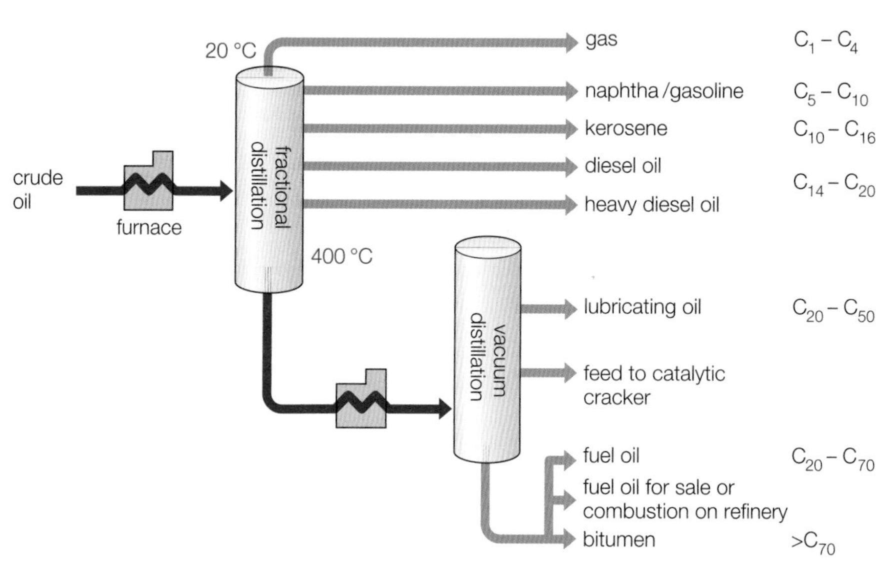

Figure 6.8 ▶ The fractional distillation of crude oil

A furnace heats the crude oil to about 400 °C. It then flows into a fractionating tower containing 40 or so horizontal 'trays' pierced with small holes.

The column is hotter at the bottom and cooler at the top. Rising vapour condenses when it reaches the tray with liquid at a temperature just below its boiling temperature. Condensing vapour releases energy. This heats the liquid on the tray and evaporates the more volatile compounds in the mixture on the tray.

With a series of trays, the outcome is that hydrocarbons with small molecules and low boiling temperatures rise to the top of the column, while larger molecules stay at the bottom. Fractions are drawn off from the column at various levels.

Some components of crude oil have boiling temperatures too high for them to vaporise at the furnace temperature and atmospheric pressure. Lowering the pressure in a separate vacuum distillation column reduces the boiling temperatures of these hydrocarbons and this makes it possible to separate them.

Fuel fractions

Petrol is a blend of hydrocarbons based on the gasoline fraction (hydrocarbons with 5–10 carbon atoms), whereas jet fuel is produced from the kerosene (or paraffin) fraction (hydrocarbons with 10–16 carbon atoms). Fuel for diesel engines is made from diesel oil (hydrocarbons with 14–20 carbon atoms).

All these fuels must be refined to remove sulfur compounds, which would cause air pollution when they burn. In addition, petrol must be blended carefully if modern engines are to start reliably and run smoothly. The proportion of volatile hydrocarbons added to petrol is higher in winter to help cold-starting, but lower in summer to prevent vapour forming too readily.

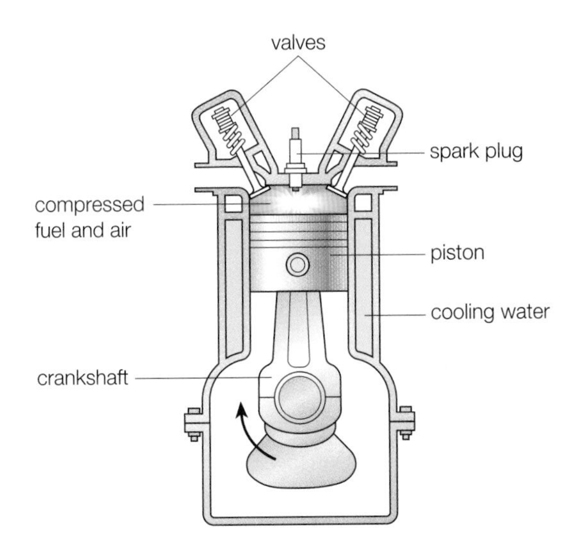

Figure 6.9 ◀ The working parts of a cylinder in an internal combustion engine that runs on petrol. Sparks from the plugs cause the compressed fuel and air to ignite. This produces more gas molecules, increasing the pressure and forcing the piston down. The product gases are then allowed to escape and, as the pressure falls, the piston rises ready for the next ignition.

For smooth running, petrol must burn smoothly in the engines of vehicles and not in fits and starts. To ensure smooth combustion, companies produce high-octane fuel by increasing the proportions of branched alkanes and arenes, or blending-in oxygen compounds. The three main approaches are:

- *cracking* – which makes smaller molecules and converts straight chain hydrocarbons to branched and cyclic hydrocarbons
- *reforming* – which turns straight chain and cyclic alkanes into arenes such as benzene and methylbenzene
- *adding ethanol and ethers* such as ETBE (initials based on its older name ethyl tertiary butyl ether. Its correct IUPAC name is 2-ethoxy-2-methylpropane). In the USA and Brazil, a mixture of 10% ethanol and 90% petrol (gasoline), known as gasohol, is commonly used.

$$CH_3—CH_2—O—C(CH_3)_2—CH_3$$

Figure 6.10 ▲ Structure of the ether ETBE. Adding ETBE is one of the methods used to raise the octane number of gasoline from as low as 70 to 120, the required level for unleaded premium petrol.

Catalytic cracking

Catalytic cracking converts heavier fractions, such as diesel oil and fuel oil, from the fractional distillation of crude oil into more useful hydrocarbon fuels by breaking up large molecules into smaller ones.

Cracking converts long-chain alkanes with twelve or more carbon atoms into smaller, more useful molecules in a mixture of branched alkanes, cycloalkanes, alkenes and branched alkenes.

The catalyst is a synthetic sodium aluminium silicate belonging to a class of compounds called 'zeolites'. A zeolite has a three-dimensional structure (Figure 6.11) in which the silicon and oxygen atoms form tunnels and cavities into which small molecules can fit. Cracking takes place on the surface of the catalyst at about 500 °C.

Synthetic zeolites make excellent catalysts because they can be developed with active sites to favour the shapes and sizes of those molecules which react to give the desired products.

Catalytic cracking is a continuous process. The finely powdered catalyst gradually becomes coated with carbon, so it circulates through a regenerator where the carbon burns away in a stream of air (Figure 6.12).

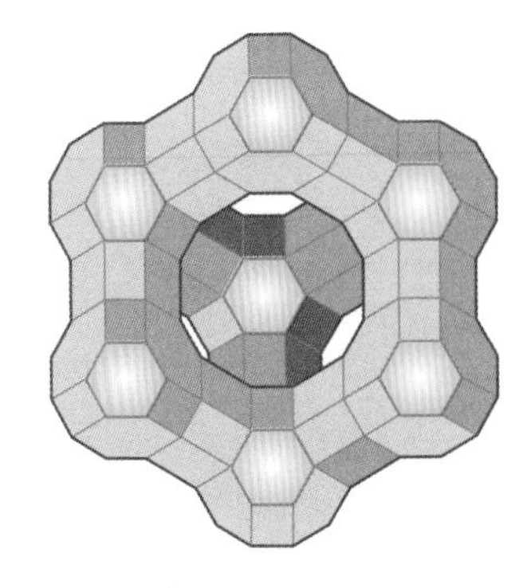

Figure 6.11 ▲ A model of the structure of a zeolite crystal

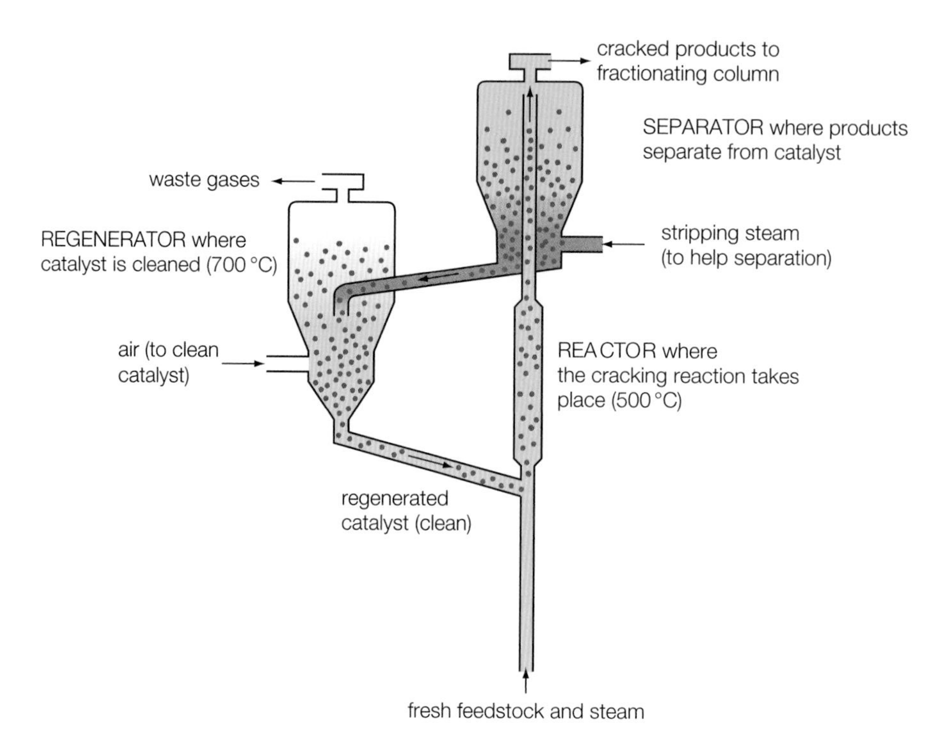

Figure 6.12 ▶ Catalytic cracking – fresh feedstock and catalyst powder flow to the vertical reactor where cracking takes place. The cracked products pass to a fractionating column, while the catalyst flows to the regenerator.

Reforming

Reforming converts straight-chain and cyclic alkanes into arenes (aromatic hydrocarbons) such as benzene and methylbenzene (Section 6.1). Hydrogen is a valuable by-product of the process – it can be used in processes elsewhere at the refinery.

The catalyst for this process is often one or more of the precious metals such as platinum and rhodium supported on an inert material such as aluminium oxide. The process operates at about 500 °C.

Figure 6.13 ▶ Examples of reforming

$$CH_3CH_2CH_2CH_2CH_2CH_2CH_3 \xrightarrow[\text{500 °C, high pressure}]{\text{Pt catalyst}} C_6H_5CH_3 + 4H_2$$

heptane → methylbenzene

$$C_6H_{12} \xrightarrow[\text{500 °C, high pressure}]{\text{Pt catalyst}} C_6H_6 + 3H_2$$

cyclohexane → benzene

Test yourself

9 a) Why is crude oil so important?
b) Why should we try to conserve our reserves of crude oil?

10 a) Why do you think that ethanol and ETBE can raise the octane number of petrol?
b) What other methods are used to increase the octane number of petrol?

11 a) Why is cracking important?
b) What conditions are used for cracking?

12 Write structural formulae to show how:
a) catalytic cracking converts hexane to butane and ethene
b) catalytic cracking converts decane, $C_{10}H_{22}$, to 2,3-dimethylpentane and propene.
c) reforming converts hexane to cyclohexane.

6.5 Alternative fuels

Crude oil is not the only source of organic raw materials and fuels. Industries are turning increasingly to plants as the source of chemicals and fuels because of the impact on the environment from our use of fossil fuels.

The concentrations of greenhouse gases, particularly carbon dioxide, in the air are rising. This is primarily due to human activity, especially the burning of fossil fuels in power stations, in aviation, in industries, in our homes and in our vehicles. The carbon dioxide produced is enhancing the greenhouse effect and there is mounting evidence that this is responsible for global warming and climate change (Section 19.3).

Concern about the enhanced greenhouse effect is encouraging people and governments to reduce their CO_2 emissions and seek a more sustainable lifestyle. This involves planning to live within the means of the environment in order that the Earth's natural resources are not destroyed, but remain available for future generations.

> **Definition**
>
> A **sustainable lifestyle** involves living within the means of the environment in order that the Earth's natural resources are available for future generations.

Any attempt to reduce CO_2 emissions means seeking alternatives to fossil fuels, such as biofuels – bioethanol and biodiesel.

Crops take in carbon dioxide from the air as they photosynthesise, making sugars and vegetable oils that can be used to produce biofuels. When the biofuels burn, the carbon dioxide taken up during photosynthesis is returned to the air. This analysis suggests that the use of biofuels should have no overall effect on the level of carbon dioxide in the atmosphere. Because of this, biofuels are sometimes described as 'carbon neutral' (Section 19.4).

However, this ignores the carbon dioxide released from fossil fuels during the mechanical planting, harvesting and processing of the crop, and in the manufacture of fertilisers applied to the crop during the growing season.

The principal biofuels now being used are bioethanol and biodiesel. Bioethanol is manufactured by fermenting carbohydrates such as starch and sugar in crops like sugar cane.

Fermentation converts starch to glucose, and then glucose to ethanol and carbon dioxide. The process is catalysed by enzymes in yeast.

$$C_6H_{12}O_6(aq) \rightarrow 2CO_2(g) + 2C_2H_5OH(aq)$$

glucose   carbon dioxide   ethanol

Biodiesel is produced by extracting and processing the oils from crops such as rapeseed.

Figure 6.14 ▶
Brazil is the leading country to use bioethanol. This photo shows a petrol station in São Paulo with separate pumps for ethanol ('álcool'), petrol ('gasolina') and an ethanol/petrol mixture.

Test yourself

13 Why are governments around the world becoming increasingly concerned about our use of fossil fuels?

14 Suggest three ways in which our use of fossil fuels might be reduced.

15 Biofuels are sometimes described as 'carbon neutral'. Why is this?

16 Explain what is meant by the term 'sustainable development', using examples from the manufacture of fuels.

Definition

Unsaturated compounds contain one or more double or triple bonds between atoms in their molecules. The term is often applied to alkenes and to describe unsaturated fats which have C=C double bonds in their hydrocarbon chains.

6.6 Alkenes

Alkenes, such as ethene and propene, are invaluable to chemists because they react with a wide range of chemicals to make useful products. Both ethene and propene are produced during the cracking of heavier fractions from crude oil. They are important starting points for the synthesis of other chemicals because of the reactivity of their double bonds.

The alkenes form a series of organic compounds in which the functional group is a carbon–carbon double bond, C=C. Because of this, their molecules have two atoms of hydrogen fewer than the corresponding alkane, and their general formula is C_nH_{2n}. As they have less than the maximum content of hydrogen, they are described as being 'unsaturated'.

Names and structures

The name of an alkene is based on the name of the corresponding alkane with the ending -ene in place of -ane (Figure 6.15).

Figure 6.15 ▶
The names and structures of the four simplest alkenes

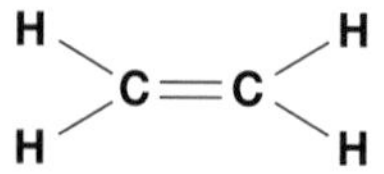

$H_2C{=}CH_2$

ethene

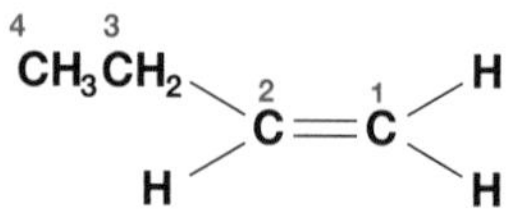

$\overset{4}{C}H_3\overset{3}{C}H_2(H)\overset{2}{C}{=}\overset{1}{C}H_2$

but-1-ene

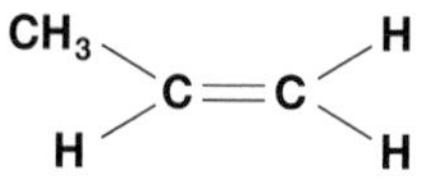

$CH_3(H)C{=}CH_2$

propene

$\overset{4}{C}H_3(H)\overset{3}{C}{=}\overset{2}{C}(H)\overset{1}{C}H_3$

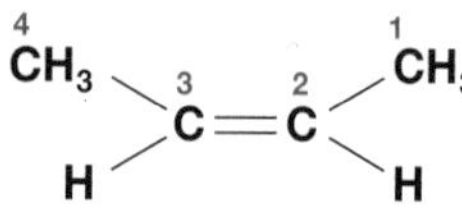

but-2-ene

Where necessary, a number in the name shows the position of the double bond, as in the structural isomers but-1-ene and but-2-ene. Counting starts from the end of the chain that gives the lowest possible number in the name. This number indicates the first of the two atoms connected by the double bond. In but-1-ene, for example, the double bond is between carbon atoms numbered 1 and 2.

Physical properties

Like the alkanes, the melting and boiling temperatures of alkenes increase as the number of carbon atoms in the molecules increase. Ethene, propene and the butenes are gases at room temperature. Alkenes, like other hydrocarbons, do not mix with or dissolve in water.

www Data

The double bond in alkenes

Chemists have extended the theory of atomic orbitals (Section 3.6) to describe the distribution of electrons in molecules. This molecular orbital theory is helpful in discussing the bonding and reactivity of alkenes.

Molecular orbitals result when atomic orbitals overlap forming bonds between atoms. The shape of a molecular orbital shows the regions in space where there is a high probability of finding electrons. A sigma (σ) bond is a single covalent bond formed by a pair of electrons in an orbital in a molecule with the electron density concentrated between two nuclei. Free rotation is possible around single bonds.

Sigma bonds can form by overlap of two s orbitals, an s orbital and a p orbital, or two p orbitals (Figure 6.16).

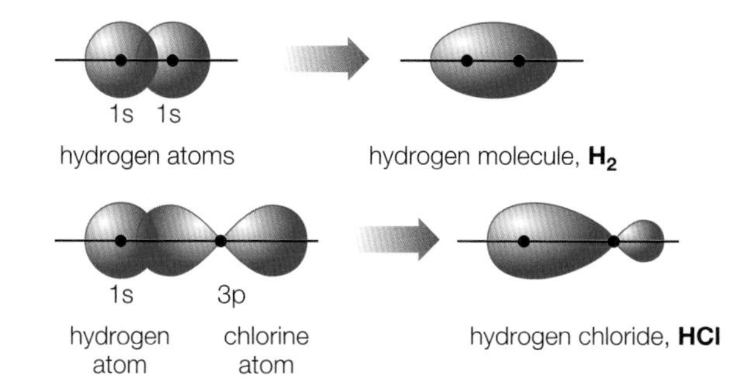

Figure 6.16 ◀ Examples of sigma bonds in molecules

A pi (π) bond is the type of bond found in molecules with double and triple bonds. The bonding electrons are in a π orbital formed by the *sideways overlap* of two atomic p orbitals. In a π bond, the electron density is concentrated in two regions – one above and the other below the plane of the molecule, on either side of the line between the nuclei of the two atoms joined by the bond (Figure 6.17).

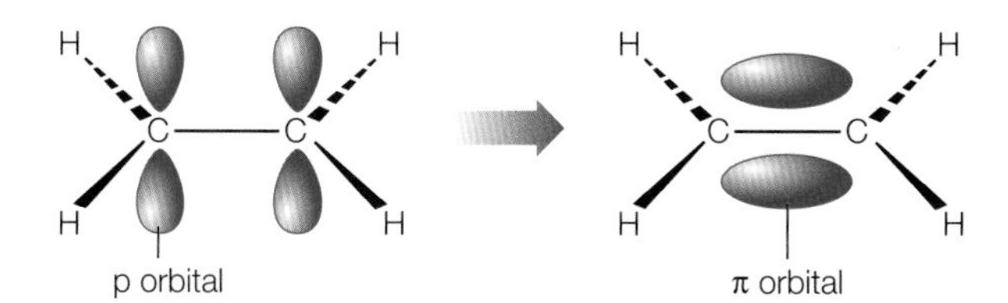

Figure 6.17 ◀ The π bond in ethene

6.7 *Cis–trans* isomerism

X-ray diffraction studies show that the ethene molecule is planar (Figure 6.17) with the three atoms around each carbon atom arranged trigonally at approximately 120°. However, the CH_2 groups in ethene cannot be rotated around the carbon–carbon double bond. In ethane and other alkanes, it is possible to rotate the whole molecule around a single σ bond because this does not affect the overlap of orbitals (Figure 6.18). But with a double bond, rotation would involve breaking the π bond and this requires more energy than the molecules possess at normal temperatures. So, free rotation is not possible around the carbon–carbon double bond in alkenes and this gives rise to *cis* and *trans* isomers.

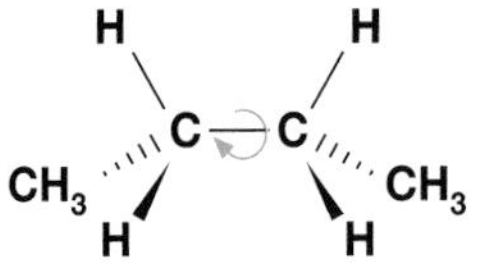

These two structures are the same compound because the ends of the molecule can rotate freely around the single bond.

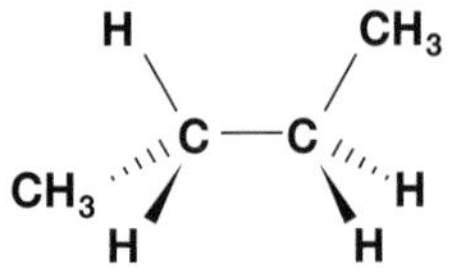

Figure 6.18 ▶
Free rotation can occur around a single σ bond

Cis and *trans* isomers have different shapes and therefore different displayed formulae. Alkenes and other compounds with a C=C bond have *cis* and *trans* isomers if there are two different groups on each carbon atom in the double bond (Figure 6.19).

In the *cis* isomer, similar groups are on the same side of the double bond (in Latin, *cis* means 'on the same side'). In the *trans* isomer, similar groups are on opposite sides of the double bond (in Latin, *trans* means 'opposite' or 'across').

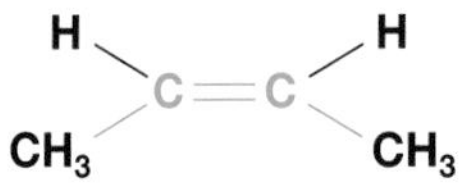

Definition

Cis **and** ***trans*** **isomers** are molecules with the same molecular formula, the same structural formula, but different displayed formulae.

Note

At one time, *cis* and *trans* isomers were often called geometric isomers.

These two structures are different compounds because the double bond stops rotation.

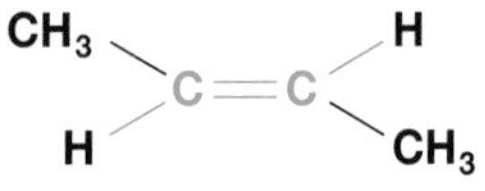

melting temperature = – 139 °C
cis-but-2-ene

melting temperature = – 106 °C
trans-but-2-ene

Figure 6.19 ▲
Cis and *trans* isomers of but-2-ene – different compounds with different melting temperatures, boiling temperatures and densities

Test yourself

17 Why do you think the bond angles around each carbon atom in ethene are approximately 120°?

18 Why is rotation about a carbon–carbon double bond restricted?

19 Which of the following unsaturated compounds have *cis* and *trans* isomers: but-1-ene, 1,1-dichloropropene, pent-2-ene, but-1,3-diene?

20 a) Draw and name the displayed structures of the three isomers of $C_2H_2Br_2$.
b) What type of structural isomerism do they show?
c) Which two are *cis–trans* isomers?

Activity

Renaming *cis–trans* as E–Z isomers

Cis–trans isomerism arises in compounds containing carbon–carbon double bonds when there are two different groups on both carbon atoms of the double bond. Different isomers result because of restricted rotation around the double bond.

The existence of a ring structure in a molecule can also restrict rotation and give rise to *cis* and *trans* isomers. Look at the structure of *trans*-1,2-dichlorocyclobutane in Figure 6.20. Its *cis* isomer is *cis*-1,2-dichlorocyclobutane.

1 Draw the displayed formula of *cis*-1,2-dichlorocyclobutane.

2 There is another pair of *cis–trans* isomers named dichlorocyclobutane and a separate structural isomer. Name and draw displayed formulae for these three molecules.

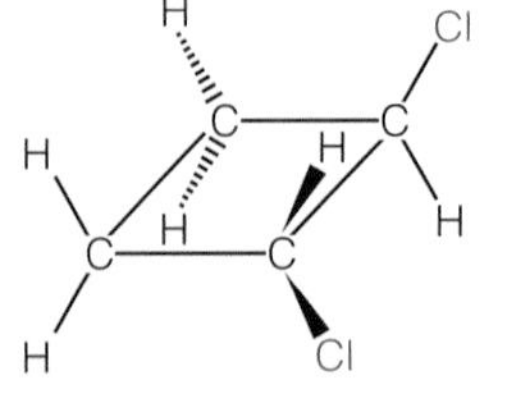

Figure 6.20 ▲
The structure of *trans*-1,2-dichlorocyclobutane

Now look at the isomer of 2-bromo-1-chloroprop-1-ene in Figure 6.21.

3 Draw the other *cis–trans* isomer of 2-bromo-1-chloroprop-1-ene.

But, which of these isomers is *cis* and which is *trans*? The rule normally used to name the isomers as *cis* or *trans* is:

- in the *cis* isomer, similar groups are on the same side of the double bond
- in the *trans* isomer, similar groups are on opposite sides of the double bond.

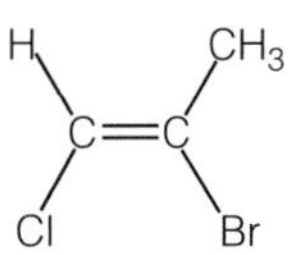

Figure 6.21 ▲
An isomer of 2-bromo-1-chloroprop-1-ene

In 2-bromo-1-chloroprop-1-ene, there are four different groups on the atoms joined by the double bond. So the normal rule, which requires one group to be the same on both carbon atoms, cannot be used.

This is where the *cis–trans* naming system breaks down and it becomes necessary to use the E–Z naming system. In fact, the E–Z naming system was developed in order to name complex alkenes in naturally occurring materials such as the red pigment in tomatoes. One molecule of this pigment has 13 C=C bonds.

The E–Z naming system

- Look at the atoms bonded to the first carbon atom in the C=C bond. The atom with the highest atomic number takes priority.
- If two atoms with the same atomic number, but in different groups, are attached to the first carbon atom, then the next bonded atom is taken into account. Thus, CH_3CH_2– has priority over CH_3–.
- This consideration is then repeated with the second carbon atom in the C=C bond.

4 a) What is the order of priority among Br, C in CH_3, Cl and H?
b) What is the priority among CH_3–, CH_3CH_2–, CH_3O– and $HOCH_2$–?

5 Look again at 2-bromo-1-chloroprop-1-ene in Figure 6.21.
a) What is the priority between H– and Cl– attached to the first carbon atom in the double bond?
b) What is the priority between Br– and CH_3– attached to the second carbon atom in the double bond?

- If the two groups of highest priority are on the same side of the double bond, the isomer is designated Z- (from the German 'zusammen' meaning 'together'), and if the two groups of highest priority are on opposite sides of the double bond, the isomer is designated E- (from the German 'entgegen' meaning 'opposite').

6 Draw the displayed structure of Z-2-bromo-1-chloroprop-1-ene.

7 Use the E–Z system to name the *cis* and *trans* isomers of:
a) but-2-ene
b) 1,2-dichlorocyclobutane.

Definition

An **addition reaction** is a reaction in which two molecules add together to form a single product.

6.8 Chemical reactions of the alkenes

The characteristic reactions of alkenes are addition reactions, in which small molecules such as H_2, Cl_2 and HBr add across the double bond to form a single product.

Addition of hydrogen

Hydrogen adds to C=C double bonds at room temperature in the presence of a platinum or palladium catalyst, or on heating to 150 °C in the presence of a nickel catalyst.

Figure 6.22 ▶
Addition of hydrogen to propene

$$CH_3(H)C{=}CH_2 + H_2 \xrightarrow[150\,°C]{\text{Ni catalyst}} CH_3{-}CH_2{-}CH_3$$

propane

Note

When describing an organic reaction, always write an equation, name the reagents and the products, and state the conditions (temperature, pressure, catalysts).

This process is known as 'catalytic hydrogenation'. The advantage of using a solid metal catalyst is that it can be held in the reaction vessel as the reactants flow in and the products flow out. There is no difficulty in separating the products from the catalyst.

At one time, catalytic hydrogenation was important in the manufacture of margarine from unsaturated vegetable oils in palm seeds and sunflower seeds. Vegetable oils are liquids containing carbon–carbon double bonds, nearly all of which are *cis*-double bonds. During hydrogenation, some of these double bonds are converted to carbon–carbon single bonds by addition of hydrogen. The change in structure turns oily unsaturated liquids into soft saturated fatty solids like margarine. However, research in the 1960s started to show that saturated fats contributed to heart disease.

Fats that have been partially saturated by hydrogenation often contain *trans*-fats and, during the 1990s, evidence began to suggest that these *trans*-fats could also lead to heart disease. However, complete hydrogenation eliminates the double bonds, and hence the *trans*-fats. This has led to a new strategy by some manufacturers of margarine and vegetable fat spreads. This involves completely hydrogenating part of the vegetable oil and then blending this with untreated oil to make a spread of the correct texture, but with no *trans*-fats.

Figure 6.23 ▲
Manufacturers of vegetable fat spreads, such as Bertolli, often claim that they contain virtually no *trans*-fats, that they are rich in unsaturated fat and low in saturated fats compared to butter. These spreads are a healthier option than butter.

Addition of halogens

Chlorine and bromine add rapidly to alkenes at room temperature. The products are dichloroalkanes and dibromoalkanes. Fluorine reacts explosively with small alkenes, such as ethene and propene, but the reaction with iodine is slow.

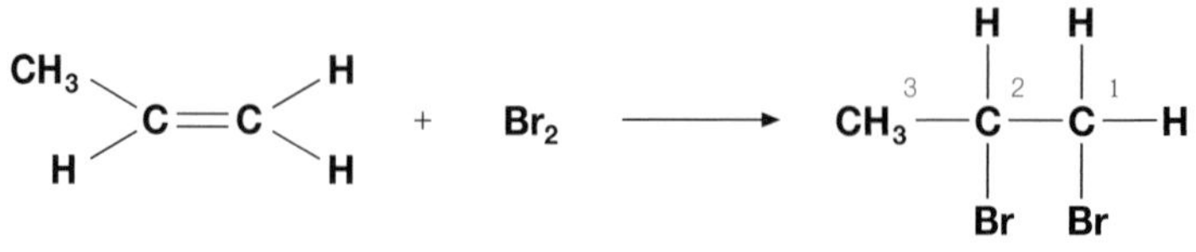

Figure 6.24 ▶
Addition of bromine to propene

$$CH_3(H)C{=}CH_2 + Br_2 \longrightarrow \overset{3}{C}H_3{-}\overset{2}{C}HBr{-}\overset{1}{C}H_2Br$$

propene → 1,2-dibromopropane

Addition of hydrogen halides

Hydrogen halides react readily with alkenes at room temperature forming halogenoalkanes. For example, hydrogen bromide reacts with ethene in the gas phase to form bromoethane. The reaction happens at room temperature.

The other hydrogen halides, HCl and HI, react in a similar way.

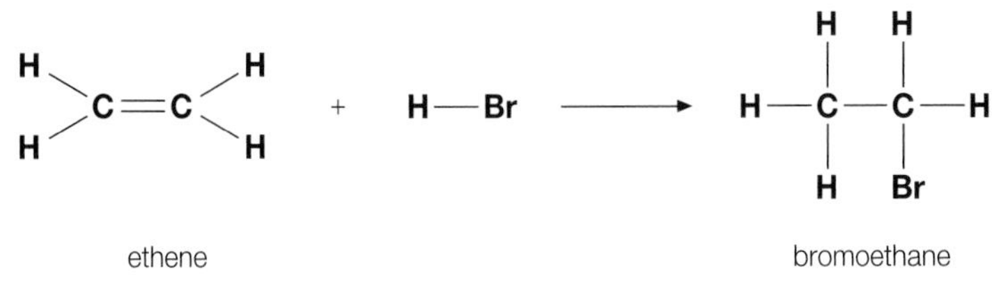

Figure 6.25 ◀
Addition of hydrogen bromide to ethene

Oxidation by potassium manganate(VII)

Potassium manganate(VII) oxidises alkenes – the products depend on the conditions. A dilute, acidified solution of potassium manganate(VII) converts an alkene to a diol at room temperature. At the same time, purple manganate(VII) ions, MnO_4^-, are reduced to very pale pink Mn^{2+} ions, and the purple colour disappears if there is excess alkene. So, like the reaction with aqueous bromine (Section 6.9), this reaction with dilute acidified potassium manganate(VII) solution can be used to distinguish between unsaturated and saturated hydrocarbons.

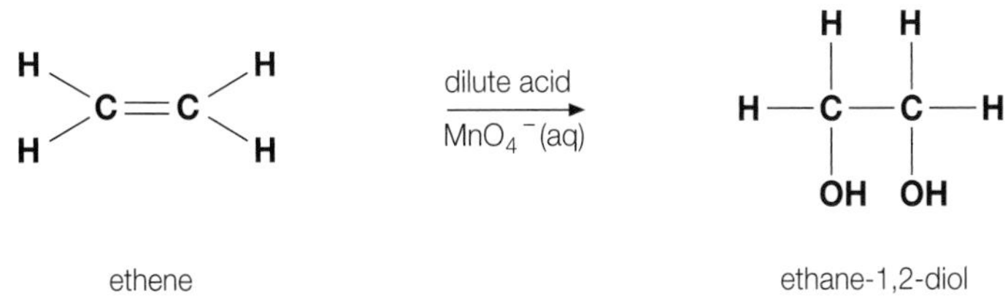

Figure 6.26 ◀
The reaction of ethene with dilute acidified manganate(VII) ions producing ethane-1,2-diol

6.9 Electrophilic addition to alkenes

Most of the reactions of alkenes are electrophilic addition reactions. Electrophiles (Section 5.6) attack the electron-rich region of the double bond in alkenes, and in particular the exposed π bond. Electrophiles that add to alkenes include hydrogen bromide, bromine and water in the presence of H^+ ions.

The addition of hydrogen bromide to ethene

Hydrogen bromide molecules are polar (Section 8.1). The hydrogen atom, with its δ+ charge at one end of the molecule, acts as an electrophile (Figure 6.27).

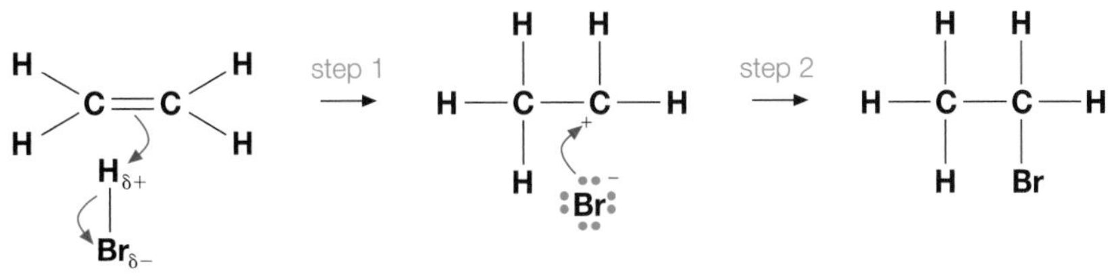

Figure 6.27 ◀
Electrophilic addition of hydrogen bromide to ethene. The reaction takes place in two steps.

In the first step of the reaction, an HBr molecule approaches an ethene molecule. The δ+ hydrogen end of the HBr is attracted towards the electron-rich double bond. As the HBr molecule gets even closer, heterolytic fission of the π bond occurs. The electrons in the π bond form a covalent bond to the hydrogen atom and, at the same time, heterolytric fission of the HBr bond also occurs. Electrons in the H–Br bond are taken over by the bromine atom, producing a Br^- ion.

The other product of step 1 is the highly reactive cation $CH_3CH_2^+$. This is so reactive that it reacts immediately with the Br^- in the second step to form bromoethane, CH_3CH_2Br.

The addition of bromine to ethene

The addition of bromine to ethene is also an electrophilic addition. Bromine molecules are not polar, but they become polarised as they approach the electron-rich region of the double bond. Electrons in the double bond repel electrons in the bromine molecule, and the $\delta+$ end of the bromine molecule (nearer the double bond) becomes electrophilic (Figure 6.28). In this case, the product is 1,2-dibromoethane.

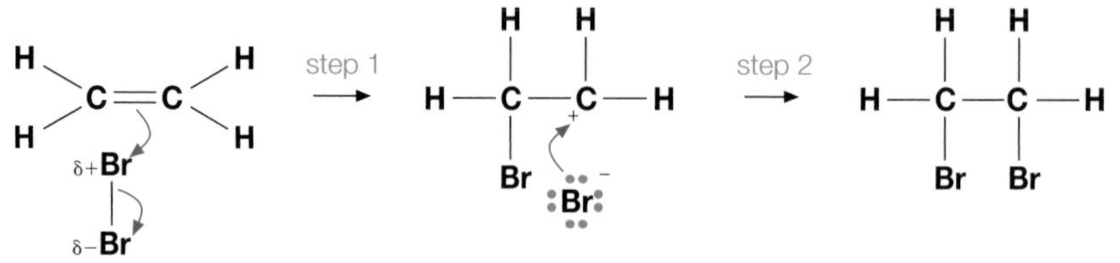

Figure 6.28 ▶
Electrophilic addition of bromine to ethene

These mechanisms for electrophilic additions are not simply hypotheses or good ideas. They are supported by significant pieces of experimental evidence to help our understanding of the reactions at a molecular level.

Firstly, it is found that the rate of addition with hydrogen halides is in the order HCl > HBr > HI. This is, of course, the order of their strengths as acids, which supports the first step of the mechanism in which an H^+ ion adds to the alkene.

Secondly, if HBr addition is carried out in the presence of NaCl then chloroethane, CH_3CH_2Cl, is obtained in addition to the expected product, CH_3CH_2Br. Similarly, if Br_2 addition is carried out in the presence of NaCl then 1-bromo-2-chloroethane, CH_2BrCH_2Cl, is obtained in addition to CH_2BrCH_2Br.

Tutorial

This supports the second step of the mechanism in which the highly reactive cations $CH_3CH_2^+$ and $CH_2BrCH_2^+$ can be attacked by either Br^- or Cl^-.

Testing for alkenes and the C=C bond

The reaction of bromine water (aqueous bromine, $Br_2(aq)$) with hydrocarbons is a useful test for alkenes and the carbon–carbon double bond.

Unsaturated hydrocarbons with a C=C bond, such as ethene and cyclohexene, quickly decolorise yellow/orange bromine water producing a colourless mixture.

Saturated hydrocarbons, such as ethane and cyclohexane, have no reaction with bromine water and the yellow/orange colour of bromine remains.

When bromine is added to water, some of it reacts with the water to form a mixture of hydrobromic acid and bromic(I) acid. Hydrobromic acid is a relatively strong acid, which ionises immediately. In comparison, bromic(I) acid is a weak acid which remains un-ionised as polar $HO^{\delta-}–Br^{\delta+}$ molecules.

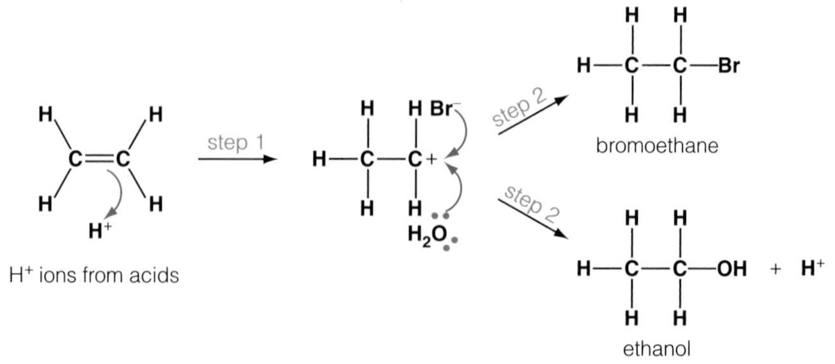

Figure 6.29 ▶
Reaction of bromine water with ethene producing bromoethane and ethanol

$Br_2(l) + H_2O(l) \rightarrow H^+(aq) + Br^-(aq) + HOBr(aq)$

So, when bromine water is shaken with ethene, H^+ ions from the HBr react with the electron-rich region of the C=C bond forming an intermediate cation. This cation then reacts rapidly with Br^- and H_2O in the bromine water to form a mixture of bromoethane and ethanol (Figure 6.29).

In addition to these products, unreacted bromine molecules react with ethene to form dibromoethane (Figure 6.28) and the polarised HO–Br reacts with ethene to form 2-bromoethanol, CH_2OHCH_2Br (Figure 6.30).

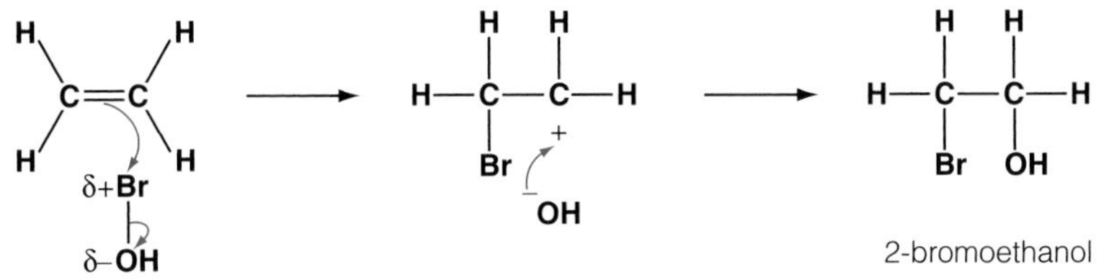

Figure 6.30 ◀ Formation of 2-bromoethanol when bromine water reacts with ethene

6.10 Addition to unsymmetrical alkenes

When a molecule such as HBr or HCl adds to an unsymmetrical alkene, such as propene, there are two possible products, but much more of one product is usually produced (Figure 6.31).

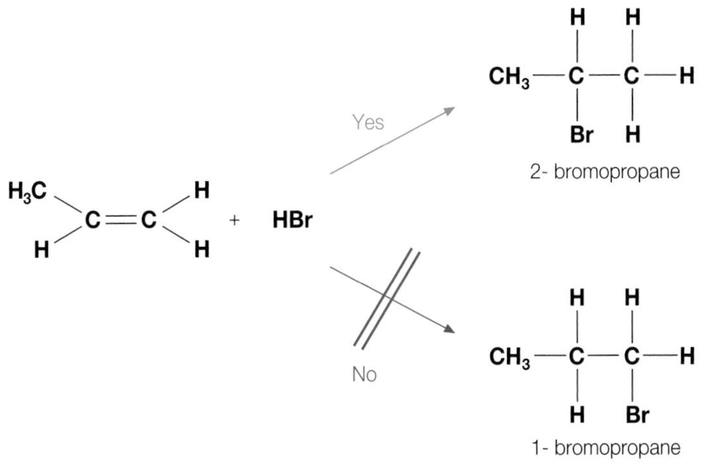

Figure 6.31 ◀ Possible products when hydrogen bromide adds to propene

Analysis of the products of the reaction between hydrogen bromide and propene shows that much more 2-bromopropane is produced than 1-bromopropane. This suggests that the hydrogen atom from HBr adds mainly to the carbon atom of the double bond which already has more hydrogen atoms attached to it. This pattern is usually called Markovnikov's rule because it was first reported by the Russian chemist Vladimir Markovnikov who studied many alkene addition reactions during the 1860s.

The mechanism for electrophilic addition helps to account for Markovnikov's rule. On adding HBr to propene, there are two possible cation intermediates (Figure 6.32).

The cation with its positive charge in the middle of the carbon chain is slightly more stable than the cation with its charge at the end of the chain. The more stable cation has two alkyl groups pushing electrons towards the positively charged carbon atom. This helps to stabilise the ion by reducing its

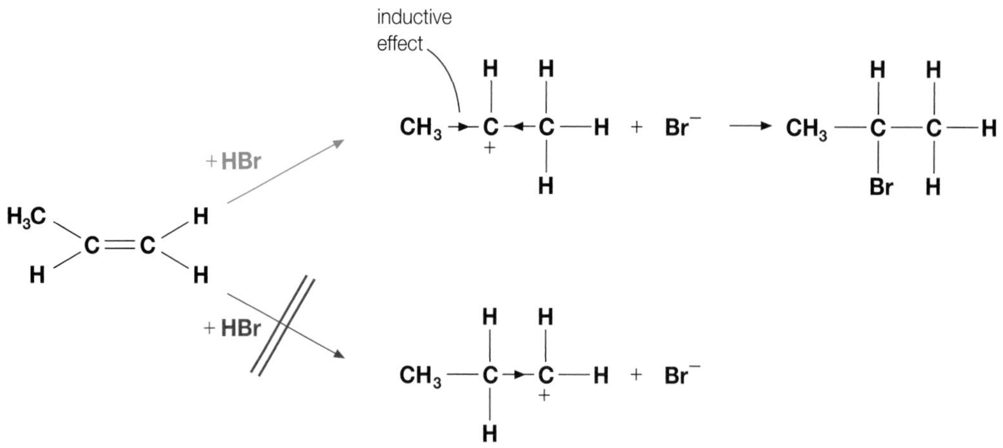

Figure 6.32 ▲
Explaining Markovnikov's rule

Definitions

Intermediates are atoms, molecules, ions or free-radicals which do not appear in the overall equation for a reaction, but which are formed during one step of a reaction and then used up in the next.

The **inductive effect** describes the way in which electrons are either pushed towards or pulled away from a carbon atom by the atoms or groups to which it is bonded. Alkyl groups have a small tendency to push electrons towards any carbon atom to which they are bonded. One of the consequences of this is that any carbon atom with a positive charge becomes more stable as more alkyl groups are attached to it.

positive charge, which is an example of the inductive effect. Owing to this inductive effect, the more stable cation persists longer and is more likely to combine with Br^- ions forming the main product, 2-bromopropane.

Test yourself

21 Write the structures and names of the products and the conditions for the reactions when propene reacts with:
   a) hydrogen
   b) chlorine
   c) potassium manganate(VII).

22 Catalytic hydrogenation is sometimes used in the manufacture of spreads, such as 'Flora™', from vegetable oils.
   a) What is meant by the term 'catalytic hydrogenation'?
   b) Explain the terms 'saturated' and 'unsaturated' as applied to organic compounds such as those in low-fat spreads and vegetable oils.
   c) Why are unsaturated fats, such as olive oil and sunflower oil, thought to be healthier foods than more saturated fats, such as cream?

23 For the reaction of hydrogen chloride with but-1-ene:
   a) draw the structure of the two possible cation intermediates
   b) say which cation is the more stable
   c) write the displayed formula and name of the main product.

Definition

Addition polymerisation is an addition reaction in which small molecules, called monomers, join together forming a giant molecule, called a polymer.

6.11 Addition polymers from alkenes

If the conditions are right, molecules of ethene will undergo addition reactions with each other to form polythene – or more correctly, poly(ethene). Two different kinds of polythene are manufactured – low-density polythene and high-density polythene.

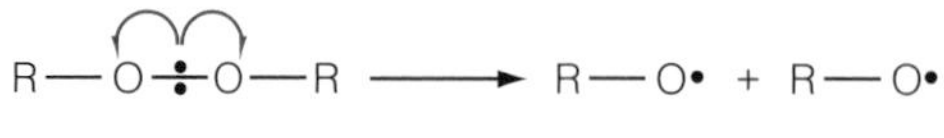

$$R—O \vdots O—R \longrightarrow R—O\bullet + R—O\bullet$$

peroxide initiator   free radicals

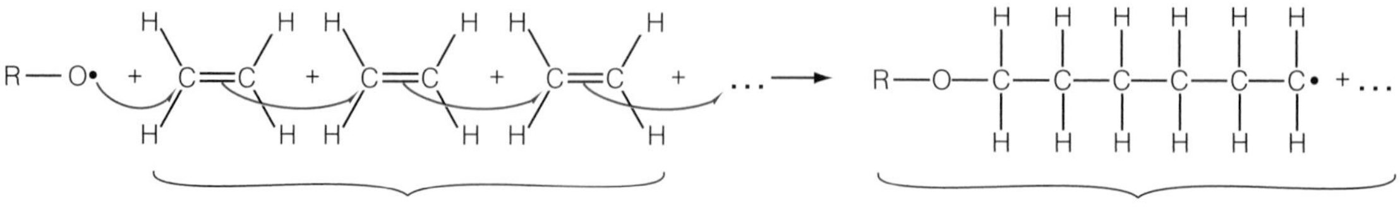

Figure 6.33 ▲
Molecules of ethene can undergo addition reactions with each other.

Low-density polythene is manufactured by heating ethene at high pressures and high temperatures with special substances called initiators. These initiators are often peroxides, which break apart to form free radicals that initiate (start) the reaction (Figure 6.33). The polyethene produced has very long chains with lots of branches. The branches prevent the molecules from packing closely and this results in low-density material.

High-density polythene is manufactured at relatively low pressure and low temperature with a special catalyst. This produces extra long chains with very little branching, so the chains pack closer.

Processes like these are called 'addition polymerisations'. During addition polymerisation, small molecules like ethene, known as monomers, add to each other to form a giant molecule called a polymer.

Polythene is by far the most important polymer at present. After polythene, the two most useful polymers are poly(propene) and poly(chloroethene) or pvc. These are also manufactured by addition polymerisation. Polythene, polypropene and pvc are soft, flexible and slightly elastic – because of this, they are often called plastics.

The monomers, polymer structures, properties and uses of polythene, polypropene and pvc are shown in Table 6.1.

Figure 6.34 ▲
Clingfilm is just a thin film of pvc. Its correct name is poly(chloroethene).

Monomer	Polymer repeat unit	Properties	Uses
Ethene $H_2C{=}CH_2$	Polythene (poly(ethene)) $-[CH_2-CH_2]_n-$	• Light, flexible • Easily moulded • Transparent • Good insulator • Resistant to water, acids and alkalis	• Plastic bags and bottles • Beakers • Insulation for cables
Propene $H_2C{=}CH(CH_3)$	Polypropene (poly(propene)) $-[CH_2-CH(CH_3)]_n-$	• Tough • Easily moulded • Easily coloured • Very resistant to water, acids and alkalis	• Fibre for ropes and carpets • Crates • Toys
Chloroethene $H_2C{=}CHCl$	pvc (poly(chloroethene)) $-[CH_2-CHCl]_n-$	• Tough • Rigid or flexible • Very resistant to water, acids and alkalis	• Guttering and window frames • Insulation for cables • Waterproof clothing • Flooring • Clingfilm

Table 6.1 ▲
Our most important polymers

Test yourself

24 $CF_2{=}CF_2$ is the monomer for the polymer ptfe.
  a) What is the systematic name for $CF_2{=}CF_2$?
  b) Draw a section of the ptfe polymer composed of three monomer units.
  c) The name 'ptfe' contains four key letters from the full name of the polymer. What is the full name?

25 a) What conditions are used to manufacture low-density polythene?
  b) Why is polythene so useful?
  c) What is the major disadvantage of plastics?

Figure 6.35 ▲
This little boy is wearing a pvc waterproof jacket.

Activity

A more sustainable future for polymers

Manufactured goods such as metals and polymers are produced at a heavy cost to society and the environment. In general, their production uses up scarce natural resources, it consumes non-renewable fuels and energy, and it disrupts wildlife and the countryside.

Today's industrialists and manufacturers are expected to assess the life cycles of their products more closely in order to find ways in which industry and society can contribute to a more sustainable use of materials. This life cycle assessment (LCA) is part of the legislation designed to protect the environment.

Figure 6.36 ▲
Oil companies extract millions of tonnes of crude oil every day.

In terms of energy and materials, manufactured goods such as computers, clothes and cars go through a life cycle with three distinct phases:

- *birth* – raw materials and energy are used to make the goods
- *life* – chemicals and energy are needed to maintain and use the goods
- *death* – energy and possibly space are needed to dispose of the goods.

Here are some issues and questions that are being raised about the life cycle of polymers from the extraction of crude oil, through the production and use of commercial polymer products to their eventual disposal.

- Only about 4% of crude oil is used to make polymers. Most crude oil is used to provide fuel for transport, heating homes and industry. If we continue to use crude oil at our present rate, our known reserves are unlikely to last more than a few decades.

1 Suggest three significant steps that should be taken to reduce our consumption of crude oil.

- Vast quantities of non-renewable fossil fuels are used in extracting crude oil, in transporting it to refineries, in processing at the refineries and then in the production of specific products such as plastic bags, bottles, toys and fibres.
- After production, the various polymer products move to the shops where we buy them, use them and then throw them away in landfill sites. This is very wasteful.

2 What are local councils and the government now doing to reduce landfill waste?

3 Why is this approach so important?

- Most plastic waste can be melted and then remoulded. Recycling would seem an obvious way forward, but there are problems in sorting and separating different types of polymer, particularly when some products have more than one type of polymer, and plastic containers often need cleaning.

- In some cases, it is possible to convert polymers back to their simple monomers with a recovery rate of 80–90%.

4 What greater efforts should be made to recycle plastics?

- Some polymers that are difficult to recycle can be burned and used as fuel. For full combustion, the process requires very high temperatures and special incinerators.

5 a) Is incineration better than landfill? Explain your answer.
b) Why are people often opposed to incinerators in their area?
c) What environmental problems does incineration make worse?

Figure 6.37 ◀
Most of our plastic waste ends up on landfill sites.

WWW Extension questions

REVIEW QUESTIONS

1 Figure 6.38 shows three important reactions of ethene.

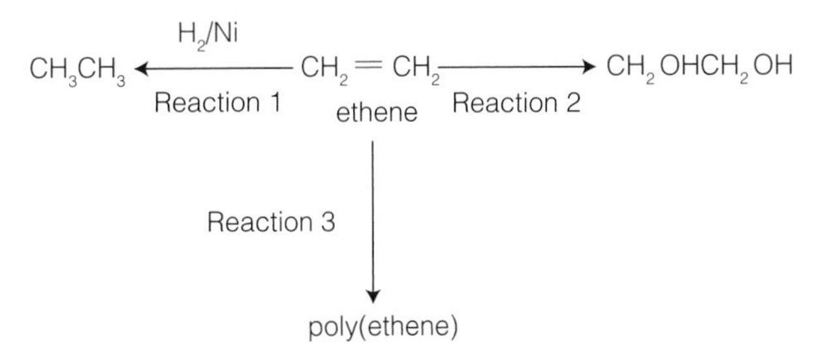

Figure 6.38 ▲

a) i) What conditions of temperature and pressure are used in Reaction 1? (2)

ii) Reaction 1 is used to convert unsaturated alkenes to saturated alkanes. What is meant by the terms 'unsaturated' and 'saturated' in this context? (3)

iii) Why are saturated and unsaturated chemicals important to dieticians and nutritionists? (2)

b) i) What chemicals are used in Reaction 2 to produce CH_2OHCH_2OH? (2)

ii) Common names for CH_2OHCH_2OH are ethylene glycol and antifreeze. What is its systematic name? (1)

iii) What is the common use of CH_2OHCH_2OH? (1)

c) i) What conditions are used in Reaction 3 to convert ethene to high-density poly(ethene)? (2)

ii) Draw a section of the polythene structure consisting of three monomer units. (1)

d) State four properties of poly(ethene) which make it particularly suitable for making plastic bags. (2)

2 Crude oil is an important source of materials for the petrochemical industry. Various products are obtained from crude oil by fractional distillation, followed by processes involving cracking and reforming.

a) How is crude oil separated into different fractions by fractional distillation? (6)

b) i) What is meant by 'cracking'? (3)

ii) Dodecane, $C_{12}H_{26}$, can be cracked into ethene and a straight-chain alkane so that the molar ratio of ethene to straight-chain alkane is 2 : 1.

Write a balanced equation for this reaction and name the straight-chain alkane. (2)

iii) Heat alone can be used to crack alkanes, but oil companies normally use catalysts as well. Suggest two reasons why oil companies use catalysts. (2)

c) Straight-chain alkanes such as heptane can be reformed into cyclic compounds.

Write a balanced equation to show how heptane can be reformed into methylcyclohexane. (2)

d) Oxygen-containing compounds are added to some brands of petrol to improve their performance. In Formula One racing cars, 2-methylpropan-2-ol is usually added to the petrol.

i) Draw the structural formula of 2-methylpropan-2-ol. (1)

ii) Why do compounds like 2-methylpropan-2-ol improve a racing car's performance on the track? (1)

3 a) Propene and but-2-ene are used in the petrochemical industry to produce important polymers.

i) Explain the term 'polymer'. (3)

ii) Poly(propene) does not have a sharp melting temperature, but softens over a wide temperature range. Why is this? (2)

iii) Draw a section of the polymer from but-2-ene showing two repeat units. (1)

b) But-2-ene can be converted to buta-1,3-diene by a process called dehydrogenation. Buta-1, 3-diene is used to make synthetic rubber.

i) Explain the term 'dehydrogenation'. (1)

ii) Draw the structure of buta-1,3-diene. (1)

iii) Write an equation for the dehydrogenation of but-2-ene to form buta-1,3-diene. (2)

4 The reaction between propene, $CH_3CH{=}CH_2$, and hydrogen bromide involves electrophilic addition producing $CH_3CHBrCH_3$ as the major product.

a) Explain the term 'electrophilic addition'. (3)

b) What is the name of the compound $CH_3CHBrCH_3$? (1)

c) Draw displayed structures to show the mechanism of the reaction between propene and hydrogen bromide. (3)

d) Why is the major product $CH_3CHBrCH_3$ rather than $CH_3CH_2CH_2Br$? (3)

Unit 2

Application of Core Principles of Chemistry

7 The shapes of molecules and ions

Ionic and covalent bonding both depend on electrostatic attractions to hold their ions and atoms together, but covalent bonding differs from ionic bonding in that it is directional. Some ions – such as OH^-, SO_4^{2-} and NH_4^+ – contain more than one atom. The atoms within these ions are held together by covalent bonds, but the overall bonding between these ions and those of opposite charge is, of course, ionic.

7.1 Directional bonds

In covalent bonds, two atoms share a pair of electrons between their nuclei, and these two atoms may share pairs of electrons with other atoms (Section 4.9). This means that atoms always stay in the same positions relative to each other with bonds in the same direction and therefore, molecules have a fixed shape. This constant three-dimensional spatial relationship between the atoms in a molecule is important because it can affect the physical and chemical properties of covalently bonded compounds.

Figure 7.1 ▶ The three-dimensional shape of molecules is crucial in the structure of enzymes, which control the rate of chemical reactions in living things. This photo shows the use of an insecticide which has just the right shape to interfere with an enzyme in the insects and prevent their growth. Fortunately, the insecticide has little or no effect on the adjacent plants, which thrive once the insects have been destroyed.

In contrast to covalent bonds, ionic bonding involves the attraction between ions around each other in every direction. The ions are held together because their arrangement in the lattice results in stronger attractive forces between ions of opposite charge than repulsive forces between ions with the same charge (Section 4.4).

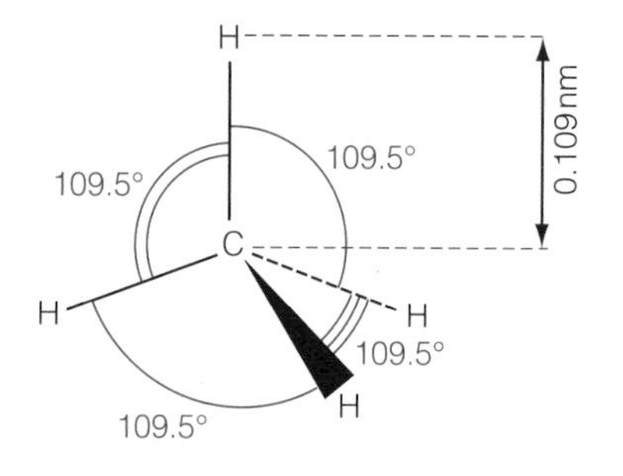

Figure 7.2 ▲ All the bond angles in methane are 109.5° and all the C–H bond lengths are 0.109 nm.

7.2 The shapes of molecules and ions

X-ray diffraction studies, and other instrumental methods, enable chemists to measure bond lengths and bond angles very accurately in molecules and ions such as NH_4^+ which have covalent bonds. The results show that covalent bonds have a definite direction and a definite length. For example, X-ray diffraction studies show that all the C–H bond lengths in methane, CH_4, are 0.109 nm and all the H–C–H bond angles are 109.5° (Figure 7.2).

Chemists have developed a very simple theory to explain and predict the shapes and bond angles of simple molecules and ions containing covalently bonded atoms. The theory is based on the repulsion of electron pairs in the outermost shell of the central atom. It is called the electron-pair repulsion theory (EPRT). The theory says that electron pairs in the outer shell of atoms and ions repel each other and get as far apart as possible.

Note

Sometimes the abbreviation VSEPR is used for the electron-pair repulsion theory. This is short for 'valence shell electron-pair repulsion'. The valence shell of an atom is its outer shell.

Look at the molecules of beryllium chloride and boron trifluoride in Figure 7.3. Beryllium chloride, $BeCl_2$, provides the simplest example of the electron-pair repulsion theory. In the $BeCl_2$ molecule, beryllium has only two pairs of electrons in its outer shell. In order to get as far apart as possible, these two pairs of electrons must be on opposite sides of the beryllium atom. The shape of the molecule is described as linear and the Cl–Be–Cl bond angle is 180°.

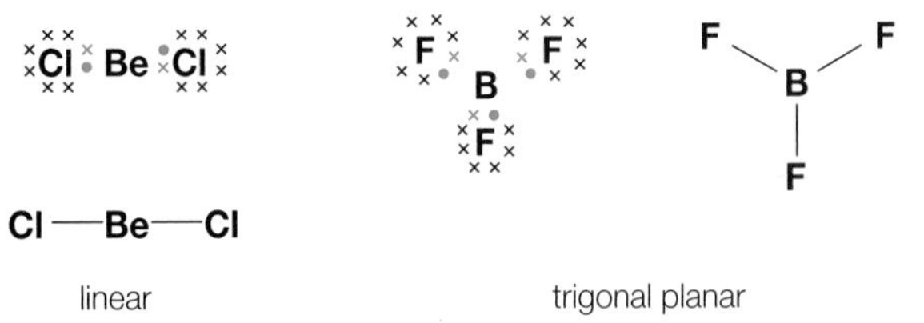

Cl—Be—Cl

linear

trigonal planar

Figure 7.3 ▲
The shapes of molecules with two and three electron pairs around the central atom

The next simplest example of the electron-pair repulsion theory is shown by boron trifluoride, BF_3. In the BF_3 molecule, boron has three electron pairs in its outer shell. This time, to get as far apart as possible, the three pairs must occupy the corners of a triangle around the boron atom. The shape of this molecule is described as trigonal planar and the F–B–F bond angles are 120°.

Number of electron pairs	Shape	Bond angle ∠XMX	Example
2	Linear	180°	Cl—Be—Cl $BeCl_2$
3	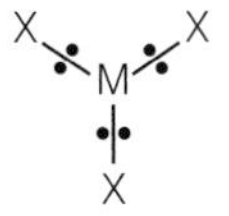 Trigonal planar	120°	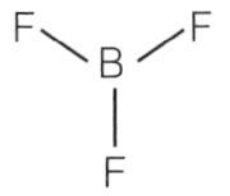 BF_3
4	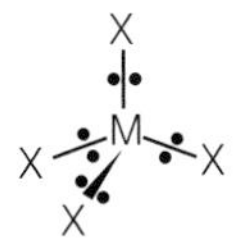 Tetrahedral	190.5°	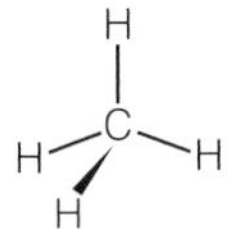 CH_4
5	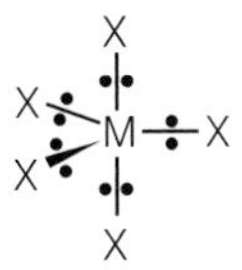 Trigonal bipyramidal (Two triangle-based pyramids joined at the bases.)	90° and 120°	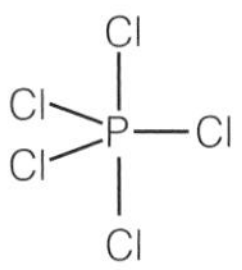 gaseous PCl_5
6	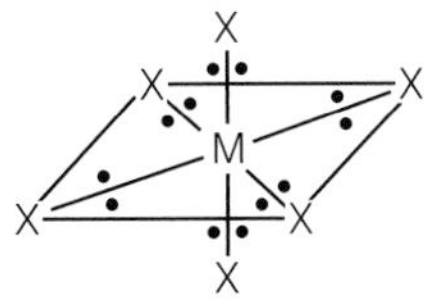 Octahedral (Two square-based pyramids joined at the bases.)	90°	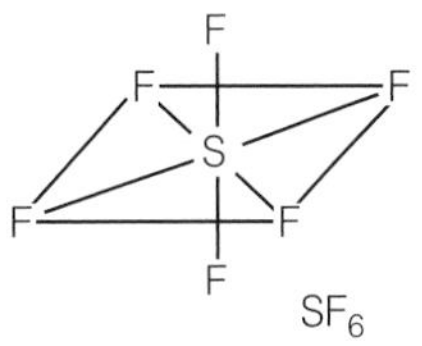 SF_6

Table 7.2 ▲
The shapes of molecules with two to six pairs of bonding electrons around the central atom

Definitions

A **bond angle** is the angle between two covalent bonds in a molecule or giant covalent structure.

Bond length is the distance between the nuclei of two atoms linked by one or more covalent bonds. For the same two atoms, triple bonds are shorter than double bonds which, in turn, are shorter than single bonds, as shown in Table 7.1.

Bond	Bond length/nm
C—C	0.154
C=C	0.134
C≡C	0.120

Table 7.1 ▲

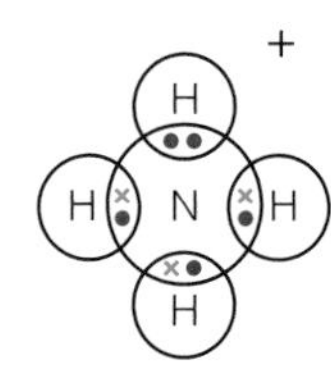

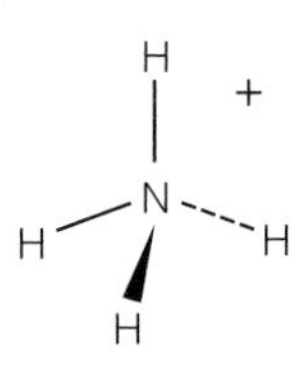

tetrahedral

Figure 7.4 ▲
The shape of the NH_4^+ ion

Now look at the shape of the ammonium ion, NH_4^+, in Figure 7.4. In the NH_4^+ ion, nitrogen has four electron pairs in its outer shell, each bonded to a hydrogen atom. To get as far apart as possible, the four pairs are in positions which would be at the corners of a tetrahedron. The four hydrogen atoms are in similar positions further away from the nitrogen atom. The shape of the NH_4^+ ion is described as tetrahedral and all the H–N–H bond angles are 109.5°.

Table 7.2 summarises the shapes of molecules with two, three, four, five and six pairs of electrons based on the electron pair repulsion theory. In each case, the electron pairs are repelled as far apart as possible. The table also shows the predicted bond angles for each molecule.

The electron-pair repulsion theory shows how chemists can make generalisations from their results and use these generalisations to make predictions.

Test yourself

1 Why are covalent bonds described as 'directional bonds'?

2 Why do you think the carbon–carbon double bond, C=C, is shorter than the carbon–carbon single bond, C–C?

3 Aluminium chloride sublimes at quite low temperatures forming simple molecules of $AlCl_3$, in which the Al atom is bonded to each Cl atom by a single covalent bond.

a) Draw a dot-and-cross diagram for an $AlCl_3$ molecule showing only electrons in the outer shell of all four atoms.

b) Describe the shape of $AlCl_3$ molecules and predict the Cl–Al–Cl bond angle.

Activity

Predicting the shapes and bond angles of molecules and ions

In this activity you can use the electron-pair repulsion theory to predict the shapes and bond angles in various molecules and ions.

1 Draw dot-and-cross diagrams of the following simple molecules, showing only electrons in the outer shell of all atoms:

a) PF_5
b) SiH_4
c) BCl_3

2 Predict the shape and bond angle in:

a) PF_5
b) SiH_4
c) BCl_3

3 a) Draw a dot-and-cross diagram for a molecule of methanal, H_2CO.

b) Methanal molecules have a trigonal planar shape, but the H–C–H bond angle is smaller than the H–C–O bond angles. Explain this.

4 a) Draw dot-and-cross diagrams for CH_4 and H_2O.

b) In each of these molecules, there are four pairs of electrons around the central atoms (carbon and oxygen, respectively). In CH_4, the H–C–H bond angle is 109.5°; in H_2O, the H–O–H bond angle is 104.5°.

i) Explain why the bond angle in CH_4 is 109.5°.
ii) Explain why the bond angle in H_2O is 104.5°, and not 180° with the hydrogen atoms as far apart as possible.
iii) Both CH_4 and H_2O have four pairs of electrons around their central atoms. Why do you think the bond angle in H_2O is less than 109.5°?
iv) Suggest a name for the shape of the H_2O molecule.

5 a) Draw a dot-and-cross diagram for the H_3O^+ ion.
b) How many pairs of electrons are there around the central oxygen atom?
c) Suggest a possible name for the shape of the H_3O^+ ion.

7.3 Molecules and ions with lone pairs and multiple bonds

Lone pairs

Some molecules, such as ammonia and water, contain non-bonding pairs (or lone pairs) of electrons as well as bonding pairs (Figure 7.5).

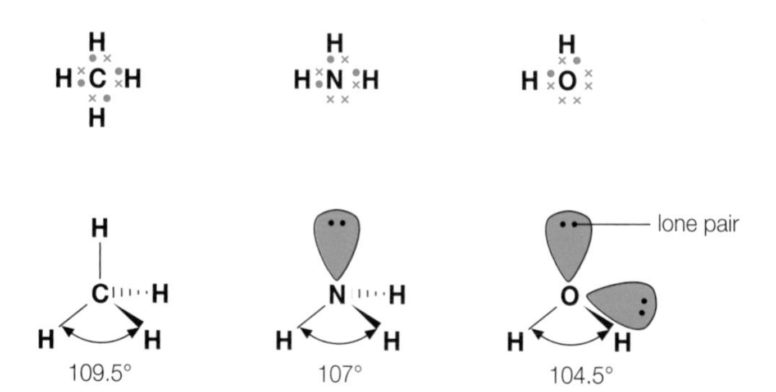

Figure 7.5 Shapes and bond angles in molecules with bonding pairs and lone pairs of electrons

Ammonia and water have exactly the same electron structure as methane (Figure 7.5) with four pairs of electrons in the outer shell of the central atom.

In methane, all four pairs of electrons are bonding pairs between the central carbon atom and a hydrogen atom. In ammonia, three of the four pairs make up N–H bonds as bonding pairs, but the fourth is a lone pair. Each of these four electron pairs repels the others, so they form a tetrahedral shape around the nitrogen atom. But the positions of atoms in the NH_3 molecule make a shape which is pyramidal – a triangle-based pyramid – with a nitrogen atom at the top and hydrogen atoms at the three corners of its base.

In water, there are also four pairs of electrons around the central atom – two bonding pairs and two lone pairs. The shape formed by these electron pairs is tetrahedral again, but the shape of the water molecule, H–O–H, is described as V-shaped or bent.

Lone pairs of electrons are held closer to the central atom than the bonding pairs. This means that they have a stronger repelling effect than bonding pairs. Therefore, the strength of repulsion between electron pairs is:

lone pair–lone pair > lone pair–bonding pair > bonding pair–bonding pair

This explains why the bond angle in ammonia, with one lone pair, is less than that in methane; and why the bond angle in water, with two lone pairs, is less than that in ammonia.

Similar predictions about shapes and bond angles can be made for ions such as H_3O^+, BF_4^- and NH_2^- (Figure 7.6).

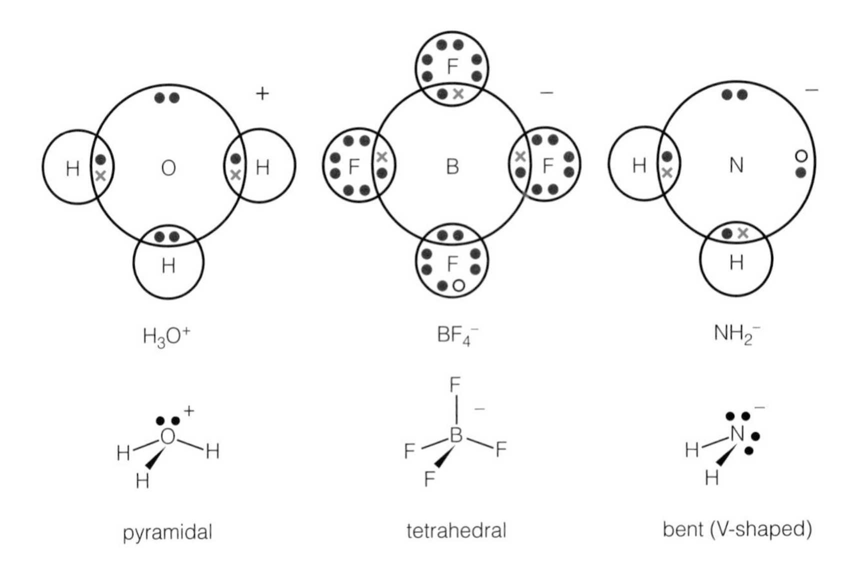

Figure 7.6 ▶
Dot-and-cross diagrams and shapes of some ions

Multiple bonds

The electrons in double bonds and triple bonds are located more or less between the nuclei of the two atoms they join. So, double bonds and triple bonds count as just one centre of negative charge (electron-pair repulsion axis) when it comes to predicting molecular shapes (Figure 7.7).

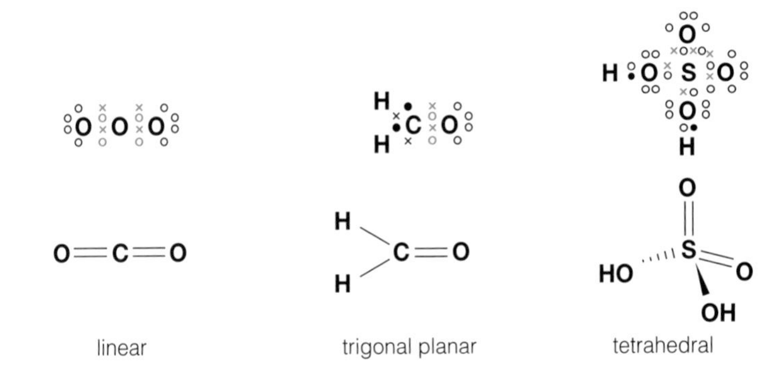

Figure 7.7 ▶
The shapes of some molecules with multiple bonds: CO_2 – carbon dioxide; H_2CO – methanal; and H_2SO_4 – sulfuric acid

Test yourself

4 Draw dot-and-cross diagrams for the following simple molecules and predict their shapes:
   a) PH_3
   b) HCN
   c) N_2
   d) CCl_4

5 Non-metals beyond silicon in the periodic table can sometimes have more than eight electrons in their outer shell, like phosphorus in PCl_5 and sulfur in SF_6 (Table 7.2). With this in mind, draw dot-and-cross diagrams for the following molecules and ions and predict their shapes:
   a) PCl_6^-
   b) SO_2
   c) XeF_2
   d) SO_4^{2-}

6 a) Draw a dot-and-cross diagram for CH_3Cl.
   b) What is the shape of the CH_3Cl molecule?
   c) The H–C–H bond angles in CH_3Cl are 111°, whereas the Cl–C–H bond angles are 108°. Why do you think the Cl–C–H bond angles are smaller than the H–C–H bond angles? (Hint: The C–H bond is much shorter than the C–Cl bond.)

From our knowledge of lone pairs of electrons and multiple bonds, it is clear that lone pairs, double bonds and triple bonds affect the shapes of molecules and ions in a similar way to electrons in single bonds. So all of these (single bonds, lone pairs, double bonds and triple bonds) can be regarded as separate centres of negative charge in predicting the overall shapes of molecules and ions.

7.4 Structure and bonding in different forms of carbon

Carbon can exist in different solid forms – diamond, graphite and various fullerenes. These solid forms of carbon are called allotropes – different forms of the same element in the same state.

These forms of carbon illustrate the important connections between the structure and bonding of materials, their properties and hence their uses (see Section 4.1).

All these forms of carbon are held together by strong covalent bonds with a definite length and direction. Diamond and graphite are giant molecules with giant covalent structures, whereas fullerenes, in comparison, are relatively simple molecules.

Definition

Allotropes are different forms of the same element in the same physical state.

Diamond

Strong covalent bonds with a definite length and fixed direction help to account for the rigid covalent structure of diamond – the hardest naturally occurring substance (Section 4.8). People have always valued diamonds for their brilliance as gemstones. But diamonds are also used industrially as abrasives for cutting and grinding hard materials such as glass and stone.

Diamond conducts thermal energy very well – five times better than copper. This important property means that diamond-tipped cutting tools don't overheat. The rigidity of the strong covalent bonds in diamond means that as the atoms close to the tip of a cutting tool get hotter and move faster, the vibrations move rapidly throughout the giant structure.

Figure 7.8 Diamonds that cannot be sold as gemstones are used in glass cutters and diamond-studded saws. This photo shows an engraver using a diamond-studded wheel to make patterns in a glass vase.

Graphite

Graphite is used to make crucibles for molten metals because it melts at the extremely high temperature of 3650 °C. For the same reason, graphite blocks are used to line the walls of industrial furnaces.

This high melting temperature also suggests that graphite has a giant structure with strong covalent bonds. This is confirmed by X-ray diffraction studies, which show that the atoms are held together in extended sheets (layers) of atoms. Each layer contains billions and billions of carbon atoms arranged in hexagons (Figure 7.9). Each carbon atom is held strongly in its layer by strong covalent bonds to three other carbon atoms. So every layer is a giant molecule with a giant covalent structure. The distance between neighbouring carbon atoms in the same layer is only 0.14 nm, but the distance between layers is 0.34 nm.

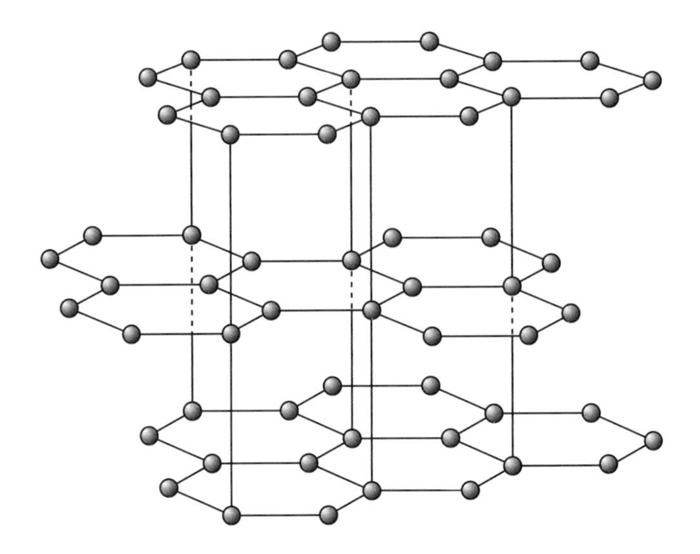

Figure 7.9 ▶
The giant covalent structure of graphite. The layers are vast sheets of carbon atoms piled on top of each other. The bonding between atoms within the layers is strong, but the bonding between layers is relatively weak.

Figure 7.10 ▲
Graphite fibres are used to reinforce the shafts of broken bones, badminton rackets and golf clubs like the one being used by Tiger Woods in this photo.

Each carbon atom in graphite uses three of its outer shell electrons to form three normal covalent bonds with other carbon atoms – this accounts for the hexagonal arrangement of the atoms within a layer.

The fourth outer shell electron on each carbon atom forms part of a cloud of delocalised electrons smeared out over each layer. Because of these delocalised electrons, graphite conducts electricity well. This explains why graphite is used for electrodes in industry and as the positive terminal in dry cells.

The covalent bonds between carbon atoms within the layers of graphite are so strong that many modern composites incorporate graphite fibres for greater tensile strength (Figure 7.10).

Unlike diamond, graphite is soft and feels greasy – this leads to the use of graphite as a lubricant. Graphite has lubricating properties because the bonds between the well-separated layers are relatively weak, allowing the layers to slide over each other.

Fullerenes

At one time, chemists believed there were only two forms of crystalline carbon. Then, in 1985, Harry Kroto and his research team at the University of Sussex discovered buckminsterfullerene, C_{60} – a black solid with a simple molecular structure. Since 1985, a whole group of similar compounds have been prepared and the group are now known as 'fullerenes'.

At the molecular level, the fullerenes mimic the geodesic (football-like) dome invented by the American engineer Robert Buckminster Fuller (Figure 7.11). Hence, the original name 'buckminsterfullerene' and the nicknames 'bucky balls' and 'footballene'.

Fullerenes are fundamentally different from diamond and graphite because they are molecular forms of carbon, rather than infinite giant structures.

Fullerenes are black solids which are soluble in various solvents because of their molecular structure. This has already led to the use of C_{60} in mascara and printing ink.

The bonding at each carbon atom in fullerenes resembles that in graphite. Three of the outer shell electrons are combined in covalent bonds with other atoms, while the fourth electron is delocalised over the whole molecule. But,

Definition

Composites combine two or more materials to create a new material which has the desirable properties of both its constituents. For example, plastic reinforced with graphite fibres combines the flexibility of the plastic with the high tensile strength of graphite.

unlike graphite which conducts, the fullerenes are good electrical insulators because the delocalised electrons cannot move between molecules. However, metals in groups 1 and 2 can react with C_{60} to form superconducting systems at very low temperatures. The reaction produces a rare type of salt in which electrons transferred to the C_{60} move around the whole salt like electrons in a metal.

$$3Rb(s) + C_{60}(s) \rightarrow (Rb^+)_3C_{60}^{3-}(s)$$

At present, these superconducting fullerene–metal systems can conduct only relatively small currents at very low temperatures, below −190 °C. The priority is to develop superconductors that can work at higher temperatures and carry much larger currents.

Now that chemists understand the structure of fullerenes, they are able to produce fullerenes in the form of tubes as well as spheres. These 'buckytubes' or 'carbon nanotubes' are not only the narrowest tubes ever made, but also the strongest and toughest weight for weight.

These carbon nanotubes have enormous potential in very diverse applications from the replacement of graphite fibres in golf clubs and fishing rods to their use in medicine as vehicles for carrying drugs into specific body cells.

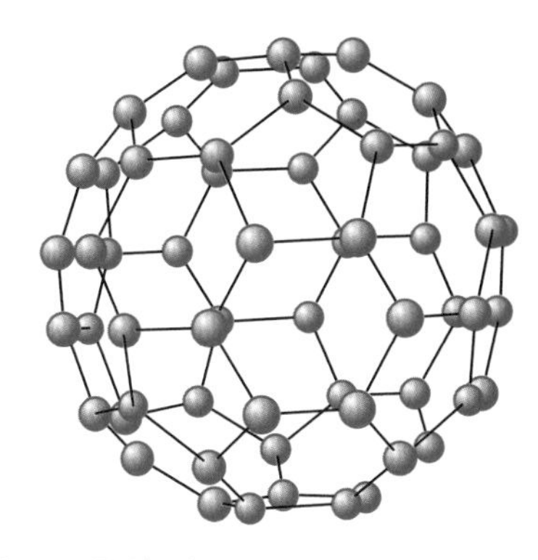

Figure 7.11 ▲ The structure of C_{60} is roughly spherical with each carbon atom bonded to three nearest neighbours. Look carefully and see if you can count all 60 carbon atoms. Other fullerenes have the formulae C_{32}, C_{50} and C_{240}.

Figure 7.12 ◄ Scientists at Berkeley National Laboratory in the USA have recently used carbon nanotubes as conveyor belts to move atom-sized particles from one site to another. By applying a small electric current to a carbon nanotube, they moved indium particles (shown here) along the tube like parts on an assembly line. Their research opens up the possibility of mass produced devices on the atomic scale.

Test yourself

7 a) Why will a zip fastener move more freely if it is rubbed with a soft pencil?
   b) Why should you use a pencil for this rather than oil?
   c) Why is graphite better than oil for lubricating the moving parts of machinery?

8 a) Name one element, other than carbon, which exists as different allotropes and name the allotropes.
   b) Explain what is meant by a 'composite'. Illustrate your answer using either reinforced concrete or wattle-and-daub.

9 Why is diamond such a poor conductor of electrical energy, but a good conductor of thermal energy?

10 a) Why do diamond, graphite and the fullerenes have different properties?
   b) Why is diamond described as a three-dimensional structure, graphite as a two-dimensional layer lattice and fullerenes as simple molecules?

REVIEW QUESTIONS

WW
Exter
quest

1 a) Phosphorus forms the chloride PCl_3. Draw a dot-and-cross diagram for PCl_3. (2)

b) Draw and name the shape of the PCl_3 molecule. (2)

c) Explain why PCl_3 has this shape. (2)

d) Why does PCl_3 form a stable compound with BCl_3? (3)

2 The covalently bonded compound urea has the formula $(NH_2)_2C{=}O$. Urea is commonly used as a fertiliser in most of Europe, whereas ionic ammonium nitrate, NH_4NO_3, is the most popular fertiliser in the UK.

a) Draw a dot-and-cross diagram for urea. (2)

b) Describe the arrangement of atoms

i) around the carbon atom in urea

ii) around a nitrogen atom in urea. (2)

c) Suggest two advantages of using urea as a fertiliser compared with ammonium nitrate. (2)

3 When space travel was being pioneered, one of the first rocket fuels was hydrazine, H_2NNH_2.

a) Draw a dot-and-cross diagram to show the electron structure of a hydrazine molecule. (2)

b) Predict the size of the H–N–H bond angle in a hydrazine molecule and explain your reasoning. (3)

c) Suggest a possible equation for the reaction which occurs when hydrazine vapour burns in oxygen. (2)

d) When 1.00 g of hydrazine burns in excess oxygen, 18.3 kJ of thermal energy is released. Calculate the enthalpy change of combustion of hydrazine. (2)

4 Diamond and graphite are described as allotropes.

a) Explain what is meant by the term 'allotropes'. (2)

b) Are fullerenes also allotropes with diamond and graphite? Explain your answer. (1)

c) Graphite fibres are often used for the brushes (contacts) in electric motors.

i) Give two reasons why graphite fibres are used in this way. (2)

ii) Give three reasons why diamonds would be unsuitable for this use. (3)

8 Polar bonds and polar molecules

All the evidence suggests that the bonding in caesium fluoride, Cs^+F^-, is ionic. This is a compound of a highly reactive metal and the most reactive non-metal. In contrast, the bonding in a molecule such as chlorine, in which both atoms are the same, is purely covalent. In most compounds, however, the bonding is neither purely ionic nor purely covalent. So how do chemists decide on the type of bonding to expect in different compounds?

8.1 A spectrum of bonding

The electron pair in a covalent bond is not shared equally if the two atoms joined by the bond are different. The nucleus of one atom attracts the electrons more strongly than the nucleus of the other. This means that one end of the bond has a slight excess of negative charge (δ−). The other end of the bond has a slight deficit of negative charge and the charge cloud of electrons does not cancel the positive charge on the nucleus. So this end of the bond has a partial positive charge (δ+).

The partial charges at the ends of some covalent bonds results in a spectrum of bonding – from pure covalent bonding (in molecules such as bromine and oxygen) through an increasing separation of partial charges (in molecules such as hydrogen chloride and hydrogen fluoride) to pure ionic bonding (in compounds such as potassium fluoride and sodium chloride).

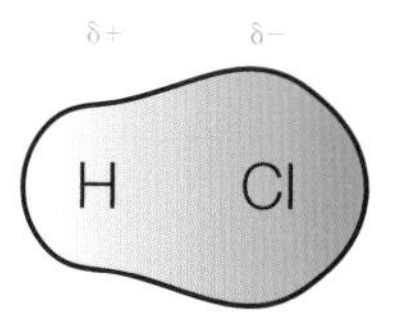

Figure 8.1 ▲
A polar covalent bond in hydrogen chloride. Overall the molecule is uncharged – it is not an ion. The uneven distribution of electrons leads to partial charges at the ends of the covalent bond.

Definition

Polar covalent bonds are bonds between atoms of different elements. The shared electrons are drawn towards the atom with the stronger pull on the electrons. The bonds have a positive pole at one end and a negative pole at the other.

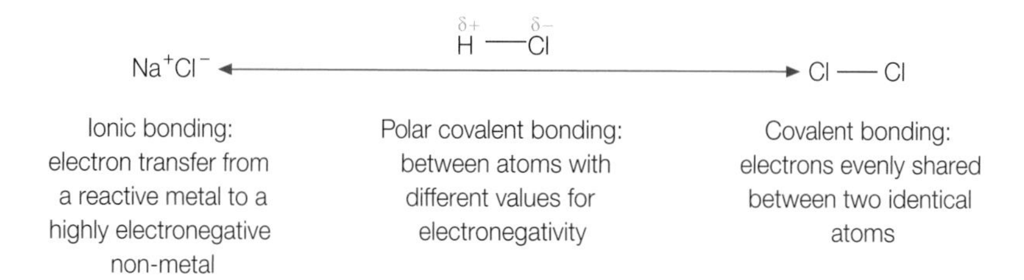

Figure 8.2 ▲
A bonding spectrum from purely ionic to purely covalent

Note

The δ symbol is the Greek letter 'delta'. Chemists use this symbol for a small quantity or change. They use the symbols δ+ and δ− for the small charges at the ends of a polar bond. They use the capital Greek 'delta', Δ, for larger changes or differences (Section 2.2).

8.2 Electronegativity

Chemists use electronegativity values to predict the extent to which the bonds between different atoms are polar. The stronger the pull of an atom on the electrons it shares with other atoms, the higher its electronegativity. Oxygen is more electronegative than hydrogen, so an O–H bond is polar with a slight negative charge on the oxygen atoms and a slight positive charge on the hydrogen atom.

There are two quantitative scales of electronegativity, one devised by Linus Pauling (1901–1994) and the other by Robert Mulliken (1896–1986).

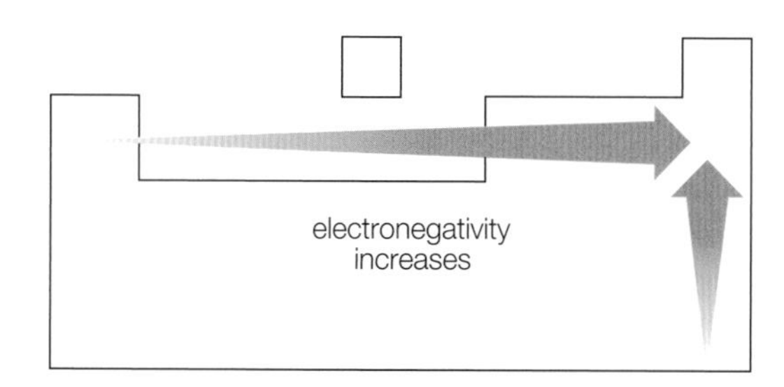

Figure 8.3 ▲
Trends in electronegativity for s and p block elements

Definition

Electronegativity is a measure of the pull of an atom of an element on the shared electrons in a covalent bond. In a polar bond the shared electrons are drawn towards the more electronegative atom.

Activity

Interpreting electronegativity values

The version of the periodic table in Table 8.1 shows electronegativity values for selected elements. The values are those from the Pauling scale.

		H 2.1				
Li 1.0	Be 1.5	B 2.0	C 2.5	N 3.0	O 3.5	F 4.0
Na 0.9	Mg 1.2	Al 1.5	Si 1.8	P 2.1	S 2.5	Cl 3.0
K 0.8	Ca 1.0					Br 2.8
Rb 0.8	Sr 1.0					I 2.5

Table 8.1 ▲
Pauling electronegativity values

1 Why are electronegativity values for He and Ne not included in the table?

2 Identify two elements in the table that would combine to form a compound with covalent bonds that are not polar.

3 Identify the two elements in the table that would form the most purely ionic compound.

4 What is the trend in the values of electronegativity from left to right across a period?

5 a) Draw diagrams showing the electrons in the main shells for lithium and fluorine.

b) Use your diagrams and the concept of shielding to explain why fluorine is much more electronegative than lithium.

6 What is the trend in the values of electronegativity down a group?

7 a) Draw diagrams showing the electrons in the main shells for fluorine and chlorine.

b) Use your diagrams and the concept of shielding to explain why fluorine is more electronegative than chlorine.

In each of these scales, the term electronegativity is used to compare one element with another qualitatively, so to compare elements it is enough to know the trends in electronegativity values across and down the periodic table.

Highly electronegative elements, such as fluorine and oxygen, are at the top right of the periodic table. The least electronegative elements, such as caesium, are at the bottom left.

The bigger the difference in the electronegativity of the elements forming a bond, the more polar, and possibly more ionic, the bond. The bonding in a compound becomes ionic when the difference in electronegativity is large enough for the more electronegative element to remove electrons completely from the other element. This happens in compounds such as sodium chloride, magnesium oxide and calcium fluoride.

Test yourself

1. Use Figure 8.3 on page 127 and the electronegativity values in Table 8.1 to predict the polarity of the bonds in these molecules: H_2S, NO, CCl_4, ICl.
2. Put these bonds in order of polarity, with the most polar first: C–I, C–H, C–Cl, C–O, C–F, C–Br
3. Put these sets of compounds in order of the character of the bonding, with the most ionic on the left and the most covalent on the right:
   a) Al_2O_3, Na_2O, MgO, SiO_2
   b) LiI, NaI, KI, CsI

8.3 Polar molecules

The covalent bonds in hydrogen chloride are polar and, as there is only one bond in each molecule, the molecules are also polar. There are, however, molecules with polar bonds that are not polar. One example is tetrachloromethane (Figure 8.4). The four polar bonds in CCl_4 are arranged symmetrically around the central carbon atom so that overall they cancel each other out.

Polar molecules are little electrical dipoles – they have a positive electric pole and a negative electric pole. These two poles of opposite charge in a molecule are called dipoles. Dipoles tend to line up in an electric field (Figure 8.5).

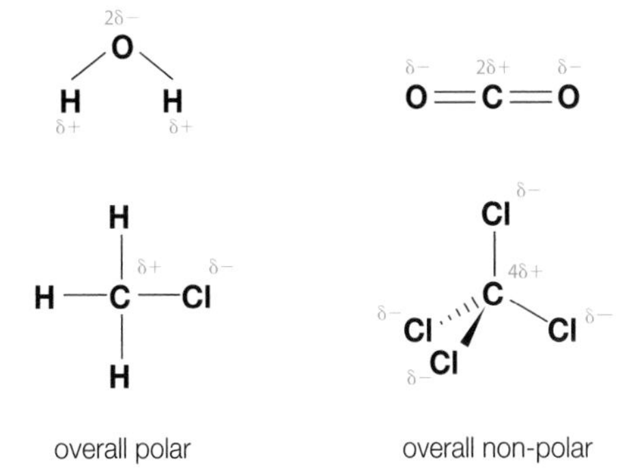

Figure 8.4 ▲
Molecules with polar bonds. Note that in the examples on the left the net effect of all the polar bonds is a polar molecule; in the examples on the right the overall effect is a non-polar molecule.

Definition

Polar molecules have polar bonds which do not cancel each other out, so that the whole molecule is polar.

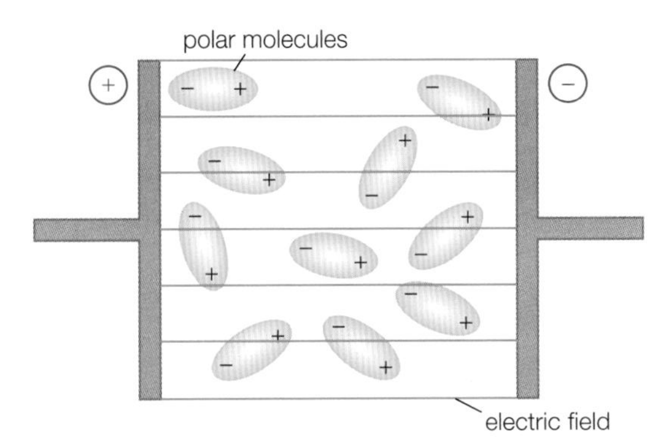

Figure 8.5 ▲
Polar molecules in an electric field. The electrostatic forces tend to line up the molecules with the field. Random movements due to the kinetic energy of the molecules tend to disrupt the alignment of the molecules.

The bigger the dipole, the bigger the twisting effect – or dipole moment – on a molecule in an electric field. By making measurements with a polar

Molecule	Dipole moment /debye units
HCl	1.08
H_2O	1.94
CH_3Cl	1.86
$CHCl_3$	1.02
CCl_4	0
CO_2	0

Table 8.2 ▲
Measures of the polarities of some molecules

substance between two electrodes it is possible to calculate dipole moments. The units are debye units, named after the physical chemist Peter Debye (1884–1966).

A thin stream of a polar liquid is attracted towards an object with an electrostatic charge. This is because the polar molecules tend to move and rotate because the charge on one side of the molecules is attracted to the opposite charge on the object.

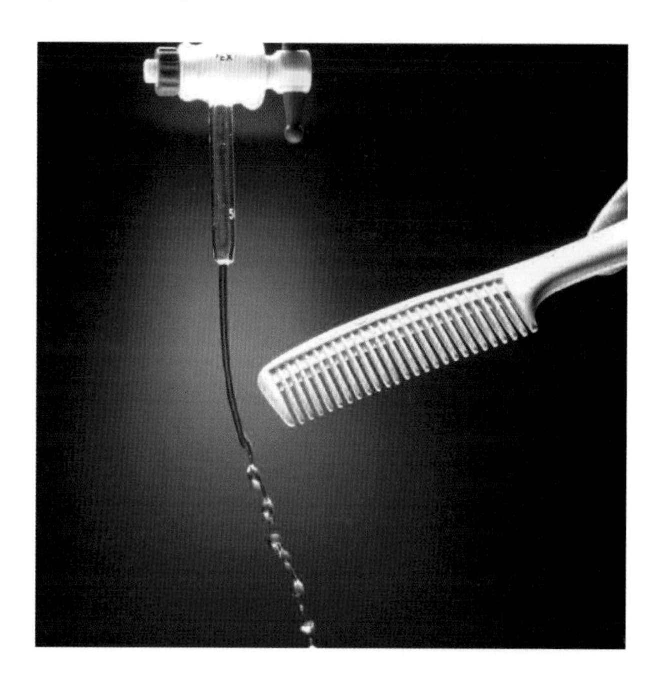

Figure 8.6 ▶
A thin stream of water is bent by a nearby comb carrying an electrostatic charge.

Tutorial

Test yourself

4 Consider the shape of the following molecules and the polarity of their bonds. Then, divide the molecules into two groups – polar and non-polar: HBr, $CHBr_3$, CBr_4, CO_2, SO_2, C_2H_6.

5 Account for the relative values of the dipole moments of the molecules in Table 8.2.

REVIEW QUESTIONS

www Extension questions

1 Three of the hydrides of the group 6 elements are H_2O, H_2S and H_2Se.

a) Explain why all three hydrides have polar molecules. (2)

b) State and explain the trend in electronegativity from O to Se down the group. (3)

c) In which of the hydrides of group 6 are the bonds most polar? (1)

2 Explain why:

a) FCl is polar but F_2 is not (2)

b) SO_2 is polar but CO_2 is not (2)

c) NCl_3 is polar but BCl_3 is not. (2)

3 Thin streams of some liquids are attracted towards a charged rod, but with other liquids there is no effect.

a) Explain why some liquids are attracted while others are not. (2)

b) Predict which of the following liquids are deflected towards a charged rod and explain your predictions: water, hexane, bromoethane, tetrachloromethane. (4)

c) Why are the affected liquids always attracted towards the charged rod and not repelled? (2)

9 Intermolecular forces

Covalent, ionic and metallic bonds are strong bonds that account for the properties of materials such as diamond, glass and metals. Equally important are the weak attractive forces between molecules. Without these intermolecular forces there would be no rivers or oceans; with no intermolecular forces the genetic code would be unreadable and our bodies would fall apart.

Figure 9.1 Geckos can climb smooth walls and hang from ceilings thanks to the intricate design of their feet. Each of a gecko's toes is lined with microscopic hairs, and each hair is further branched into finer structures. Weak intermolecular forces over the large surface area of the hairs are strong enough to grip on any surface, but weak enough to break as the gecko moves by peeling its feet away.

9.1 London forces

The Dutch physicist Johannes van der Waals (1837–1923) developed a theory of intermolecular forces to explain why real gases behave in the way that they do. If there were no attractions between molecules, it would be impossible to turn a gas into a liquid by cooling. For some gases, the attractive forces are so weak that they do not liquefy until very low temperatures are reached. The boiling temperature of hydrogen, for example, is −253 °C, just 20 degrees above absolute zero.

It is not obvious why there are weak attractions between uncharged non-polar molecules, such as those of iodine, hydrocarbons and the noble gases. The German physicist who developed the theory to explain these forces was Fritz London (1900–1954), so they are sometimes called London forces.

When non-polar atoms or molecules meet, there are fleeting repulsions and attractions between the nuclei of the atoms and the surrounding clouds of electrons. Temporary displacements of the electrons lead to temporary dipoles. These instantaneous dipoles can induce dipoles in neighbouring molecules –

Definitions

Intermolecular forces are weak attractive forces between molecules.

London forces are the intermolecular forces that exist between all molecules. They arise from the attractions between temporary dipoles and the fleeting dipoles they induce in neighbouring molecules.

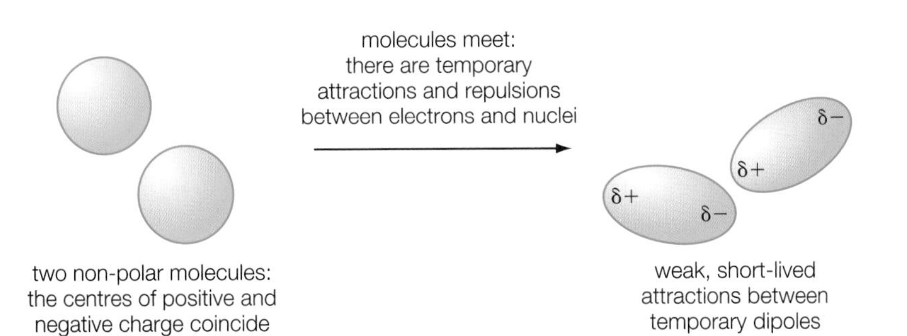

Figure 9.2 The origins of temporary induced dipoles

positive dipoles induce negative dipoles and vice versa. The attractions between these instantaneous and induced dipoles are the weakest kind of intermolecular forces, but their presence gives molecules a tendency to cohere. Intermolecular forces of this kind are roughly a hundred times weaker than covalent bonds.

The greater the number of electrons in a molecule, the more polarisable it is and the greater the possibility for temporary, induced dipoles. This explains why the boiling temperatures rise down group 7 (the halogens) and group 0 (the noble gases). For the same reason, the boiling temperatures of alkanes increase with the increasing number of carbon atoms.

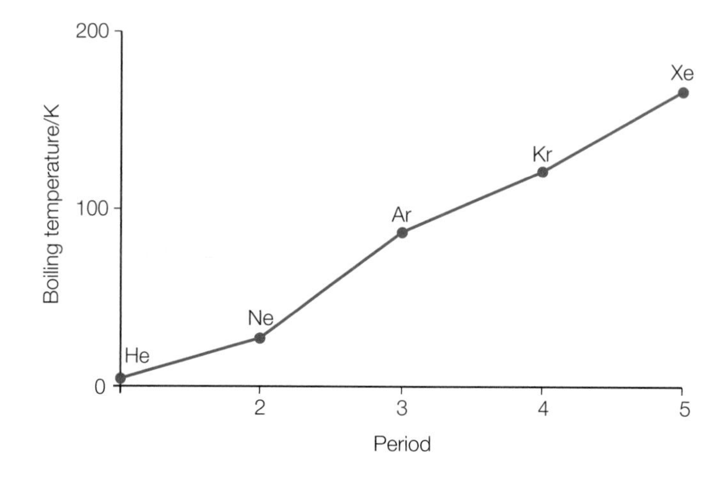

Figure 9.3 ▶ The boiling temperatures of noble gases plotted against atomic number

Definition

Polarisability is an indication of the extent to which the electron cloud in a molecule can be distorted by a nearby electric charge.

The shapes of molecules can also affect the overall size of London forces. The attractions between long thin molecules are stronger than those between short, fat molecules. Long thin molecules can lie close to each other and the attractions can take effect over a larger surface area.

Test yourself

1 Explain how Figure 9.3 illustrates the fact that the strength of intermolecular forces varies with the number of electrons in the molecules of monatomic gases.

2 Account for the states of the halogens at room temperatures – chlorine is a gas; bromine is a liquid; while iodine is a solid.

Activity

Intermolecular forces and the properties of alkanes

Refer to the data sheet for alkanes on the Dynamic Learning Student website as you answer the questions in this activity. Explain the patterns you find in terms of intermolecular forces.

Boiling temperatures of the unbranched alkanes

Plot the boiling temperatures of unbranched alkanes against the number of carbon atoms in the molecules for the range C_1 to C_{10}.

1 Which of these alkanes are gases at room temperature and pressure and which are liquids?

2 What is the approximate increase in boiling temperature for each $-CH_2-$ added to an alkane chain?

3 Estimate the boiling temperature for dodecane.

4 What type of intermolecular forces act between alkane molecules?

5 What two features of alkane molecules account for the trend in values shown by your graph?

Boiling temperatures of branched alkanes

Add to your graph the points for three 2-methyl alkanes, and also one for a 2,2-dimethyl alkane.

6 What is the effect of chain branching on the boiling temperatures of alkanes?

7 How do you account for this trend?

Melting temperatures of the unbranched alkanes

On the same axes as your other graphs, plot the melting temperatures of unbranched alkanes.

8 Identify one similarity and one difference between the plots of melting temperatures and boiling temperatures.

9 Suggest an explanation for the pattern of melting temperatures for alkanes with an odd number of carbon atoms compared to the alkanes with an even number of carbon atoms.

10 Polythene can be regarded as a long chain polymer of ethane, $-(CH_2)_n-$. The value of n can be around 100 000. How do you account for the strength of this material, which softens and melts in the range 100–150 °C?

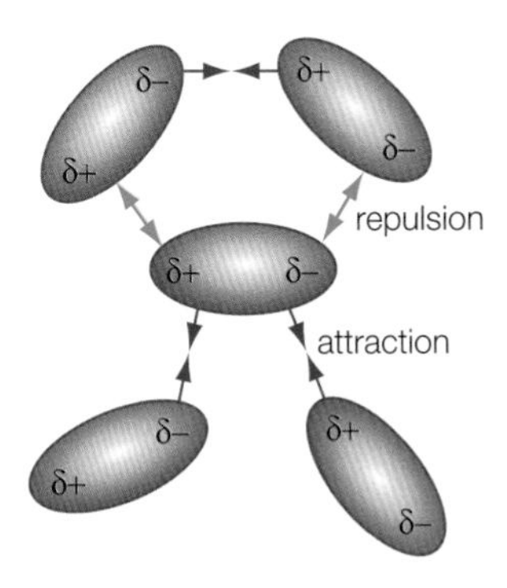

Figure 9.4 ▲
Attractions between molecules with permanent dipoles

9.2 Dipole–dipole interactions

Molecules with permanent dipoles attract each other a little more strongly than non-polar molecules. The positive dipole of one molecule tends to attract the negative dipoles of others and vice versa (Section 8.3).

This contribution to intermolecular forces from permanent dipoles occurs in addition to the London forces which act between all molecules.

Test yourself

3 Account for the difference in boiling temperature between the following pairs of molecules by considering both London forces and dipole–dipole attractions:
- **a)** ethane which boils at −88 °C and fluoromethane which boils at −78 °C
- **b)** butane which boils at −0.5 °C and propanone, CH_3COCH_3, which boils at 56 °C.

9.3 Hydrogen bonding

Hydrogen bonding is a type of dipole–dipole attraction between molecules which is much stronger than other types of intermolecular force, but still at least 10 times weaker than covalent bonds.

Hydrogen bonding affects molecules in which hydrogen is covalently bonded to one of the three highly electronegative elements – fluorine, oxygen and nitrogen. This stronger type of intermolecular force acts in addition to London forces.

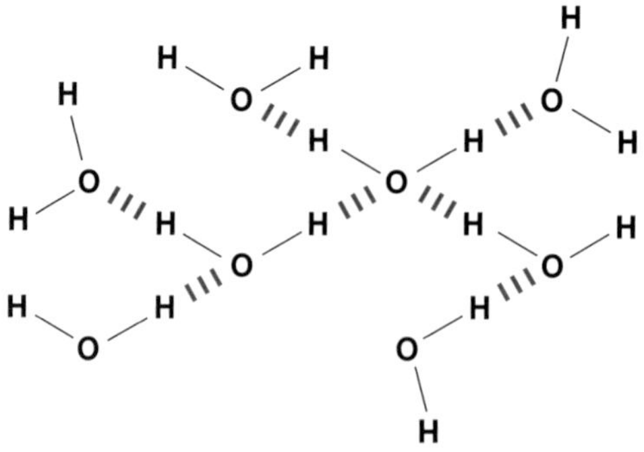

Figure 9.5 ▶
Hydrogen bonding in water

In a hydrogen bond, the hydrogen atom lies between two highly electronegative atoms. It is hydrogen bonded to one of them and covalently bonded to the other. The covalent bond is highly polar – the small hydrogen atom, with its δ+ dipole, has no inner shells of electrons to shield its nucleus. So, it gets close to a lone pair of electrons on another electronegative atom (δ−) in a neighbouring molecule. The three atoms associated with a hydrogen bond are always in a straight line.

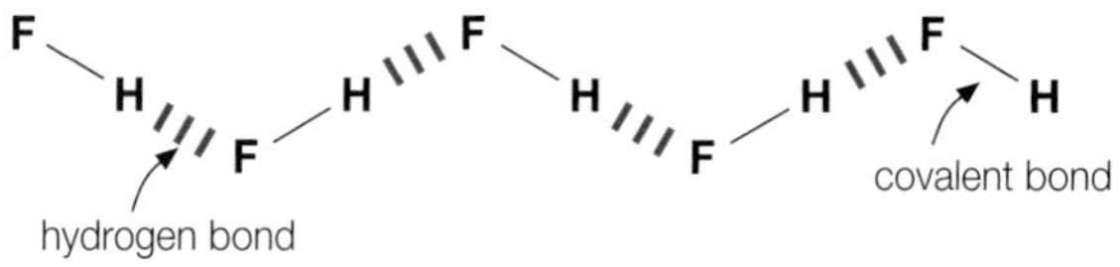

Figure 9.6 ▶
Hydrogen bonding in hydrogen fluoride

The essential requirements for a hydrogen bond are:

- a hydrogen atom covalently bonded to a highly electronegative atom
- an unshared pair of electrons on a second electronegative atom.

In a water molecule there are two O–H bonds and two unshared electron pairs on the oxygen atom. This means that each water molecule can form two hydrogen bonds. This helps to explain the three-dimensional structure of ice.

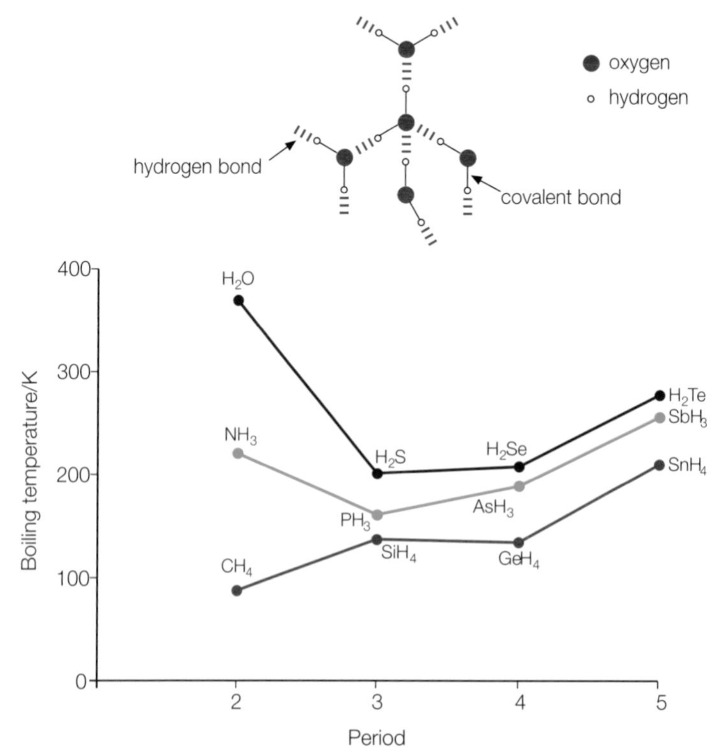

Figure 9.7 ▶
Molecules in ice are held together by hydrogen bonding. The molecules form a giant lattice structure in which each oxygen atom is bonded to two hydrogen atoms by covalent bonds and two others by hydrogen bonds.

Figure 9.8 ▲
Boiling temperatures for the hydrides of the elements in groups 4, 5 and 6

Definition

Hydrogen bonding is a type of attraction between molecules which is much stronger than other types of intermolecular force, but much weaker than covalent bonding.

Hydrogen bonding accounts for:

- the relatively high boiling temperatures of ammonia, water and hydrogen fluoride, which are out of line with those of the other hydrides in groups 5, 6 and 7
- the open structure and low density of ice
- the pairing of bases in a DNA double helix.

Figure 9.9 An iceberg in Antarctica. About 10% of the ice is above the surface of the sea because ice is less dense than water at 0 °C. This is explained by the way the hydrogen bonds in the solid hold the water molecules in an open structure. This structure collapses as the ice melts and the molecules get closer together.

Test yourself

4 State and explain the maximum number of hydrogen bonds per molecule formed in:
  a) hydrogen fluoride, HF
  b) ammonia, NH_3.

5 Draw diagrams to show hydrogen bonding between water molecules and:
  a) ammonia molecules in a solution of ammonia, NH_3
  b) ethanol molecules in a solution of ethanol, CH_3CH_2OH.

6 a) Use the data sheet on Dynamic Learning Student website to plot a graph showing how the boiling temperatures of the hydrogen halides vary with the atomic number of the halogen.
  b) Describe and explain the pattern shown by the graph with reference to the types of intermolecular forces which act between the molecules.

Data

7 Explain the differences in Figure 9.8 between the plot for the hydrides of group 6 and the plot for the hydrides of group 4.

8 Which types of intermolecular force hold together the molecules in:
  a) hydrogen bromide, HBr
  b) propane, $CH_3CH_2CH_3$
  c) methanol, CH_3OH?

9 Explain the difference in the boiling temperatures between the two isomers – ethanol which boils at 78 °C and methoxymethane which boils at −25 °C.

9.4 Solutions and solubility

Patterns of solubility

As a rough-and-ready rule, 'like dissolves like'. Water, which is highly polar, dissolves many ionic compounds and compounds with –OH groups such as alcohols and sugars. Non-polar solvents, such as cyclohexane, dissolve hydrocarbons, molecular elements and molecular compounds.

But there is a limit to the quantity of a chemical that can dissolve in a solvent. A solution is saturated when it contains as much of the dissolved solute as possible at a particular temperature.

Definitions

A **solute** is a chemical which dissolves in a **solvent** to make a **solution**.

A **saturated solution** contains as much of the solute as possible at a particular temperature.

Solubility is a measure of the concentration of a saturated solution of a solute at a specified temperature. Solubilities are commonly recorded as 'mol per 100 g water' or 'g per 100 g water' at 25 °C (298 K).

Soluble or insoluble?

No chemicals are completely soluble and none are completely insoluble in water. Even so, chemists find it useful to use a rough classification of solubility based on what they see on shaking a little of the solid with water in a test tube:

- very soluble, like potassium nitrate – plenty of the solid dissolves quickly
- soluble, like copper(II) sulfate – crystals visibly dissolve to a significant extent
- sparingly or slightly soluble, like calcium hydroxide – little solid seems to dissolve but, in this case, the pH of the solution changes
- insoluble, like iron(III) oxide – no sign that any of the material dissolves.

A similar rough classification applies to gases dissolving in water. Ammonia and hydrogen chloride are very soluble; sulfur dioxide is soluble; carbon dioxide is slightly soluble; helium is insoluble.

Solubility and intermolecular forces

Patterns of solubility for molecular solids are determined by intermolecular forces. Three interactions are involved:

- the intermolecular forces between solute molecules
- the intermolecular forces between solvent molecules
- the intermolecular forces between solute and solvent molecules.

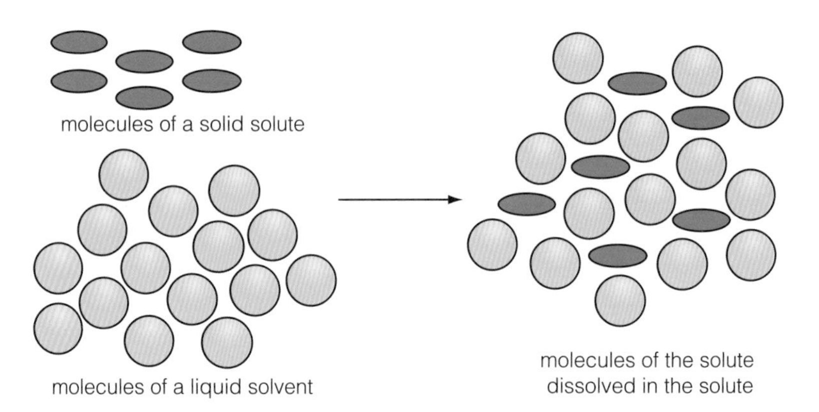

Figure 9.10 ▶ A molecular chemical dissolves if the energy needed to break intermolecular forces and to separate the molecules in the solute and in the solvent is about the same as the energy released as the solute forms new intermolecular forces with the solvent.

Definitions

A **non-aqueous solvent** is any solvent other than water.

Miscible liquids are those which mix with each other – water and alcohol are miscible; oil and water are immiscible.

When all three types of force are about the same strength, the solute dissolves freely in the solvent. So non-polar molecules, such as those of a hydrocarbon wax, dissolve and mix freely with non-polar liquids such as cyclohexane. All the intermolecular forces involved are London forces.

However, non-polar molecules, such as hydrocarbons, do not dissolve in water. The non-polar molecules can separate easily because their intermolecular forces are weak. But the strong hydrogen bonding between water molecules acts as a barrier which keeps out molecules that cannot themselves form hydrogen bonds.

Organic molecules that can form hydrogen bonds, such as alcohols, do dissolve and mix with water. Ethanol molecules, for example, can break into the hydrogen-bonded structure of water by forming new hydrogen bonds between ethanol and water molecules.

The two liquids ethanol (C_2H_5OH) and water are miscible. Alcohols with longer hydrocarbon chains do not mix with water so easily. The longer the chain, the less the miscibility of the alcohol with water.

Test yourself

10 Explain why methane gas is insoluble in water, but ammonia is freely soluble.

11 Explain why iodine is soluble in a non-aqueous solvent such as cyclohexane, but almost insoluble in water.

12 Explain why methanol is miscible with water.

Solutions of ionic salts in water

It is not obvious why the charged ions in a crystal of sodium chloride can separate and dissolve in water with only a small energy change. The value of the lattice energy for a salt such as sodium chloride (Section 4.5) shows that a large amount of energy is needed to separate the ions from a crystal.

The explanation of the solubility of some ionic salts in water is that the ions are strongly hydrated by polar water molecules. The water molecules cluster round the ions and bind to them. In the case of sodium chloride, the energy released when the water molecules bind to the ions is enough to compensate for the energy needed to overcome the ionic bonding between the ions.

Other salts are insoluble in water because the energy that would be released when the ions are hydrated is not large enough to balance the lattice energy of the crystals.

> **Definition**
>
> **Hydration** takes places when water molecules bond to ions or add to molecules. Water molecules are polar so they are attracted to both positive ions and negative ions.

Data

> **Test yourself**
>
> 13 With the help of data sheets on the Dynamic Learning Student website, classify the following salts as very soluble, soluble, slightly soluble or insoluble according to their solubility in water: barium sulfate, caesium fluoride, calcium hydroxide, calcium sulfate, lithium fluoride, lithium chloride, magnesium chloride, magnesium sulfate, potassium iodide.

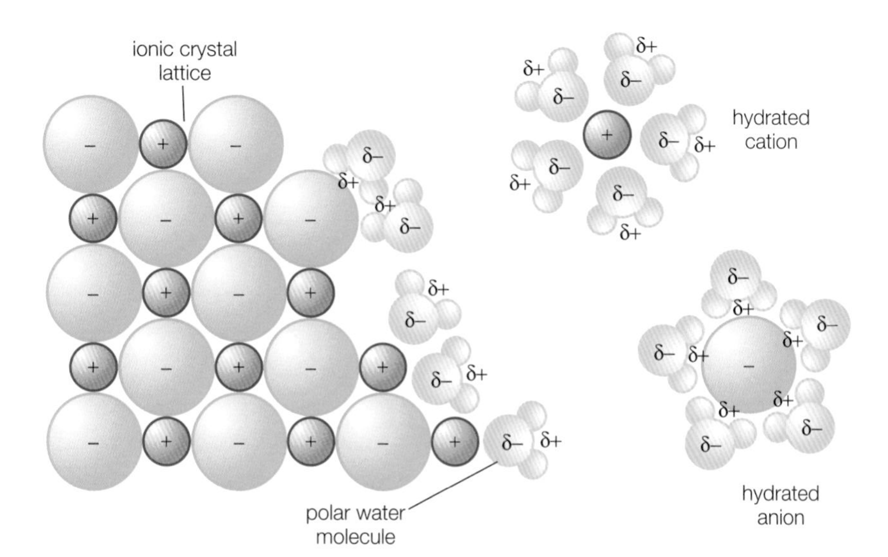

Figure 9.11 ◄ Sodium ions and chloride ions leaving a crystal lattice and becoming hydrated as they dissolve in water. Here the bond between the ions and the polar water molecules is electrostatic attraction.

Extension questions

REVIEW QUESTIONS

1 a) State the types of intermolecular forces present in:

i) propane

ii) ethanol. (2)

b) Why is the boiling temperature of propane (−42.2 °C) lower than the boiling temperature of ethanol (−78.5 °C)? (2)

c) Glycerol is a type of alcohol (propan-1,2,3-triol).

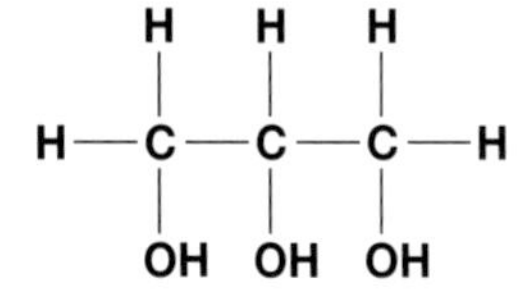

Figure 9.12 ▲

Predict whether the boiling temperature of glycerol is higher or lower than that of ethanol, giving your reasons. (2)

2 Explain the following statements about oil refining in terms of intermolecular forces.

a) It is possible to separate the hydrocarbons in crude oil into fractions by distillation. (3)

b) Cracking turns a mixture of liquids from crude oil into a mixture of gases. (2)

c) Isomerisation turns straight chain alkanes into alkanes with lower boiling temperatures. (2)

3 a) Name the strongest of the intermolecular forces between water molecules, and describe the bonding with the help of a diagram. (3)

b) Explain why:

i) the boiling temperature of water is higher than the boiling temperatures of the other hydrides of group 6 elements

ii) ice is less dense than water at 0 °C

iii) water and pentane are immiscible liquids. (6)

10 Redox

Oxidation and reduction reactions are very common. Chemists have devised a number of ways of recognising and describing what happens during changes of this kind.

10.1 Oxidation and reduction

Burning is perhaps the commonest example of oxidation; another example is rusting, which converts iron to a form of iron oxide. At its simplest, oxidation involves adding oxygen to an element or compound.

Reduction is the opposite of oxidation. Metal oxides are reduced during the extraction of metals from their ores – in a blast furnace, for example, carbon monoxide takes the oxygen away from iron oxide to leave metallic iron (Figure 10.1).

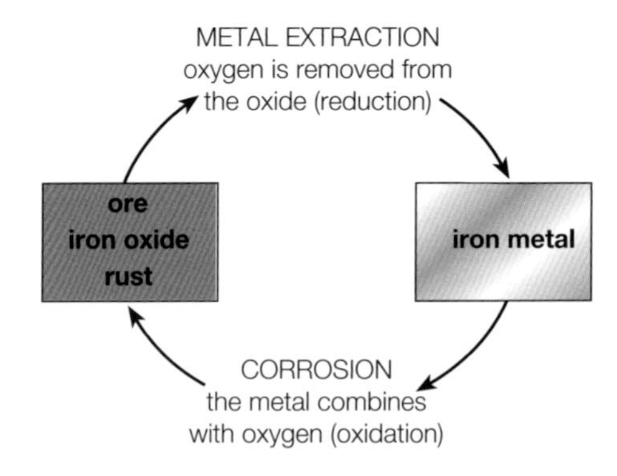

Figure 10.1 ▶
The cycle of extraction and corrosion for iron

Test yourself

1 Which element or compound is oxidised and which is reduced in the reaction of:
- a) steam with hot magnesium
- b) copper(II) oxide with hydrogen
- c) aluminium with iron(III) oxide
- d) carbon dioxide with carbon to form carbon monoxide?

Electron transfer

Magnesium burns brightly in air. The product is a white solid – the ionic compound magnesium oxide, $Mg^{2+}O^{2-}$.

$$2Mg(s) + O_2(g) \rightarrow 2Mg^{2+}O^{2-}(s)$$

During the reaction, each magnesium atom gives up two electrons, turning into a magnesium ion:

$$2Mg \rightarrow 2Mg^{2+} + 4e^-$$

Oxygen takes up the electrons from the magnesium producing oxide ions:

$$O_2 + 4e^- \rightarrow 2O^{2-}$$

In this way, electrons transfer from magnesium atoms to oxygen atoms, forming ions from atoms.

Magnesium atoms also turn into ions when they react with other non-metals such as chlorine, bromine and sulfur.

In all its reactions with non-metals, magnesium loses electrons and its atoms become positive ions. The non-metal's atoms gain electrons and become negative ions. All the reactions involve electron transfer.

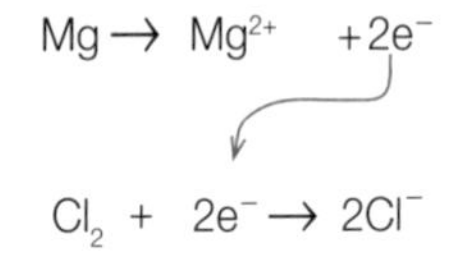

Figure 10.2 ▲
Electron transfer in the reaction of magnesium with chlorine

Magnesium is oxidised as it loses electrons; the non-metal is reduced as it gains electrons.

Ionic half-equations

A half-equation is an ionic equation used to describe either the gain or the loss of electrons during a redox process. Half-equations help to show what is happening during a reaction. Two half-equations combine to give the overall balanced equation.

Zinc metal can reduce copper ions to copper. This can be shown as two half-equations:

electron gain (reduction): $Cu^{2+}(aq) + 2e^- \rightarrow Cu(s)$

electron loss (oxidation): $Zn(s) \rightarrow Zn^{2+}(aq) + 2e^-$

Adding together the two half-equations leads to the full ionic equation. The electrons must cancel in order that the electrons gained on one side of the equation equal the electrons lost on the other side.

$$Cu^{2+}(aq) + 2e^- \rightarrow Cu(s)$$

$$Zn(s) \rightarrow Zn^{2+}(aq) + 2e^-$$

$$Cu^{2+}(aq) + Zn(s) \rightarrow Cu(s) + Zn^{2+}(s)$$

Definition

Reduction and **ox**idation always go together, hence the term **redox** reaction.

Note

You may find the mnemonic, **oil rig** helpful.

Oxidation	**R**eduction
Is	**I**s
Loss	**G**ain

Definitions

Ionic equations describe chemical changes by showing only the reacting ions, leaving out spectator ions.

A **half-equation** is an ionic equation used to describe either the gain, or the loss, of electrons during a redox reaction.

Test yourself

2 Write the separate ionic half-equations for the reactions of:
  a) sodium with chlorine
  b) zinc with oxygen
  c) calcium with bromine.

3 Write the ionic half-equations and the full ionic equation for the reaction of zinc with silver nitrate solution.

Data

10.2 Oxidation numbers

Chemists use oxidation numbers to keep track of the electrons transferred or shared during chemical changes. With the help of oxidation numbers it becomes much easier to recognise redox reactions. Oxidation numbers provide a useful way of organising the chemistry of elements such as chlorine, which can be oxidised or reduced to varying degrees. Chemists have also chosen to base the names of inorganic compounds on oxidation numbers.

Oxidation numbers and ions

Oxidation numbers show how many electrons are gained or lost by an element when atoms turn into ions and vice versa. In Figure 10.3, movement up the diagram involves the loss of electrons and a shift to more positive oxidation numbers – this is oxidation. Movement down the diagram involves the gain of electrons and a shift to less positive, or more negative, oxidation numbers – this is reduction.

The oxidation number of all uncombined elements is zero. In a simple ion, the oxidation number of the element is the charge on the ion. For example, in calcium chloride the metal is present as the Ca^{2+} ion and the oxidation number of calcium is +2.

Oxidation numbers distinguish between the compounds of elements such as iron that can exist in more than one oxidation state. In iron(II) chloride the Roman number 'II' shows that iron is in oxidation state +2. Iron atoms lose two electrons when they react with hydrogen chloride to make iron(II) chloride.

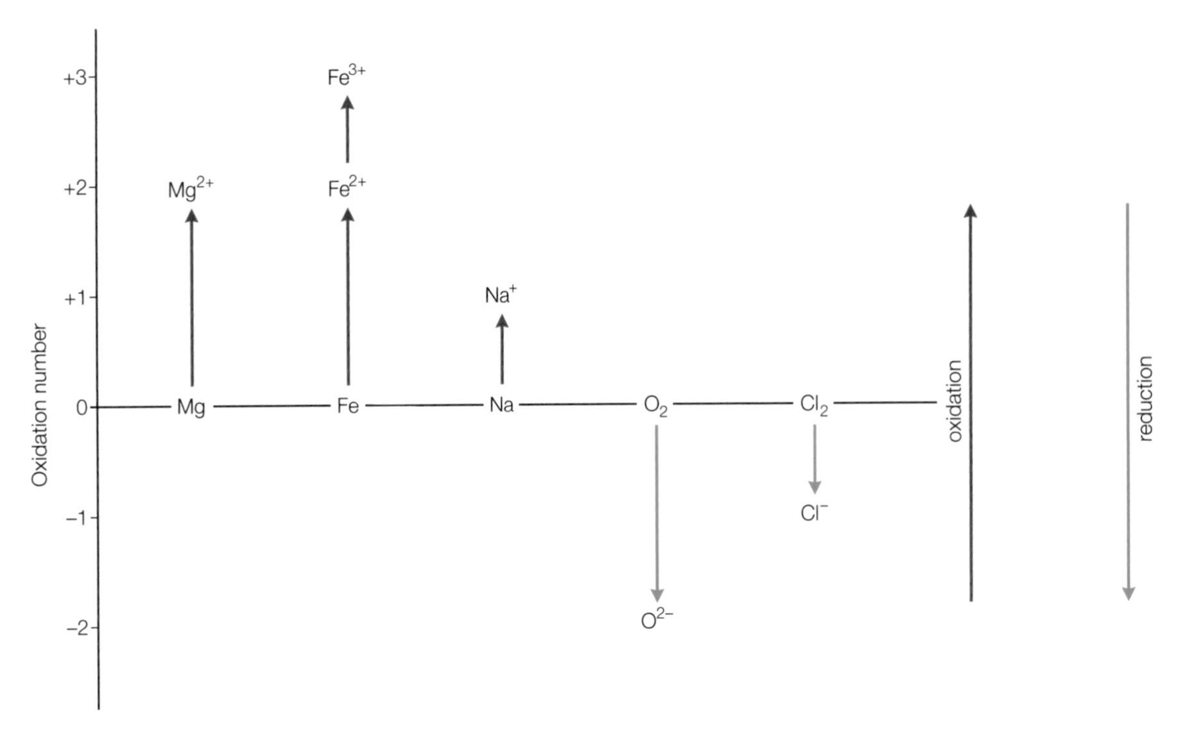

Figure 10.3 ▲
Oxidation numbers of atoms and ions

Figure 10.4 ►
Oxidation number rules

Oxidation number rules

1. The oxidation number of uncombined elements is zero.
2. In simple ions the oxidation number of the element is the charge on the ion.
3. The sum of the oxidation numbers in a neutral compound is zero.
4. The sum of the oxidation numbers for an ion is the charge on the ion.
5. Some elements have fixed oxidation numbers in all their compounds.

Metals

group 1 metals (e.g. Li, Na, K) +1
group 2 metals (e.g. Mg, Ca, Ba) +2
aluminium +3

Non-metals

hydrogen (except in metal hydrides, H^-) +1
fluorine −1
oxygen (except in peroxides, O_2^{2-}, and compounds with fluorine) −2
chlorine (except in compounds with oxygen and fluorine) −1

NH_4^+
−3 +1

MnO_4^-
+7 −2

SO_4^{2-}
+6 −2

$Cr_2O_7^{2-}$
+6 −2

Figure 10.5 ▲
Oxidation numbers in ions with more than one atom. Note the use of 2− for the electric charge on a sulfate ion (number first for ionic charges) but the use of −2 to refer to the oxidation state of oxygen in the ion (charge first for oxidation states in ions and molecules).

With the help of the rules in Figure 10.4, it is possible to extend the use of oxidation numbers to ions consisting of more than one atom. The charge on an ion, such as the sulfate ion, is the sum of the oxidation numbers of the atoms. The normal oxidation state of oxygen is −2. There are four oxygen atoms (four at −2) in the sulfate ion so the oxidation state of sulfur must be +6 to give an overall charge on the ion of −2 (Figure 10.5).

Definitions

Oxidation originally meant combination with oxygen, but the term now covers all reactions in which atoms, molecules or ions lose electrons. The definition is extended to cover molecules, as well as atoms and ions, by defining oxidation as a change which makes the oxidation number of an element more positive, or less negative.

Reduction originally meant removal of oxygen or addition of hydrogen, but the term now covers all reactions in which atoms, molecules or ions gain electrons. The definition is extended to cover molecules, as well as atoms and ions, by defining oxidation as a change which makes the oxidation number of an element more negative, or less positive.

Test yourself

4 What is the oxidation number of:
a) aluminium in Al_2O_3
b) nitrogen in magnesium nitride, Mg_3N_2
c) nitrogen in barium nitrate, $Ba(NO_3)_2$
d) nitrogen in the ammonium ion, NH_4^+?

Oxidation numbers and molecules

The rules in Figure 10.4 make it possible to apply the definitions of oxidation and reduction to molecules. In most molecules, the oxidation state of an atom corresponds to the number of electrons from that atom which are shared in covalent bonds.

Where two atoms are linked by covalent bonds, the more electronegative atom (Section 8.2) has the negative oxidation state. Fluorine always has a negative oxidation state of −1 because it is the most electronegative of all atoms. Oxygen normally has a negative oxidation state (−2) but it has a positive oxidation state (+1) when combined with fluorine.

The reason for writing oxidation numbers as +1, +2 and so on is to make quite clear that when dealing with molecules they do not refer to electric charges. Molecules are not charged and the sum of the oxidation states for all the atoms in a molecule is zero.

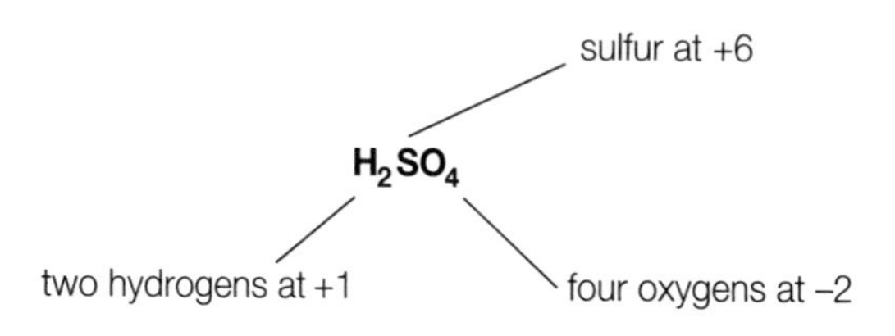

Figure 10.6 ◀
Oxidation numbers of the elements in sulfuric acid

Test yourself

5 Are these elements oxidised or reduced in these conversions?
a) calcium to calcium bromide
b) chlorine to lithium chloride
c) chlorine to chlorine dioxide
d) sulfur to hydrogen sulfide
e) sulfur to sulfuric acid.

Oxidation numbers and the chemistry of elements

Oxidation numbers help to make sense of the chemistry of an element such as bromine (see Figure 10.7).

The oxidation numbers of the elements lithium to chlorine in their oxides reveal a periodic pattern when plotted against atomic number (Figure 10.8). The most positive oxidation number for each element corresponds to the number of electrons in the outer shell of the atoms.

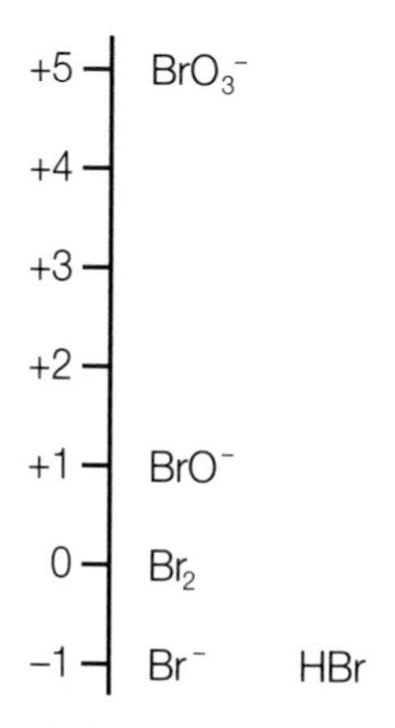

Figure 10.7 ▲
A reaction turning bromine into BrO^- ions involves oxidation of bromine. The conversion of BrO^- ions to BrO_3^- ions involves further oxidation of bromine.

Definition

The degrees of oxidation or reduction shown by an element are its **oxidation states**. The states are labelled with the oxidation number of the element in each state.

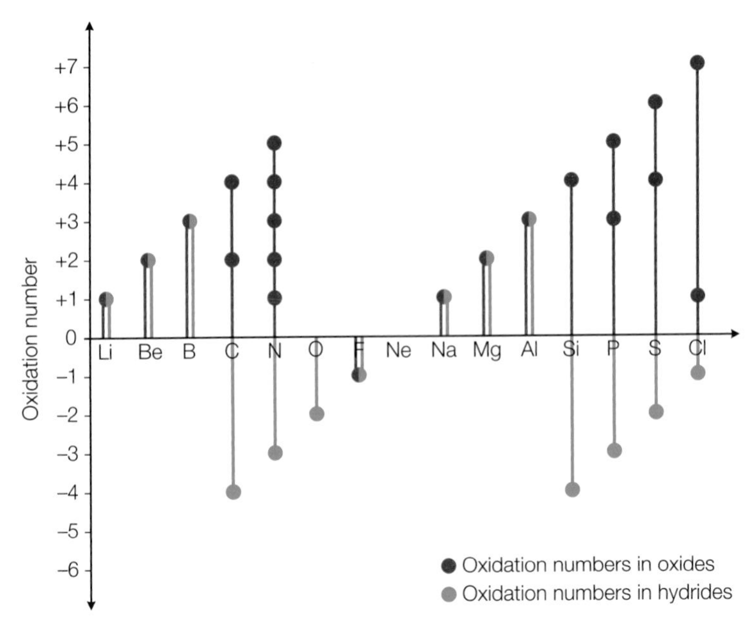

Figure 10.8 ▲
Oxidation numbers of elements in their oxides and hydrides. Note that some elements form oxides in a variety of oxidation states.

Test yourself

6 This question refers to Figure 10.8.
   a) Give the formula of the oxide of lithium.
   b) Give the formulae of the two oxides of carbon.
   c) What are the oxidation states of nitrogen in these oxides: NO, N_2O, NO_2, N_2O_3, N_2O_5?
   d) Give the formulae of the hydrides of nitrogen and phosphorus.
   e) Why is there only one element with a negative oxidation number in an oxide?

Oxidation numbers and the names of compounds

The names of inorganic compounds are becoming increasingly systematic, but chemists still use a mixture of names. Most prefer the name 'copper sulfate' for the blue crystals with the formula $CuSO_4.5H_2O$. This is hydrated copper(II) sulfate. Its fully systematic name, tetraaquocopper(II) tetraoxosulfate(VI)-1-water, is rarely used. The systematic name has much more to say about the arrangement of atoms, molecules and ions in the blue crystals but it is too cumbersome for normal use. The systematic name also shows the oxidation states of copper and sulfur in the compound.

Here are some of the basic rules for naming common inorganic compounds:

- the ending '-ide' shows that a compound contains just the two elements mentioned in the name. The more electronegative element comes second – for example, sodium sulfide (Na_2S), carbon dioxide (CO_2) and magnesium nitride (Mg_3N_2)
- the Roman numbers in names indicate the oxidation numbers of the elements – for example iron(II) sulfate ($FeSO_4$) and iron(III) sulfate ($Fe_2(SO_4)_3$)
- the traditional names of oxoacids end in '-ic' or '-ous' as in sulfuric (H_2SO_4) and sulfurous (H_2SO_3) acids, and nitric (HNO_3) and nitrous (HNO_2) acids. The '-ic' ending is for the acid in which the central atom has the higher oxidation number

- the corresponding traditional endings for the salts of oxoacids are '-ate' and '-ite' as in sulfate (SO_4^{2-}) and sulfite (SO_3^{2-}), and in nitrate (NO_3^-) and nitrite (NO_2^-)
- the more systematic names for oxoacids and oxosalts use oxidation numbers as in sulfate(VI) for sulfate (SO_4^{2-}) and sulfate(IV) for sulfite (SO_3^{2-}).

When in doubt, chemists give the name and the formula. In some cases, they may give two names – the systematic name and the traditional name.

Test yourself

7 Write the formulae of the compounds:
 a) tin(II) oxide
 b) tin(IV) oxide
 c) iron(III) nitrate(V)
 d) potassium chromate(VI).

10.3 Recognising redox reactions

www Tutorial

Oxidation numbers help to identify redox reactions. In the equation for any redox reaction, at least one element changes to a more positive oxidation state, while another changes to a less positive oxidation state. A reaction is not a redox reaction if there are no changes of oxidation state.

Oxidising and reducing agents

An agent is someone or something which gets things done. In spy stories, the main players are secret agents with a mission to make a change. In redox reactions, the chemicals with a mission are the oxidising and reducing agents.

The term 'oxidising agent' (or oxidant) describes chemical reagents which can oxidise other atoms, molecules or ions by taking electrons away from them. Common oxidising agents are oxygen, chlorine, nitric acid, potassium manganate(VII), potassium dichromate(VI) and hydrogen peroxide.

The term 'reducing agent' (or reductant) describes chemical reagents which can reduce other atoms, molecules or ions by giving them electrons. Common reducing agents are hydrogen, sulfur dioxide and zinc or iron in acid.

It is easy to get into a mental tangle when using these terms. When an oxidising agent reacts it is reduced; when a reducing agent reacts, it is oxidised. This is illustrated by the reaction of magnesium with chlorine (Figure 10.9).

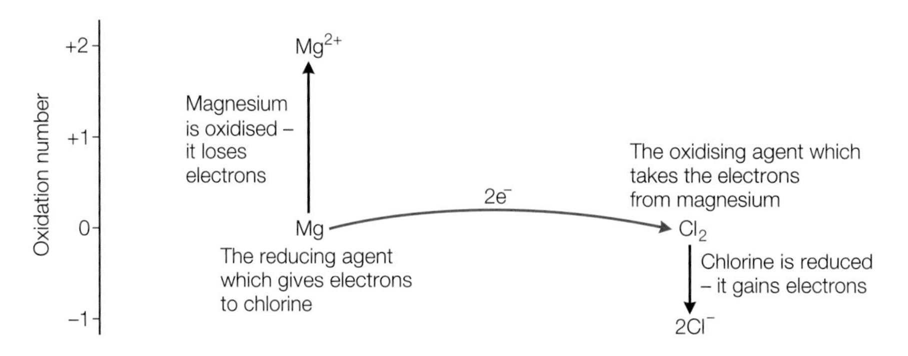

Figure 10.9 Magnesium is oxidised by loss of electrons. It is oxidised by chlorine, so chlorine is the oxidising agent. At the same time chlorine gains electrons and is reduced by magnesium. Magnesium is the reducing agent.

Disproportionation reactions

In some reactions, the same element both increases and decreases its oxidation number. In other words, some of the element is oxidised while the rest of it is reduced.

Warming copper(I) oxide with dilute sulfuric acid does not produce a solution of copper(I) sulfate. Instead, the products are a solution of copper(II) sulfate and a precipitate of copper metal.

$$\overset{+1}{Cu_2O(s)} + H_2SO_4(aq) \rightarrow \overset{+2}{CuSO_4(aq)} + \overset{0}{Cu(s)} + H_2O(l)$$

Some of the copper(I) is oxidised to copper(II), while the rest is reduced to copper(0). This is an example of a disproportionation reaction. Reactions of this kind are important in the chemistry of the halogens (Section 11.9).

Test yourself

8 Classify these reactions as redox, acid–base, precipitation or thermal decomposition. Identify the disproportionation reaction.

a) $CaCl_2(aq) + K_2SO_4(aq) \rightarrow CaSO_4(s) + 2KCl(aq)$
b) $CaCO_3(s) \rightarrow CaO(s) + CO_2(g)$
c) $Ca(s) + 2H_2O(l) \rightarrow Ca(OH)_2(aq) + H_2(g)$
d) $Ca(OH)_2(s) + 2HCl(aq) \rightarrow CaCl_2(aq) + 2H_2O(l)$
e) $2Ca(OH)_2(s) + 2Cl_2(aq) \rightarrow CaCl_2(aq) + Ca(ClO)_2(aq) + 2H_2O(l)$

Tutorial

10.4 Balancing redox equations

Oxidation numbers help when balancing redox equations because the total decrease in oxidation number for the element reduced must equal the total increase in oxidation number for the element oxidised.

This is illustrated below by the oxidation of hydrogen bromide by concentrated sulfuric acid. The main products are bromine, sulfur dioxide and water.

Worked example

Step 1: Write down the formulae for the atoms, molecules and ions involved in the reaction.

$$HBr + H_2SO_4 \rightarrow Br_2 + SO_2 + H_2O$$

Step 2: Identify the elements which change in oxidation number and the extent of change.

In this example only bromine and sulfur show changes of oxidation state.

Step 3: Balance so that the total increase in oxidation number of one element equals the total decrease of the other element.

In this example, the increase of +1 in the oxidation number of two bromine atoms balances the −2 decrease of one sulfur atom.

$$2HBr + H_2SO_4 \rightarrow Br_2 + SO_2 + H_2O$$

Step 4: Balance for oxygen and hydrogen.

In this example, the four hydrogen atoms on the left of the equation join with the two remaining oxygen atoms to form two water molecules.

$$2HBr + H_2SO_4 \rightarrow Br_2 + SO_2 + 2H_2O$$

Step 5: Add the state symbols.

$$2HBr(g) + H_2SO_4(l) \rightarrow Br_2(l) + SO_2(g) + 2H_2O(l)$$

Test yourself

9 Write balanced equations for these redox reactions. State which element is oxidised and which is reduced in each example.

a) Fe with Br_2 to give $FeBr_3$
b) F_2 with H_2O to give HF and O_2
c) IO_3^- and H^+ with I^- to give I_2, and H_2O
d) $S_2O_3^{2-}$ and I_2 to give $S_4O_6^{2-}$ and I^-.

Activity

Preparing a sample of an oxide of nitrogen

Lead(II) nitrate decomposes on heating to form three products. These are lead(II) oxide, nitrogen dioxide and oxygen. Nitrogen dioxide, NO_2, is a brown gas. Some of the nitrogen dioxide molecules pair up to form N_2O_4. Cooling the gas mixture condenses the N_2O_4 as a greenish liquid.

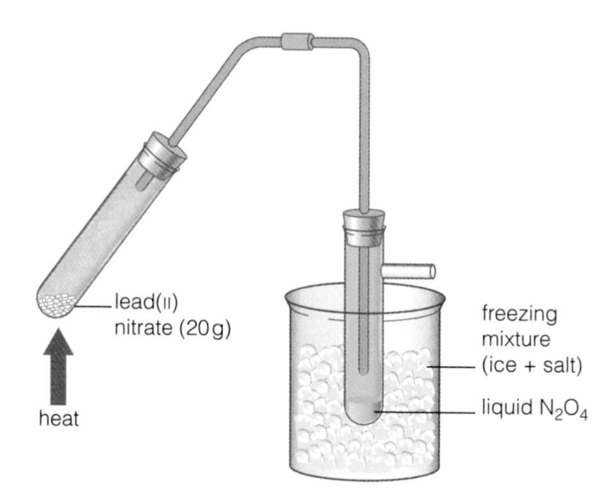

Figure 10.10 ▲
Heating lead(II) nitrate to collect a sample of N_2O_4

1 Give the oxidation states of all the elements in: **a)** lead(II) nitrate, **b)** lead(II) oxide, **c)** nitrogen dioxide, **d)** N_2O_4 and **e)** oxygen gas.

2 Identify as oxidation, reduction or neither, the formation of:
- **a)** lead(II) oxide from lead(II) nitrate
- **b)** nitrogen dioxide from lead(II) nitrate
- **c)** oxygen from lead(II) nitrate
- **d)** N_2O_4 from NO_2.

3 Write balanced equations for:
- **a)** the decomposition of lead(II) nitrate
- **b)** the formation of N_2O_4 from NO_2.

4 Calculate the theoretical mass of N_2O_4 that could be collected by condensing the gases given off when 20 g lead(II) nitrate decomposes.

5 Suggest reasons why the mass of N_2O_4 collected by heating 20 g lead(II) nitrate in the apparatus in Figure 10.10 would be less than your answer to question **4**.

REVIEW QUESTIONS

1 These are incomplete half-equations for changes involving reduction in solution. Complete and balance the half-equations – you need to add the electrons:

a) $H^+(aq) \rightarrow H_2(g)$ (1)

b) $Fe^{3+}(aq) \rightarrow Fe^{2+}(aq)$ (1)

c) $H_2O_2(aq) \rightarrow 2OH^-(aq)$ (1)

2 These are incomplete half-equations for changes involving oxidation in solution. Complete and balance the half-equations – you need to add the electrons:

a) $Mg(s) \rightarrow Mg^{2+}(aq)$ (1)

b) $Sn^{2+}(aq) \rightarrow Sn^{4+}(aq)$ (1)

c) $I^-(aq) \rightarrow I_2(aq)$ (1)

3 Pick a reduction from question **1** and an oxidation from question **2** and combine them to give the full ionic equation for these reactions:

a) iron(III) ions with tin(II) ions (2)

b) magnesium with dilute hydrochloric acid (2)

c) hydrogen peroxide with iodide ions. (2)

4 What are the oxidation numbers of chlorine in these ions: Cl^-, ClO^-, ClO_2^-, ClO_3^-, ClO_4^-? (3)

5 What are the oxidation numbers of nitrogen in these molecules: N_2, NH_3, N_2H_4, HNO_3, HNO_2, NH_2OH, NF_3? (4)

6 Write ionic half-equations for these changes in solution – in each case state whether the process is an example of oxidation or of reduction:

a) cobalt(II) ions turning into cobalt(III) ions (2)

b) hydroxide ions turning into oxygen and water molecules (2)

c) hydrogen molecules turning into hydrogen ions. (2)

7 Identify the element that disproportionates in each of these reactions by giving the oxidation states of the element before and after reaction:

a) $2H_2O_2(aq) \rightarrow 2H_2O(l) + O_2(g)$ (2)

b) $Cl_2(aq) + 2NaOH(aq) \rightarrow NaCl(aq) + NaClO(aq) + H_2O(l)$ (2)

c) $3MnO_4^{2-}(aq) + 4H^+(aq) \rightarrow 2MnO_4^-(aq) + MnO_2(s) + 2H_2O(l)$ (2)

11 Groups in the periodic table

Inorganic chemistry is the study of the hundred or so chemical elements and their compounds. The amount of information can be bewildering, hence the importance of the periodic table which helps to identify patterns in all the facts about properties and reactions.

11.1 Group 1, the alkali metals

The group 1 elements are better known as the alkali metals. These elements are more alike than the elements in any other group. Their compounds are widely used as chemical reagents.

The elements

The metals are soft and easily cut with a knife. They are shiny when freshly cut but quickly become dull in air as they react with moisture and oxygen. Laboratory specimens are kept in oil to protect them from the air.

Figure 11.1 ▲
A sample of lithium metal. Lithium compounds such as lithium carbonate are used in drugs for mental disease. Other lithium compounds are important reagents for organic synthesis.

Figure 11.2 ▲
A sample of sodium. Sodium metal is a powerful reducing agent used to extract titanium and some other metals. Sodium vapour is used in street lights. Sodium hydroxide is the most important industrial alkali. Many other sodium compounds are important commercially.

Figure 11.3 ▲
A sample of potassium. Potassium ions are an essential nutrient for plants and an ingredient of NPK fertilisers. There are many other important potassium compounds.

Atoms of the elements

The group 1 elements have similar chemical properties because their atoms have similar electron structures with one electron in the outer s orbital (Section 3.7). Even so, there are trends in properties down the group from lithium to caesium.

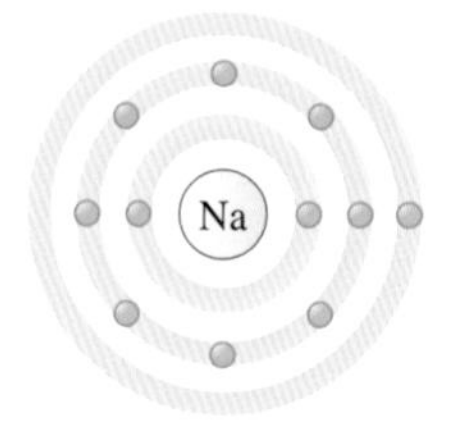

Figure 11.4 ◄
Diagrams to represent the electron configurations of lithium and sodium

The atoms change in two ways down the group – the charge on the nucleus increases and the number of filled inner shells also increases. The shielding effect of the inner electrons means that the effective nuclear charge attracting the outer electron is 1+. Down the group, the outer electrons get further and further away from the same effective nuclear charge and so they are held less strongly (Section 3.4).

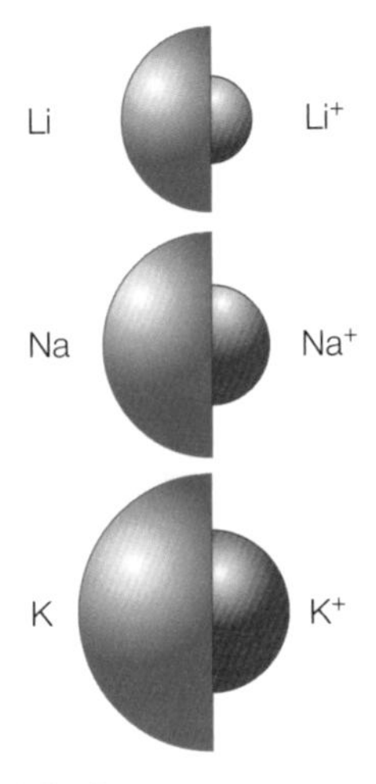

Figure 11.5 ▲
Relative sizes of the atoms and ions of group 1 elements

Test yourself

1 The electron configuration of lithium can be shortened to [He] $2s^1$ (Section 3.6). Using this style, give the electron configurations of:
a) sodium
b) potassium.

2 The first ionisation energies of the alkali metals get smaller down the group. Why is this?

3 Explain why the ions of group 1 metals are smaller than their atoms.

Oxidation states

When the atoms of alkali metals react, they lose their single s electron from the outer shell turning into ions with a single positive charge: Li^+, Na^+, K^+ and so on. So the only oxidation state in their compounds is +1.

Reactions of the elements

The alkali metals are powerful reducing agents because they react by giving up electrons to form M^+ ions.

All the metals react with water to form hydroxides, MOH, and hydrogen (where M is Li, Na, K and so on). The rate and violence of the reaction increases down the group – lithium reacts steadily with cold water; caesium reacts explosively.

$$2Li(s) + 2H_2O(l) \rightarrow 2LiOH(aq) + H_2(g)$$

All the metals react vigorously with chlorine to form colourless, ionic chlorides, M^+Cl^-. The chlorides are soluble in water, forming colourless solutions.

$$2K(s) + Cl_2(g) \rightarrow 2KCl(s)$$

Test yourself

4 Write balanced equations and use oxidation numbers to show that sodium acts as a reducing agent when it reacts with:
a) water
b) chlorine.

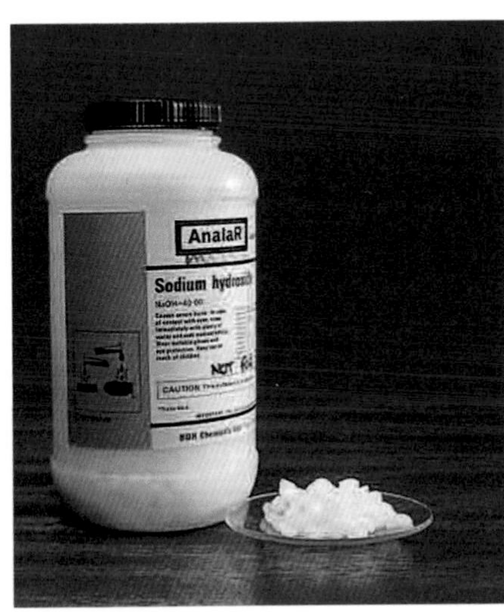

Figure 11.6 ▲
Sodium hydroxide, NaOH, is deliquescent, which means that it picks up water from moist air and then dissolves in it. Sodium hydroxide is a strong base – it dissolves in water to form a highly alkaline solution. The traditional name for the alkali is 'caustic soda'. Sodium hydroxide is highly corrosive and more hazardous to the skin and eyes than many acids.

11.2 Properties of compounds of group 1 metals

The hydroxides

The hydroxides are all white solids, commonly supplied as pellets or flakes. These are soluble in water, forming alkaline solutions, although the solubility of these hydroxides increases down the group.

The hydroxides are strong bases because they are fully ionised in water, giving solutions of hydroxide ions (Section 1.4). Note that it is the hydroxide ions which make the solutions basic and alkaline, not the metal ions.

The carbonates

The carbonates are all white with the general formula M_2CO_3. They are unusual metal carbonates in that they dissolve in water. Solutions of these carbonates are alkaline because the carbonate ion is a base. Carbonate ions remove H^+ ions from water molecules to form hydroxide ions which make the solution alkaline.

$$CO_3^{2-}(aq) + H_2O(l) \rightarrow HCO_3^-(aq) + OH^-(aq)$$

Another unusual feature of group 1 carbonates is that most of them do not decompose on heating. The exception is lithium carbonate, which breaks down to the oxide and carbon dioxide when hot.

The nitrates

The nitrates of group 1 metals are white crystalline solids with the formula MNO_3. They are very soluble in water.

The crystals of sodium and potassium nitrates are much harder to decompose on heating than most other metal nitrates. On heating, these nitrates first melt and then on stronger heating start to decompose giving off oxygen. They only decompose as far as the nitrite.

$$2KNO_3(s) \rightarrow 2KNO_2(s) + O_2(g)$$

Lithium nitrate is the exception. It behaves like most other metal nitrates, decomposing to the oxide, nitrogen dioxide and oxygen.

Test yourself

5 Write ionic equations for the reaction of:
  a) potassium hydroxide with dilute sulfuric acid
  b) aqueous sodium carbonate with dilute nitric acid.

6 Write balanced equations for the thermal decomposition of:
  a) lithium carbonate
  b) lithium nitrate.

Sodium and potassium compounds as chemical reagents

The compounds of sodium and potassium are widely used as chemical reagents. One reason for this is that the ions of alkali metals are unreactive. So they act as spectator ions which take no part in reactions when the reagents are used (Section 1.4).

A second reason is that most sodium and potassium compounds are soluble in water, including their hydroxides and carbonates. Most other metal hydroxides and carbonates are insoluble so not available in aqueous solution.

A third reason is that the ions of alkali metals are colourless in aqueous solution so they do not hide or interfere with colour changes. Sodium or potassium compounds are coloured only if the negative ion is coloured. Potassium chromate(VI), for example, is yellow because CrO_4^- ions are yellow.

Flame colours

Flame colours help to detect some metal ions. They are particularly useful in identifying group 1 metal ions, which are otherwise very similar.

Ionic compounds such as sodium chloride do not burn during a flame test. The energy from the flame excites electrons in the sodium atoms, raising them to higher energy levels. The atoms then emit the characteristic yellow light as the electrons drop back to lower energy levels (Section 3.4).

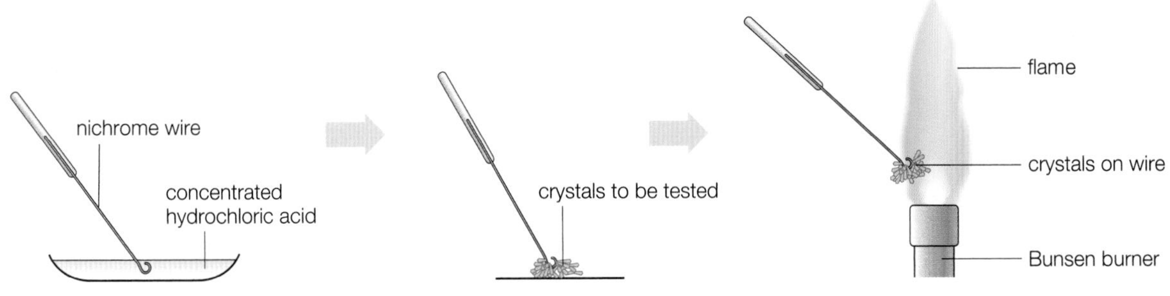

Figure 11.7 ▲
Procedure for a flame test. Chlorides evaporate more easily and so colour flames more strongly than less volatile compounds. Concentrated hydrochloric acid converts involatile compounds such as carbonates to chlorides.

Metal ion	Colour
Lithium	Bright red
Sodium	Bright yellow
Potassium	Lilac

Table 11.1 ▲
Flame colours of group 1 metal compounds

Test yourself

7 Classify these reactions as acid–base, thermal decomposition or redox:

a) $2K(s) + 2H_2O(l) \rightarrow 2KOH(aq) + H_2(g)$

b) $2KNO_3(s) \rightarrow 2KNO_2(s) + O_2(g)$

c) $2Li(s) + Br_2(l) \rightarrow 2LiBr(s)$

d) $Li_2CO_3(s) \rightarrow Li_2O(s) + CO_2(g)$

11.3 Group 2, the alkaline earth metals

Group 2 elements belong to the family of alkaline earth metals. Many of the compounds of these elements occur as minerals in rocks – hence the name 'earth metals'. Chalk, marble and limestone, for example, are forms of calcium carbonate. Dolomite consists of a mixture of calcium and magnesium carbonates. Fluorspar is a form of calcium fluoride which is mined in Derbyshire as the ornamental mineral called Blue John. These group 2 compounds are insoluble, unlike the equivalent group 1 compounds, so they do not dissolve in water.

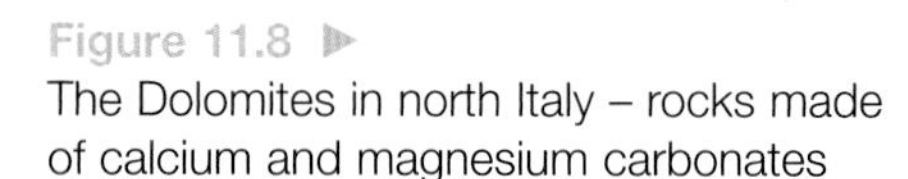

Figure 11.8 ▶
The Dolomites in north Italy – rocks made of calcium and magnesium carbonates

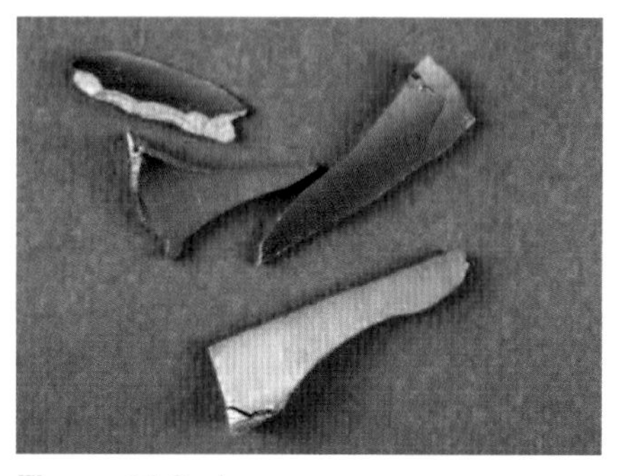

Figure 11.9 ▲
A sample of beryllium metal. Beryllium is light and strong with a high melting temperature so it is used for the construction of high-speed aircraft, missiles and space rockets. Thin sheets of the metal are transparent to X-rays and neutrons.

The elements

The group 2 metals are harder and denser than group 1 metals and they have higher melting temperatures. In air, the surface of the metals is covered with a layer of oxide.

The first member of the group, beryllium (Be) is a strong metal with a high melting temperature, but its density is much less than that of transition metals such as iron. The element makes useful alloys with other metals.

Magnesium is manufactured by the electrolysis of molten magnesium chloride from sea water or from salt deposits. The low density of the metal helps to make light alloys, especially with aluminium. These alloys, which are strong for their weight, are especially useful for car and aircraft manufacture.

Barium is a soft, silvery-white metal. It is so reactive with air and moisture that it is generally stored under oil, like the alkali metals.

Figure 11.10 ▲
Samples of the silvery metal calcium usually look grey because they are covered with a layer of calcium oxide.

Atoms and ions

Like the atoms of the alkali metals, the group 2 atoms change in two ways down the group: the charge on the nucleus increases and the number of filled inner shells also increases.

Metal	Electron configuration
Magnesium, Mg	[Ne] $3s^2$
Calcium, Ca	[Ar] $4s^2$
Strontium, Sr	[Kr] $5s^2$
Barium, Ba	[Xe] $6s^2$

Table 11.2

The shortened forms of the electron configurations of group 2 metals

The increasing number of filled inner shells means that atomic and ionic radii increase down the group. For each element, the 2+ ion is smaller than the atom because of the loss of the outer shell of electrons. The tendency to react and form ions increases down the group.

The first and second ionisation energies decrease down the group. The shielding effect of the inner electrons means that the effective nuclear charge attracting the outer electron is 2+. Down the group the outer s electrons get further and further away from the same effective nuclear charge, and so they are held less strongly and the ionisation energies decrease. This trend helps to account for the increasing reactivity of the elements down the group.

The removal of a third electron to form a 3+ ion takes much more energy because the third electron has to be removed against the attraction of a much larger effective nuclear charge. This means that it is never energetically favourable for the metals to form M^{3+} ions.

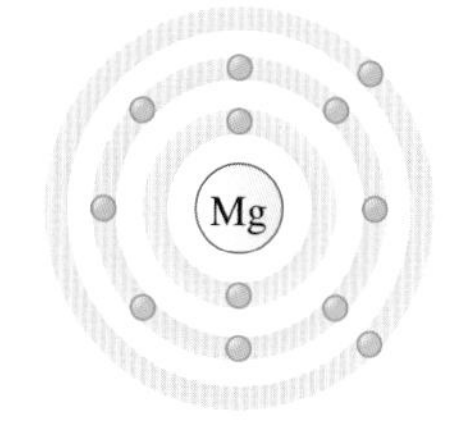

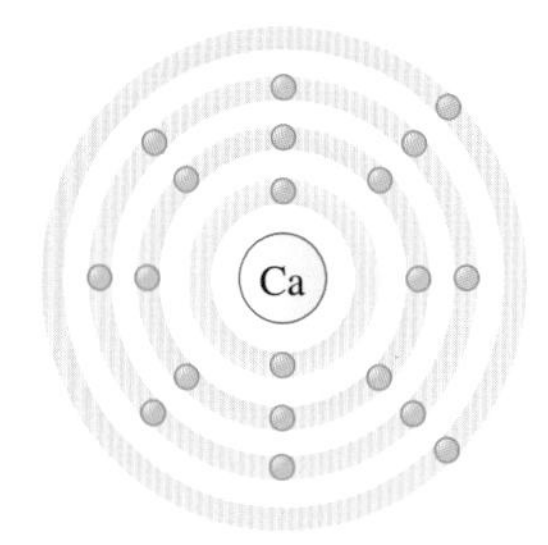

Figure 11.11

Diagrams to represent the electron configurations of magnesium and calcium atoms

Test yourself

8 Write the full electron configurations of the atoms and ions of Mg, Ca and Sr showing the numbers of s, p and d electrons (Section 3.6).

9 Explain why group 2 ions in any period are smaller than the group 1 ions in that period.

Oxidation states

All the group 2 metals have similar chemical properties because they have similar electron structures with two electrons in an outer s orbital. When the metal atoms react to form ions, they lose the two outer electrons, giving ions with a 2+ charge: Mg^{2+}, Ca^{2+}, Sr^{2+} and Ba^{2+}. So these elements exist in the +2 oxidation state in all their compounds.

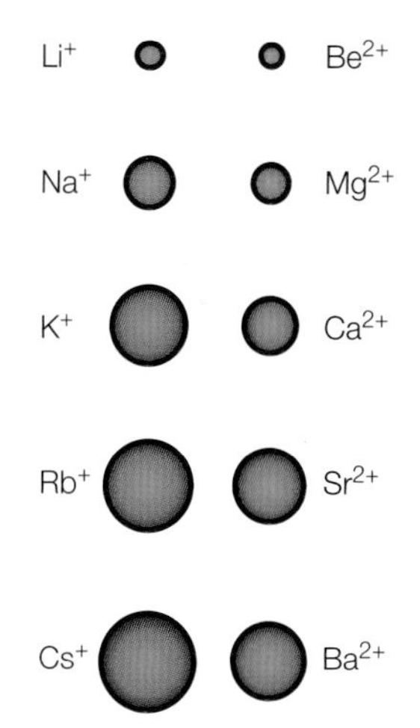

Figure 11.12

Comparison of the trend in ionic radii of group 1 and group 2 metals

11.4 Reactions of the group 2 elements

Group 2 metals are reducing agents. Apart from beryllium, they readily give up their two s electrons to form M^{2+} ions (where M represents Mg, Ca, Sr or Ba).

$$M \rightarrow M^{2+} + 2e^-$$

Reactions with oxygen

Apart from beryllium, the group 2 metals burn brightly in oxygen on heating to form white, ionic oxides, $M^{2+}O^{2-}$.

Magnesium burns very brightly in air with an intense white flame forming the white solid magnesium oxide, MgO. For this reason, magnesium powder is used in fireworks and flares.

Calcium also burns brightly in air but with a red flame forming the white solid calcium oxide, CaO. Strontium reacts in a similar way.

Barium burns in excess air or oxygen with a green flame to form a peroxide, BaO_2, which contains the peroxide ion, O_2^{2-}.

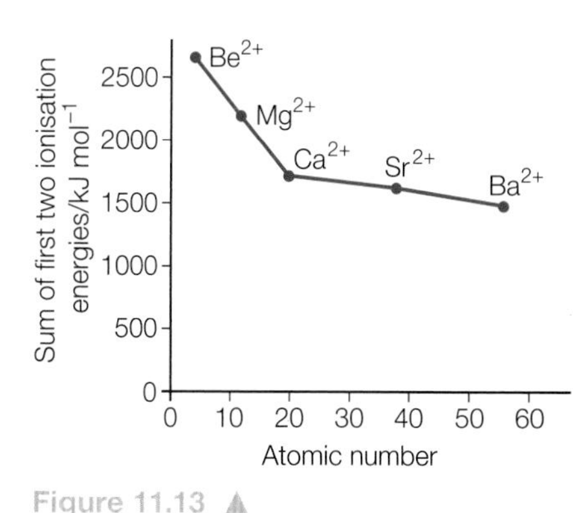

Figure 11.13

Graph showing the trend in the sum of the first two ionisation energies of group 2 metals: $M(g) \rightarrow M^{2+}(g) + 2e^-$

Test yourself

10 Write balanced equations for the reactions of:
a) strontium with oxygen
b) barium with oxygen
c) strontium with chlorine
d) barium with water.

Reactions with water

The metals Mg to Ba in group 2 react with water. The reactions are not as vigorous as the reactions of the group 1 metals. As in group 1, the rate of reaction increases down the group.

Magnesium reacts very slowly with cold water but much more rapidly on heating in steam.

$$Mg(s) + H_2O(g) \rightarrow MgO(s) + H_2(g)$$

Calcium reacts with cold water to produce hydrogen and a white precipitate of calcium hydroxide.

$$Ca(s) + 2H_2O(l) \rightarrow Ca(OH)_2(s) + H_2(g)$$

Barium reacts even faster with cold water and its hydroxide is more soluble.

Reaction with chlorine

All the metals, including beryllium, react with chlorine on heating to form white chlorides, MCl_2.

$$Mg(s) + Cl_2(g) \rightarrow MgCl_2(s)$$

11.5 Properties of the compounds of group 2 metals

Definition

A **basic oxide** is a metal oxide which reacts with acids to form salts and water. It is the oxide ion which acts as a base by taking a hydrogen ion from the acid. Basic oxides which dissolve in water are alkalis.

The oxides

Apart from beryllium oxide, the oxides of group 2 metals are basic, reacting with acids to form salts.

$$CaO(s) + 2HNO_3(aq) \rightarrow Ca(NO_3)_2(aq) + H_2O(l)$$

Magnesium oxide is a white solid made by heating magnesium carbonate. In water it turns to magnesium hydroxide, which is slightly soluble. Magnesium oxide has a high melting temperature and is used as a heat-resistant ceramic to line furnaces.

Calcium oxide is a white solid made by heating calcium carbonate. Calcium oxide reacts very vigorously with cold water, hence its traditional name 'quicklime'. The product is calcium hydroxide.

The hydroxides

The hydroxides of the elements Mg to Ba are:

- similar in that they all have the formula $M(OH)_2$, and are to some degree soluble in water forming alkaline solutions
- different in that their solubility increases down the group.

Magnesium hydroxide is the active ingredient in milk of magnesia, used as an antacid and laxative. It is only slightly soluble in water.

Calcium hydroxide is only sparingly soluble in water forming an alkaline solution, usually called limewater. The limewater test for carbon dioxide works because a solution of calcium hydroxide reacts with the gas forming a white, insoluble precipitate of calcium carbonate.

Barium hydroxide is the most soluble of the hydroxides. It is sometimes used as an alkali in chemical analysis. It has the advantage over sodium and potassium hydroxides in that it cannot be contaminated by its carbonate because barium carbonate is insoluble in water.

Test yourself

11 Write balanced equations for the reactions:
   a) magnesium oxide with dilute hydrochloric acid
   b) calcium oxide with water
   c) limewater with carbon dioxide.

The carbonates

The carbonates of group 2 metals (Mg to Ba) are:

- similar in that they all have the formula MCO_3, are insoluble in water, react with dilute acids and decompose on heating to give the oxide and carbon dioxide

  $$CaCO_3(s) \rightarrow CaO(s) + CO_2(g)$$

- different in that they become more difficult to decompose down the group – in other words, they become more thermally stable.

The nitrates

The nitrates of group 2 metals (Mg to Ba) are:

- similar in that they all have the formula $M(NO_3)_2$, are colourless crystalline solids, are very soluble in water and decompose to the oxide on heating

  $$2Mg(NO_3)_2 \rightarrow 2MgO(s) + 4NO_2(g) + O_2(g)$$

- different in that they become more difficult to decompose down the group.

The sulfates

The sulfates are:

- similar in that they are all colourless solids with the formula MSO_4
- different in that they become less soluble down the group.

Epsom salts consist of hydrated magnesium sulfate, $MgSO_4.7H_2O$, which is a laxative.

Plaster of Paris is the main ingredient of building plasters and much is used to make plasterboard. The white powder is made by heating the mineral gypsum in kilns to remove most of the water of crystallisation.

Stirring plaster of Paris with water produces a paste which soon sets as it turns back into interlocking grains of gypsum. Plaster makes good moulds because it expands slightly as it sets so that it fills every crevice.

Barium sulfate absorbs X-rays strongly so it is the main ingredient of 'barium meals' used to diagnose disorders of the stomach or intestines. Soluble barium compounds are toxic but barium sulfate is very insoluble so it is not absorbed into the bloodstream from the gut. X-rays cannot pass through the 'barium meal', which therefore creates a shadow on the X-ray film.

A soluble barium salt can be used to test for sulfate ions because barium sulfate is insoluble even when the solution is acidic. Adding a solution of barium nitrate or barium chloride to an acidified solution produces a white precipitate only if sulfate ions are present.

$$Ba^{2+}(aq) + SO_4^{2-}(aq) \rightarrow BaSO_4(s)$$

white precipitate

Figure 11.14 ▲
Crystals of gypsum – a hydrated form of calcium sulfate

Figure 11.15 ▲
Barium occurs naturally as barytes, $BaSO_4$ (above), and also as witherite, $BaCO_3$.

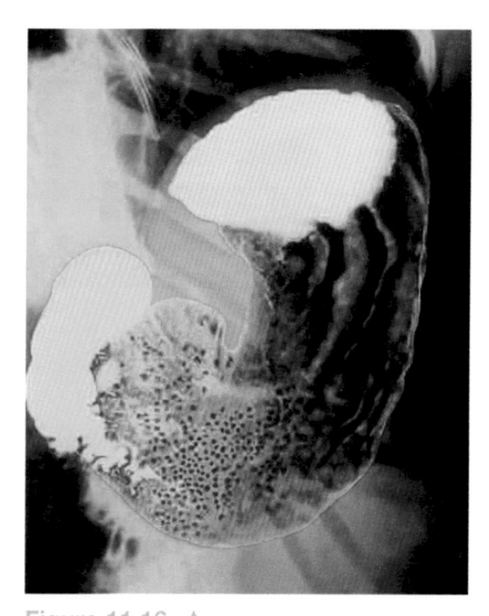

Figure 11.16 ▲
X-ray photograph of the digestive system of a patient who has taken a barium meal

Activity

Limestone and its uses

Calcium carbonate occurs naturally as limestone, chalk and marble. Limestone is an important mineral.

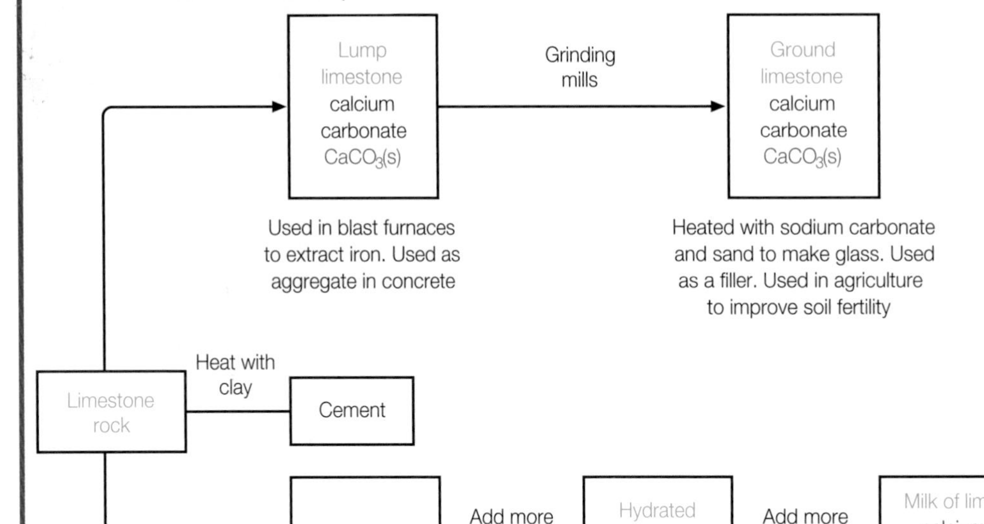

Figure 11.17 ◀ Products from limestone and their uses

Pure limestone is used in the chemical industry. Heating limestone in a furnace at 1200 K converts it to calcium oxide (quicklime). The reaction of quicklime with water produces calcium hydroxide (slaked lime).

Figure 11.18 ▲ Limestone cliff in Yorkshire. Limestone gives rise to attractive scenery.

1. Name two other minerals that consist largely of calcium carbonate.
2. Suggest a reason for grinding up limestone lumps before heating them with sodium carbonate and sand to make glass.
3. Write an equation for the decomposition of calcium carbonate on heating in a furnace.
4. Show, with the help of ionic equations, that both calcium oxide and calcium hydroxide can be used to raise the pH in soils that are too acidic.
5. Lime mortar was used in older buildings – the mortar is a mixture of slaked lime, sand and water. It sets slowly by reacting with carbon dioxide in the air. Identify the main product of the reaction of slaked lime with carbon dioxide.
6. A suspension of calcium hydroxide in water is used as an industrial alkali. Suggest why a suspension is used and not just a solution of the hydroxide.
7. What is the laboratory use of a solution of calcium hydroxide?

Thermal stability of the carbonates and nitrates

Whenever chemists use the term 'stability' they are making comparisons. For the group 2 carbonates, the question is which is more stable – the metal carbonate, or a mixture of the metal oxide and carbon dioxide?

Most of the compounds of group 1 and 2 elements are ionic. Chemists try to explain differences in the stabilities of their compounds in terms of two factors:

- the charge on the metal ions
- the size of the metal ions.

Group 2 carbonates and nitrates are generally less stable than the corresponding group 1 compounds. This suggests that the larger the charge on the metal ion and the smaller the metal ion, the less stable the compounds.

The carbonates become more stable down both group 1 and group 2. This helps to confirm that the larger the metal ion, the more stable the compounds.

Beryllium carbonate is so unstable that it does not exist. Table 11.3 shows the temperatures at which the carbonates of group 2 metals begin to decompose. The values indicate that magnesium carbonate is the least stable. It decomposes easily to the oxide and carbon dioxide when heated with a Bunsen flame. Barium carbonate is the most stable.

Chemists explain the trend in thermal stability by analysing the energy changes. Two of the energy quantities they take into account are:

- the energy needed to break up the carbonate ion into an oxide ion and carbon dioxide
- the energy given out as the 2+ and 2− charges get closer together when the larger carbonate ions break up into smaller oxide ions and carbon dioxide.

It turns out that all the group 2 carbonates (except beryllium carbonate) are thermally stable at room temperature but become unstable as the temperature rises. The key factor is the energy released as the ions get closer together. This is greater when the metal ion is small than when the metal ion is large, and this explains why $MgCO_3$ decomposes more easily than $BaCO_3$.

Carbonate	Decomposition temperature/°C
$MgCO_3$	540
$CaCO_3$	900
$SrCO_3$	1280
$BaCO_3$	1360

Table 11.3 ▲
The temperature at which group 2 carbonates begin to decompose

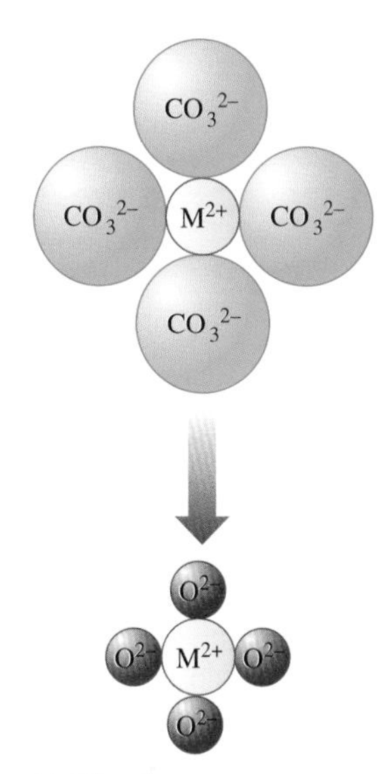

Figure 11.19 ◀
Decomposition of a group 2 carbonate to a group 2 oxide. The smaller the metal ion, the less stable the carbonate.

Flame colours

Flame tests help to identify compounds of calcium, strontium and barium.

Metal ion	Colour
Beryllium	No colour
Magnesium	No colour
Calcium	Brick red
Strontium	Bright red
Barium	Pale green

Table 11.4 ▲
Flame colours of group 2 metal compounds

Test yourself

12 Draw and label a diagram of some simple apparatus to show that magnesium carbonate decomposes on heating.

13 Write equations for:
a) the thermal decomposition of magnesium carbonate
b) the reaction of magnesium carbonate with hydrochloric acid
c) the thermal decomposition of calcium nitrate
d) the reaction of barium nitrate solution with zinc sulfate solution.

14 With the help of oxidation numbers, identify the elements that are oxidised and reduced during the thermal decomposition of magnesium nitrate.

11.6 Group 7, the halogens

Fluorine, chlorine, bromine and iodine belong to the family of halogens. All four are reactive non-metals – fluorine and chlorine extremely so. The elements are hazardous because they are so reactive. For the same reason they are never found free in nature. However, they do occur as compounds with metals. Many of the compounds of group 7 elements are salts – hence the name 'halo-gen' meaning 'salt-former'. The halogens are important economically as the ingredients of plastics, pharmaceuticals, photographic chemicals, anaesthetics and dyestuffs.

Data

The elements

Under laboratory conditions chlorine is a yellow–green gas, bromine is a dark red liquid, while iodine is a dark grey solid.

Fluorine is much too dangerous to be used in normal laboratories. Astatine, the final member of the group is the rarest naturally occurring element. It is highly radioactive – its most stable isotope has a half-life of just over 8 hours.

All the halogens consist of diatomic molecules, X_2, linked by a single covalent bond. They are all volatile. Intermolecular forces increase down the group as the numbers of electrons in the molecules increase (Section 9.1). The larger molecules are more polarisable than the smaller molecules, so melting temperatures and boiling temperatures rise down the group.

Halogen	Electron configuration
Fluorine, F	[He] $2s^2\,2p^5$
Chlorine, Cl	[Ne] $3s^2\,3p^5$
Bromine, Br	[Ar] $3d^{10}\,4s^2\,4p^5$
Iodine, I	[Kr] $4d^{10}\,5s^2\,5p^5$

Table 11.5 ▲
Shortened form of the electron configurations of the halogens

The halogens have similar chemical properties because they all have seven electrons in the outer shell – one fewer than the next noble gas in group 8.

Fluorine is the most electronegative of all elements. Its oxidation state is −1 in all its compounds. Uses of fluorine include the manufacture of a wide range of compounds consisting of only carbon and fluorine (fluorocarbons). The most familiar of these is the very slippery, non-stick polymer, poly(tetrafluorethene).

Chlorine reacts directly with most elements. In its compounds, chlorine is usually present in the −1 oxidation state but it can be oxidised to positive oxidation states by oxygen and fluorine. Most chlorine is used in the production of polymers such as pvc. Water companies use chlorine to kill bacteria in drinking water, while another important use of the element is to bleach paper and textiles.

Bromine, like the other halogens, is a reactive element but it is a less powerful oxidising agent than chlorine. Bromine is used to make a range of products including flame retardants, medicines and dyes.

Figure 11.20 ▲
Chlorine gas

Figure 11.21 ▲
Bromine is a dark red liquid at room temperature. It is very volatile, giving off a choking, orange vapour.

Figure 11.22 ▲
Iodine is a lustrous grey–black solid at room temperature. It sublimes when gently warmed to give a purple vapour.

Iodine is also an oxidising agent but a less powerful than bromine. Iodine and its compounds are used to make many products including medicines, dyes and catalysts. In many regions, sodium iodide is added to table salt to supplement iodine in the diet and to drinking water so as to prevent goitre – a swelling of the thyroid gland in the neck. Iodine is needed in our diet so that the thyroid gland can make the hormone thyroxine, which regulates growth and metabolism.

Test yourself

15 Predict the state of the following at room temperature and pressure, giving your reasons:
  a) fluorine
  b) astatine.

16 Write down the full electron configurations of:
  a) a chlorine atom
  b) a chloride ion
  c) a bromine atom
  d) a bromide ion.

17 Explain why:
  a) the atomic radii of halogen atoms increase down the group
  b) the ionic radii of halides are larger than their corresponding atomic radii.

18 Draw dot-and-cross diagrams to shown the bonding in:
  a) a fluorine molecule
  b) a molecule of hydrogen bromide
  c) a molecule of iodine monochloride, ICl.

Solutions of the halogens

The halogens dissolve freely in hydrocarbon solvents, such as hexane. When dissolved in hexane the solutions have a very similar colour to the free halogen vapours. So iodine in hexane, for example, has an attractive violet colour.

The halogens are less soluble in water than in organic solvents. Aqueous chlorine and bromine are useful reagents. Their colours are similar to the colours of their vapours. These elements also react with water (Section 11.9).

Iodine does not dissolve in water, but it will dissolve in aqueous potassium iodide. Iodine dissolves in this way because iodine molecules react with iodide ions to form triiodide ions, I_3^-. The solution is yellow when very dilute but dark orange–brown when more concentrated.

11.7 Reactions of the group 7 elements

The halogens are powerful oxidising agents. Apart from fluorine, chlorine is the strongest oxidising agent among these elements.

Halogen atoms are highly electronegative (Section 8.2), although the electronegativity decreases down the group. They form ionic compounds or compounds with polar bonding.

Reactions with metals

Chlorine and bromine react with s-block metals to form ionic halides in which the halogen atoms gain one electron to fill the outermost p sub-shell.

Iodine also reacts with metals to form iodides, but because of the polarisability of the large iodide ion, those iodides formed with small cations such as Li^+, or highly charged cations such as Al^{3+}, are essentially covalent (Section 8.1). The halogens also react with most metals in the d block. Hot iron, for example, burns brightly in chlorine forming iron(III) chloride. The reaction with bromine is similar but much less exothermic. Iodine(III) iodide does not exist because iodide ions reduce iron(III) ions to iron(II) ions. So, heating iron with iodine vapour produces iron(II) iodide.

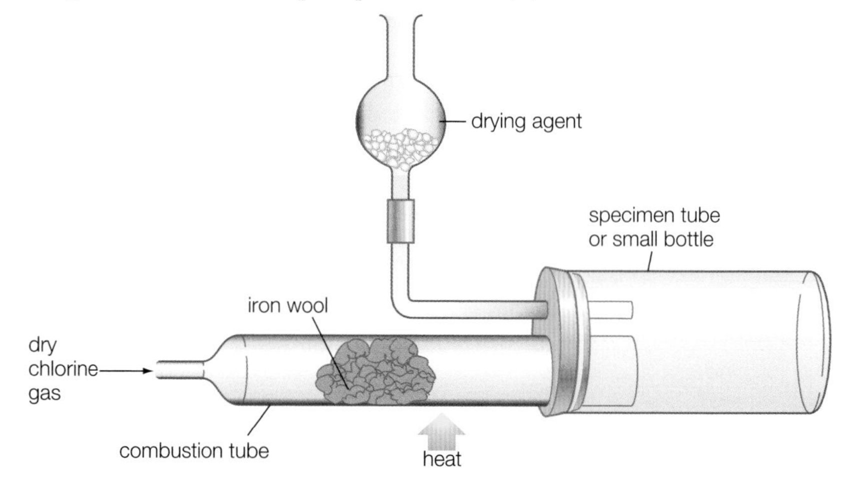

Figure 11.23 ▶ The laboratory apparatus for making anhydrous iron(III) chloride

Test yourself

19 Write balanced equations for these reactions, and show the changes in oxidation states:
- a) bromine with magnesium
- b) chlorine with iron
- c) iodine with iron.

20 In Figure 11.23, explain the reason for:
- a) drying the chlorine gas
- b) using iron wool instead of small lumps of iron
- c) collecting the product in a specimen tube
- d) allowing excess gas to escape through a tube with a drying agent.

Reactions with non-metals

Chlorine reacts with most non-metals to form molecular chlorides. Hot silicon, for example, reacts to form silicon tetrachloride, $SiCl_4(l)$, and phosphorus produces phosphorus trichloride, $PCl_3(l)$. However, chlorine does not react directly with carbon, oxygen or nitrogen.

Hydrogen burns in chlorine to produce the colourless, acidic gas hydrogen chloride, HCl. Igniting a mixture of chlorine and hydrogen gases leads to a violent explosion.

Bromine also oxidises non-metals such as sulfur and hydrogen on heating, forming molecular bromides. A mixture of bromine vapour and hydrogen gas reacts smoothly with a pale bluish flame.

$$H_2(g) + Br_2(g) \rightarrow 2HBr(g)$$

Iodine oxidises hydrogen on heating to form hydrogen iodide. Unlike the reactions of chlorine and bromine, this is a reversible reaction.

$$H_2(g) + I_2(g) \rightleftharpoons 2HI(g)$$

Reactions with aqueous Fe^{2+} ions

Chlorine and bromine can oxidise iron(II) ions in solution to iron(III) ions. Iodine is such a weak oxidising agent that it cannot oxidise iron(II) compounds.

$$Fe^{2+}(aq) + Cl_2(aq) \rightarrow Fe^{3+}(aq) + 2Cl^-(aq)$$

Test yourself

21 Show that the reactions of halogens with hydrogen illustrate a trend in reactivity down group 7.

22 Predict the formula of the product and vigour of the reaction when fluorine reacts with hydrogen.

23 Explain, in terms of structure and bonding, why silicon tetrachloride is a liquid.

24 Write equations for the reactions and use oxidation numbers to show that:
  a) phosphorus is oxidised when it reacts with chlorine
  b) chlorine is reduced when it displaces iodine from a solution of potassium iodide.

25 Write ionic half-equations for the reaction of bromine with aqueous iron(II) ions.

11.8 Halogens in oxidation state −1

Halide ions

Halide ions are the ions of the halogen elements in oxidation state −1. They include the fluoride (F^-), chloride (Cl^-), bromide (Br^-) and iodide (I^-) ions.

In group 7, a more reactive halogen oxidises the ions of a less reactive halogen. So bromine reacts with a solution of an iodide to produce iodine and a bromide. This is because bromine has a stronger tendency than iodine to gain electrons and turn into ions.

$$Br_2(aq) + 2I^-(aq) \rightarrow 2Br^-(aq) + I_2(s)$$

Figure 11.24 ▲
Adding a solution of chlorine in water to aqueous potassium iodide. Adding a hydrocarbon solvent to the mixture and shaking produces a violet colour in the organic solvent showing that iodine has been formed.

Silver nitrate solution can be used to distinguish between halides. Silver fluoride is soluble, so there is no precipitate on adding silver nitrate to a solution of fluoride ions. The other silver halides are insoluble – adding silver nitrate to a solution of one of these halide ions produces a precipitate.

$$Ag^+(aq) + Cl^-(aq) \rightarrow AgCl(s)$$

Silver chloride is white and quickly turns purple–grey in sunlight. This distinguishes it from silver bromide, which is a creamy colour, and silver iodide, which is a brighter yellow.

The colour changes are not very distinct but a further test with ammonia helps to distinguish the precipitates. Silver chloride dissolves in dilute ammonia solution easily. Silver bromide dissolves in concentrated ammonia solution, but silver iodide does not dissolve in ammonia solution at all.

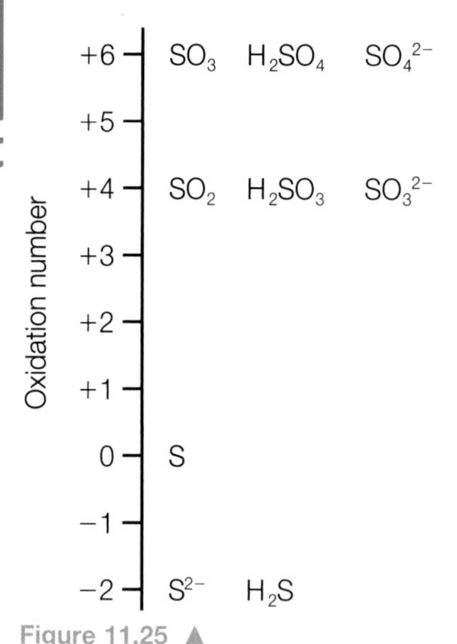

Figure 11.25 ▲
The oxidation states of sulfur

Test yourself

26 Draw a diagram of laboratory apparatus for collecting several large test tubes full of hydrogen chloride gas from the reaction of sodium chloride with concentrated sulfuric acid.

27 With the help of oxidation numbers, write a balanced equation for the reaction of sodium bromide with concentrated sulfuric acid.

Reactions of halides with concentrated sulfuric acid

Warming solid sodium chloride with concentrated sulfuric acid produces hydrogen chloride gas. This colourless gas produces white fumes in moist air. The acid–base reaction between sodium chloride and sulfuric acid can be used to make hydrogen chloride.

$$NaCl(s) + H_2SO_4(l) \rightarrow HCl(g) + NaHSO_4(s)$$

Both sulfuric acid and hydrogen chloride are strong acids. The reaction goes from left to right because hydrogen chloride is a gas and escapes from the reaction mixture. So the reverse reaction cannot happen.

This type of reaction cannot be used to make hydrogen bromide or hydrogen iodide because bromide and iodide ions are strong enough reducing agents to reduce sulfur from the +6 state to lower oxidation states.

The reactions of halide ions with sulfuric acid show that there is a trend in the strength of the halide ions as reducing agents:

- with sodium chloride the product is hydrogen chloride gas – there is no oxidation or reduction
- sodium bromide turns to orange bromine molecules as the bromide ions reduce H_2SO_4 to SO_2, together with some hydrogen bromide gas
- iodide ions are the strongest reducing agents, so with sodium iodide little or no hydrogen iodide forms – instead the products are iodine molecules mixed with sulfur and hydrogen sulfide.

So the trend as reducing agents is $I^- > Br^- > Cl^-$.

Chlorine is the strongest oxidising agent of these halogens, so it has the greatest tendency to form negative ions. Conversely, chloride ions are reluctant to give up their electrons and turn back into chlorine molecules. So the chloride ion is the weakest reducing agent.

Iodine is the weakest oxidising agent, so it has the least tendency to form negative ions. Conversely, the iodide ion is the strongest reducing agent, being most ready to give up electrons and turn back into iodine molecules.

Hydrogen halides

The hydrogen halides are compounds of hydrogen with the halogens. They are all colourless, molecular compounds with the formula HX, where X stands for Cl, Br or I. The bonds between hydrogen and the halogens are polar.

Hydrogen chloride, hydrogen bromide and hydrogen iodide are similar in that they are:

- colourless gases at room temperature which fume in moist air
- very soluble in water forming acidic solutions (hydrochloric, hydrobromic and hydriodic acids) which ionise completely in water
- strong acids, so they ionise completely in water.

H^+ transferred

$$HCl + H_2O \rightarrow Cl^-(aq) + H_3O^+(aq)$$

oxonium ion

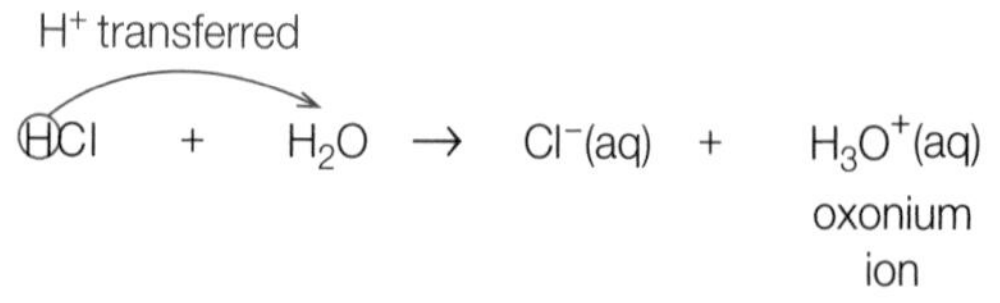

Figure 11.26 ▶
The reaction of hydrogen chloride with water. Hydrogen chloride is a strong acid which is fully ionised in aqueous solution.

Mixing any of the hydrogen halides with ammonia produces a white smoke of an ammonium salt. Ammonia molecules turn into ammonium ions, NH_4^+, in this reaction.

$$NH_3(g) + HBr(g) \rightarrow NH_4Br(s)$$

Test yourself

28 Describe the colour changes on adding:
  a) a solution of chlorine in water to aqueous sodium bromide
  b) a solution of bromine in water to aqueous potassium iodide.

29 Put the chloride, bromide and iodide ions in order of their strength as reducing agents, with the strongest reducing agent first. Explain your answer.

30 Write ionic equations for the reactions of silver nitrate solution with:
  a) potassium iodide solution
  b) sodium bromide solution.

31 Explain why the compound of hydrogen and fluorine is a liquid at room temperature on a cool day, when the other hydrogen halides are gases.

32 a) Show that the reaction of ammonia with hydrogen chloride gas involves proton transfer.
  b) Explain why the product of the reaction is a solid.

11.9 Halogens in oxidation states +1 and +5

Chlorine oxoanions form when chlorine reacts with water and alkalis.

When chlorine dissolves in water, it reacts reversibly forming a mixture of weak chloric(I) acid and strong hydrochloric acid. This is an example of a disproportionation reaction.

$$H_2O(l) + Cl_2(g) \rightleftharpoons HClO(aq) + HCl(aq)$$

Bromine reacts in a similar way but to a much lesser extent. Iodine is insoluble in water and hardly reacts at all.

When chlorine dissolves in potassium (or sodium) hydroxide solution at room temperature it produces chlorate(I) and chloride ions.

Definition

A **disproportionation** reaction is a change in which the same element both increases and decreases its oxidation number. So the element is both oxidised and reduced.

$$\overset{\text{+1 (to HOCl)}}{} \quad Cl_2(aq) + H_2O(l) \rightleftharpoons HOCl(aq) + Cl^-(aq) + H^+(aq) \quad \underset{\text{−1 (to Cl}^-\text{)}}{}$$

The active ingredient in household bleach is sodium chlorate(I), made by dissolving chlorine in sodium hydroxide solution.

On heating, the chlorate(I) ions disproportionate to chlorate(V) and chloride ions.

$$3ClO^-(aq) \rightarrow ClO_3^-(aq) + 2Cl^-(aq)$$

$$+1 \qquad\qquad +5 \qquad\qquad -1$$

Bromine and iodine react in a similar way to chlorine with alkalis. The BrO^- and IO^- ions are less stable, so they disproportionate at a lower temperature. A hot solution of iodine in potassium hydroxide produces a solution containing potassium iodate(V) and potassium iodide.

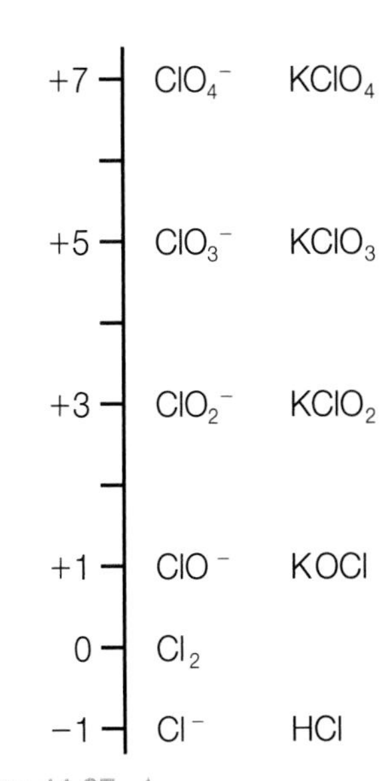

Figure 11.27 ▲ The oxidation states of chlorine

Test yourself

33 a) Use oxidation numbers to write a balanced equation for the reaction of iodine with hot aqueous hydroxide ions to form IO_3^- and I^- ions.
  b) Show that this is a disproportionation reaction.

Activity

Water treatment

Chlorine disinfects tap water by forming chloric(I) acid, HClO, on reaction with water. Chloric(I) acid is a powerful oxidising agent and a weak acid. It is an effective disinfectant because, unlike ClO^- ions, the molecule can pass through the cell walls of bacteria. Once inside the bacterium, the HClO molecules break the cell open and kill the organism by oxidising and chlorinating molecules which make up its structure.

Chloric(I) acid is a weak acid so it does not ionise easily. The concentration of un-ionised HClO in a solution depends on the pH, as shown in Figure 11.28.

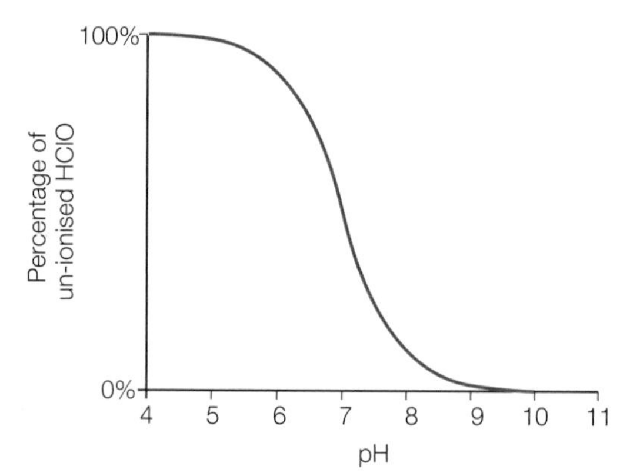

Figure 11.28 ▲
Graph to show how the concentration of chloric(I) acid, HClO, varies over a range of pH values at 20 °C

Swimming pools can be sterilised with chlorine compounds which produce chloric(I) acid when they dissolve in water. Swimming pool managers have to check the pH of the water carefully – they aim to keep the pH in the range 7.2–7.8.

1 Write an equation to show the formation of chloric(I) acid when chlorine reacts with water.

2 Swimming pools used to be treated with chlorine gas from cylinders containing the liquefied gas under pressure. Now they are usually treated with chemicals that produce chloric(I) acid when added to water. Suggest reasons for this change.

3 Explain why sodium chlorate(I) produces chloric(I) acid when added to water at pH 7–8.

4 Explain why the pH of swimming pool water must not be allowed to rise above 7.8.

5 Suggest reasons why pool water must not become acidic, even though this would increase the concentration of un-ionised HClO.

6 Nitrogen compounds, including ammonia, urea and proteins, react with HClO to form chloramines, which are irritating to skin and eyes. Chloramines formed from ammonia can react with themselves to form nitrogen and hydrogen chloride, which gets rid of the problem. However, if there is excess HClO, another reaction produces nitrogen trichloride, which is responsible for the so-called 'swimming pool smell'. Nitrogen trichloride is very irritating to skin and eyes. Write equations to show:

   a) the formation of the chloramine, NH_2Cl, from ammonia and chloric(I) acid
   b) the removal of chloramines by reaction of NH_2Cl with $NHCl_2$
   c) the formation of nitrogen trichloride from chloramine, NH_2Cl.

Figure 11.29 ▲
Chlorine compounds treat the water in swimming pools

7 Explain why swimming pools do not smell of chlorine if they are properly maintained.

REVIEW QUESTIONS

Extension questions

1 The elements Mg to Ba in group 2, and their compounds, can be used to show the trends in properties down a group of the periodic table. State and explain the trend down the group in:

a) atomic radius (3)

b) first ionisation energy (3)

c) thermal stability of the carbonates. (3)

2 Radium is a highly radioactive element. Use your knowledge of the chemistry of the elements Mg to Ba in group 2 to predict properties of radium and its compounds. Include in your predictions a description of the following changes, equations for any chemical changes and the appearance of the products:

a) the reaction of radium with oxygen (3)

b) the reaction of radium with water (3)

c) the reaction of radium oxide with water (3)

d) the reaction of radium hydroxide with dilute hydrochloric acid (3)

e) the solubility of radium sulfate in water (2)

f) the effect of heating radium nitrate. (3)

3 Astatine, At, is the element below iodine in group 7. Predict, giving your reasons:

a) the physical state of astatine at room temperature (1)

b) the effect of bubbling chorine though an aqueous solution of sodium astatide (2)

c) whether or not hydrogen astatide forms on adding concentration sulfuric acid to solid sodium astatide (2)

d) the colour of silver astatide and its solubility in concentrated ammonia solution. (2)

4 a) Write an equation for the reaction of chlorine with aqueous sodium hydroxide, and use this example to explain what is meant by disproportionation. (2)

b) On heating, chlorate(I) ions in solution disproportionate to chlorate(V) ions and chloride ions. Write an ionic equation for this reaction. (2)

c) On heating to just above its melting temperature, $KClO_3$ reacts to form $KClO_4$ and KCl. Write a balanced equation for the reaction and show that it is a disproportionation reaction. (3)

5 State the trend in the power of chloride, bromide and iodide ions as reducing agents. Describe how you could demonstrate the trend by carrying out experiments in the laboratory. Include in your description the main observations that illustrate the differences between the ions. Include equations for the reactions. (10)

12 Chemical analysis

Types of analysis

Analysis is a vital part of what chemists and other scientists do. Doctors rely on analysis for diagnosing disease; environmental scientists analyse our drinking water; food scientists check that our food is safe to eat. People rightly worry about pollution of the natural environment, but without chemical analysis we would not know anything about the causes or the scale of pollution.

12.1 Qualitative analysis

Qualitative analysis answers the question 'What is it?' In advanced chemistry courses, this question can be answered by careful observation of the changes during test tube experiments. These changes include gases bubbling off, different smells, new colours appearing, precipitates forming, solids dissolving and temperature changing.

The skill is knowing what to look for. Some visible changes are much more significant than others and a capable analyst can spot the important changes and know what they mean. Good chemists have a 'feel' for the way in which chemicals behave and recognise characteristic patterns of behaviour. With experience they know what to look for when making observations.

Success also depends on good techniques when mixing chemicals, heating mixtures and testing for gases.

Test yourself

1 Give an example of a reaction which:
   a) gives off a gas
   b) creates a smell
   c) produces a colour change
   d) forms a precipitate
   e) gets very hot.

Figure 12.1 ▲
An analyst measures the pollution levels in water from the Amur River to see if it is fit to drink. The river forms the border between China and eastern Russia.

12.2 Inorganic reactions used in tests

There are four main types of inorganic reaction which you need to know about when interpreting observations made while testing ionic compounds in solution.

Ionic precipitation reactions

The simplest examples of precipitation reactions involve mixing two solutions of soluble ionic compounds. The positive ions from one compound combine with the negative ions of the other to form an insoluble precipitate (Section 1.4).

This type of reaction is used to test for negative ions (anions). Adding a soluble barium salt (nitrate or chloride) to a solution of a sulfate, for example, produces a white precipitate of insoluble barium sulfate.

Precipitation reactions can also help to identify positive ions (cations). Adding a dilute solution of sodium hydroxide to a solution of a metal salt produces a precipitate if the metal hydroxide is insoluble in water. Some of the insoluble metal hydroxides are amphoteric. If so, they dissolve in excess of the sodium hydroxide solution (Section 1.4).

Figure 12.2 ◀ Precipitates of silver chloride, silver bromide and silver iodide formed by adding silver nitrate solution to solutions of the halide ions.

Acid–base reactions

Acids and alkalis are commonly used in chemical tests. Dilute hydrochloric acid is a convenient strong acid. Sodium hydroxide solution is often chosen as a strong base.

Adding dilute hydrochloric acid to a carbonate adds protons to the carbonate ions, CO_3^{2-}, turning them into carbonic acid molecules, H_2CO_3, which immediately decomposes into carbon dioxide and water.

$$2H^+(aq) + CO_3^{2-}(aq) \rightarrow H_2CO_3(aq) \rightarrow H_2O(l) + CO_2(g)$$

Testing with limewater can then identify the gas given off, confirming that the compound tested is a carbonate.

Redox reactions

Common oxidising agents used in inorganic tests include chlorine, bromine and acidic solutions of iron(III) ions, manganate(VII) ions and dichromate(VI) ions.

Some reagents change colour when oxidised, which makes them useful for detecting oxidising agents. In particular, a colourless solution of iodide ions turns to a yellow–brown colour when oxidised. This can be a very sensitive test if starch is present because starch gives an intense blue–black colour with low concentrations of iodine. This is the basis of using starch–iodide paper to test for chlorine and other oxidising gases.

Common inorganic reducing agents are metals (in the presence of acid or alkali), sulfur dioxide and iron(II) ions.

Some reagents change colour when reduced. In particular, dichromate(VI) ions in acid change from yellow to green. This is the basis of a test for sulfur dioxide.

Many thermal decompositions involve redox. Thermal decomposition is a fifth type of inorganic reaction that you should know about (Section 1.4).

Test yourself

2 Identify the type of reaction which happens when testing:

a) a solution of a sulfate with barium chloride

b) magnesium with dilute hydrochloric acid

c) hydrogen iodide with ammonia gas

d) a solution of an iron(II) salt with sodium hydroxide

e) potassium nitrate by heating it strongly.

Complex-forming reactions

A complex ion forms when a metal ion combines with other molecules or ions. Anhydrous copper(II) sulfate is colourless. It turns blue on adding water because the copper ions, Cu^{2+}, join with water molecules to form a complex ion $Cu(H_2O)_4^{2+}$.

The formation of a complex ion can bring about two visible changes that can be useful in analysis – there may be a colour change or an insoluble compound may dissolve. Sometimes both types of change happen at the same time.

The common test for water is based on the formation of a complex ion. Anhydrous cobalt(II) chloride is blue. It turns pink in the presence of water as the cobalt ions form a pink complex ion, $Co(H_2O)_6^{2+}$, with water molecules.

12.3 Tests and observations

www Practical guidance

Preliminary tests

Preliminary observations and tests provide a general introduction to the properties of a compound. They can include:

- the state and appearance of the compound at room temperature
- the solubility of the compound in cold and hot water, and the pH of any solution that forms
- the effect of heating the compound.

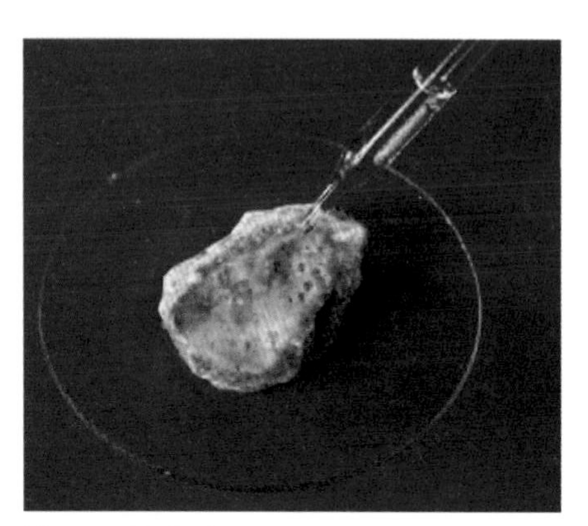

Figure 12.3 ▲ Geologists use dilute hydrochloric acid to test for carbonates in rocks

Chemists use a rough classification of solubility based on what they see when a small amount of the solid is shaken with water in a test tube:

- very soluble, for example potassium nitrate – plenty of the solid dissolves quickly
- soluble, for example sodium chloride – the crystals visibly dissolve to a significant extent and colour the solution if the compound is coloured
- slightly soluble, for example calcium hydroxide – the solid does not appear to dissolve but testing with indicator shows that enough has dissolved to affect the pH of the solution
- insoluble, for example barium sulfate – there is no sign that the solid dissolves and the pH of the water remains unchanged.

www Data

Familiarity with the patterns of solubility of common inorganic compounds helps with the interpretation of observations during test tube experiments.

Heating a small sample of a solid inorganic compound can provide clues to the identity of the compound. If a gas or vapour is given off it can be identified with the help of tests.

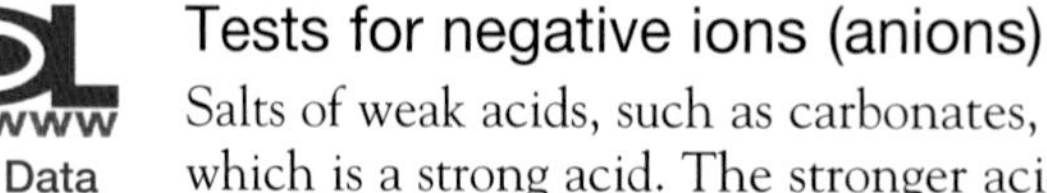

www Data

Tests for negative ions (anions)

Salts of weak acids, such as carbonates, react with dilute hydrochloric acid, which is a strong acid. The stronger acid displaces the weaker acid, which may then be identifiable itself or decompose to a recognisable gas.

Salts of some strong acids, such as halides and sulfates, can be identified by precipitation reactions.

Tests for positive ions (cations)

Aqueous sodium hydroxide can help to identify metal ions in solution. Adding sodium hydroxide solution produces a precipitate if the hydroxide of the metal is insoluble in water. The precipitate dissolves in excess of the alkali if the hydroxide is amphoteric (Section 1.4).

Mixing an ammonium salt with sodium hydroxide solution produces free ammonia molecules because hydroxide ions remove hydrogen ions from ammonium ions forming water and ammonia. Warming drives off the ammonia as a gas, which can be detected by its smell and its effect on moist red litmus paper.

$$NH_4^+(aq) + OH^-(aq) \rightarrow NH_3(aq) + H_2O(l)$$

Some metal ions do not give precipitates with alkalis. These can be identified by flame tests (Section 11.2). Flame tests can also distinguish between metal ions which behave in an identical way with solutions of sodium hydroxide.

Test yourself

3 For each of the following tests, identify the type of chemical reaction taking place, name the products and write a balanced equation for the reaction:
- a) testing for iodide ions with silver nitrate solution
- b) adding dilute hydrochloric acid to magnesium carbonate
- c) using limewater to confirm the presence of carbon dioxide
- d) adding dilute sodium hydroxide solution to a solution of iron(III) ions
- e) using concentrated ammonia solution to detect hydrogen chloride.

12.4 Quantitative analysis

Quantitative analysis involves techniques which answer the question 'How much?' In many laboratories, quantitative analysis is based on instrumental techniques such as chromatography and spectroscopy (see Topic 17).

Accurate chemical analysis generally involves preparing a solution of an unknown sample. Then it may be necessary to dilute the solution before analysing it.

Titrations are an important procedure for checking and calibrating instrumental methods. In a titration, the analyst finds the volume of the sample solution that reacts with a certain volume of a reference solution with an accurately known concentration.

Figure 12.4 ◀ A scientist in Nigeria adjusting an automatic titration device. This is being used to check that a pharmaceutical product contains the right amount of folic acid.

Test yourself

4 Why might the results of the following examples of quantitative analysis be important and why must the results be reliable:
a) the concentration of sugars in urine
b) the concentration of alcohol in blood
c) the percentage by mass of haematite (iron ore) in a rock sample
d) the concentration of nitrogen oxides in the air?

Many laboratories have automatic instruments for carrying out titrations, but the principle is exactly the same as in titrations where the volumes are measured with a traditional burette and pipette. Volumetric titrations with the kinds of glassware used in school and college laboratories are widely used in the food, pharmaceutical and other industries.

Pipettes, burettes and graduated flasks make it possible to measure out volumes of solutions very precisely during a titration. There are correct techniques for using all this glassware which must be followed carefully for accurate results.

12.5 Solutions for titrations

Any titration involves two solutions. Typically, a measured volume of one solution is run into a flask from a pipette. Then the second solution is added bit by bit from a burette until the colour change of an indicator, or the change in a signal from an instrument, shows that the reaction is complete.

The procedure only gives accurate results if the reaction between the two solutions is rapid and proceeds exactly as described by the chemical equation. So long as these conditions apply, titrations can be used to study acid–base, redox and other types of reactions.

Definitions

A standard solution is a solution with an accurately known concentration.

A primary standard is a chemical which can be weighed out accurately to make up a standard solution.

Standard solutions

Standard solutions make volumetric analysis possible. The direct way of preparing a standard solution is to dissolve a known mass of a chemical in water and then to make the volume of solution up to a definite volume in a graduated flask

This method for preparing a standard solution is only appropriate with a chemical that:

- is very pure
- does not gain or lose mass when in the air
- has a relatively high molar mass so that weighing errors are minimised.

Chemicals that meet these criteria are called primary standards. A titration with a primary standard can be used to measure the concentration of the solution to be analysed.

Test yourself

5 Explain why sodium hydroxide cannot be used as a primary standard.
6 Explain why anhydrous sodium carbonate can be used as a primary standard but hydrated sodium carbonate cannot.

Diluting a solution quantitatively

Quantitative dilution is an important procedure in analysis. Two common reasons for carrying out dilutions are:

- to make a solution with known concentration by diluting a standard solution
- to dilute an unknown sample for analysis to give a concentration suitable for titration.

The procedure for dilution is to take a measured volume of the more concentrated solution with a pipette (or burette) and run it into a graduated flask. The flask is then carefully filled to the mark with pure water.

The key to calculating the volumes to use when diluting a solution is to remember that the amount, in moles, of the chemical dissolved in the diluted solution is equal to the amount, in moles, of the chemical taken from the concentrated solution. If c is the concentration in mol dm^{-3} and V is the volume in dm^3, then we can write the following expressions.

The amount, in moles, of the chemical taken from the concentrated solution $= c_A V_A$

The amount, in moles, of the same chemical in the diluted solution $= c_B V_B$.

These two amounts are the same, so $c_A V_A = c_B V_B$.

Worked example

An analyst requires a $0.10\,\text{mol dm}^{-3}$ solution of sodium hydroxide, NaOH(aq). The analyst has a $250\,\text{cm}^3$ graduated flask and a supply of $0.50\,\text{mol dm}^{-3}$ sodium hydroxide solution. What volume of the concentrated solution should be measured into the graduated flask?

Answer

$c_A = 0.50\,\text{mol dm}^{-3}$ $\qquad c_B = 0.10\,\text{mol dm}^{-3}$

$V_A =$ to be calculated $\qquad V_B = 250\,\text{cm}^3 = 0.25\,\text{dm}^3$

$$V_A = \frac{c_B V_B}{c_A} = \frac{0.10\,\text{mol dm}^{-3} \times 0.25\,\text{dm}^3}{0.50\,\text{mol dm}^{-3}} = 0.050\,\text{dm}^3 = 50.0\,\text{cm}^3$$

Pipetting $50.0\,\text{cm}^3$ of the concentrated solution into the $250\,\text{cm}^3$ graduated flask and making up to the mark with pure water gives the required dilution after thorough mixing.

Test yourself

7 How would you prepare:
   a) a $0.05\,\text{mol dm}^{-3}$ solution of HCl(aq) given a $1000\,\text{cm}^3$ graduated flask and a $1.00\,\text{mol dm}^{-3}$ solution of the acid
   b) a $0.01\,\text{mol dm}^{-3}$ solution of NaOH(aq) given a $500\,\text{cm}^3$ graduated flask and a $0.50\,\text{mol dm}^{-3}$ solution of the alkali?

8 What is the concentration of the solution produced when making up to the mark with pure water and mixing:
   a) $10.0\,\text{cm}^{-3}$ of a $0.01\,\text{mol dm}^{-3}$ solution of $AgNO_3$(aq) in a $100\,\text{cm}^3$ graduated flask
   b) $50.0\,\text{cm}^{-3}$ of a $2.00\,\text{mol dm}^{-3}$ solution of nitric acid in a $250\,\text{cm}^3$ graduated flask?

12.6 Titration principles

A titration involves two solutions. A measured volume of one solution is run into a flask. The second solution is then added bit by bit from a burette until the reaction is complete.

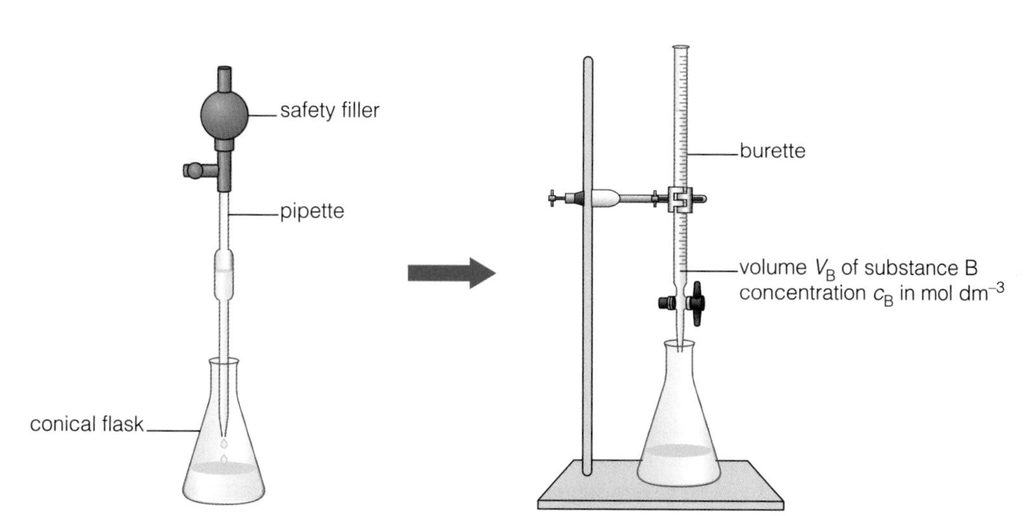

Figure 12.5 The apparatus used for a titration based on a reaction between two chemicals, A and B. The situation is essentially the same for acid–base and redox titrations, such as iodine–thiosulfate titrations.

Definition

The **end-point** is the point during a titration when enough of the solution in the burette has been added to react exactly with the amount of chemical in the flask.

Some titrations are used to investigate reactions. In these experiments the concentrations of both solutions are known and the aim is to determine the equation for the reaction.

More often, titrations are used to measure the concentration of an unknown solution knowing the equation for the reaction and using a second solution of known concentration.

In general, n_A moles of A react with n_B moles of B.

$$n_A A + n_B B \rightarrow \text{products}$$

The concentration of solution B in the flask is c_B and the concentration of solution A in the burette is c_A. Both are measured in mol dm^{-3}.

The analyst uses a pipette to run a volume V_B of solution B into the flask. Then solution A is added from the burette until an indicator shows that the reaction is complete. At the end-point, the volume added is the titre, V_A. The analyst should repeat the titration enough times to achieve consistent results.

Titration calculations

In the laboratory, volumes of solutions are normally measured in cm^3, but they should be converted to dm^3 in calculations so that they are consistent with the units used for concentrations.

The amount, in moles, of B in the flask at the start $= c_B \times V_B$

The amount, in moles, of A added from the burette $= c_A \times V_A$

The ratio of these amounts must be the same as the ratio of the reacting amounts n_A and n_B. We can therefore write:

$$\frac{c_A \times V_A}{c_B \times V_B} = \frac{n_A}{n_B}$$

In any titration, all but one of the values in this relationship are known. The one unknown is calculated from the results.

Analysing solutions

In titrations designed to analyse solutions, the equation for the reaction is given so that the ratio n_A/n_B is known. The concentration of one of the solutions is also known. The volumes V_A and V_B are measured during the titration. Substituting all the known quantities in the titration formula allows the concentration of the unknown solution to be calculated.

Investigating reactions

In titrations to investigate reactions, the problem is to determine the ratio n_A/n_B. The concentrations c_A and c_B are known and the volumes V_A and V_B are measured during the titration. So the ratio n_A/n_B can be calculated from the formula.

12.7 Acid–base titrations

Coloured indicators can be used to detect the end-point of acid–base reactions. These are chemicals which change colour as the pH varies. Typically, an indicator completes its colour change over a range of about 2 pH units as shown in Table 12.1.

In any acid–base titration there is a sudden change of pH at the end-point. The chosen indicator must therefore complete its colour change within the range of pH values spanned at the end-point.

For a titration of a strong acid with a strong alkali, the pH jumps from around pH 3 to pH 10 at the end-point. Most common indicators change colour sharply within this range.

For a strong acid titrated against a weak alkali, the pH jump is not so large and is from around pH 3 to pH 8. For a weak acid with strong base, the jump is

Indicator	Colour change low pH – high pH	pH range over which colour change occurs
Methyl orange	Red – yellow	3.2–4.2
Methyl red	Yellow – red	4.8–6.0
Bromothymol blue	Yellow – blue	6.0–7.6
Phenolphthalein	Colourless – red	8.2–10.0

Table 12.1 ▲
Some common indicators and the pH range over which they change colour

from about pH 6 to pH 10. So the indicator must be chosen with care in these two cases.

Determining the solubility of calcium hydroxide

Calcium hydroxide is an alkali that is only slightly soluble in water. Its solubility can be determined by titration of a saturated solution of the alkali with a standard solution of hydrochloric acid. The equation for the reaction is known, so the unknown to be determined is the concentration of the saturated solution of the alkali.

The results of a titration are shown in Figure 12.6.

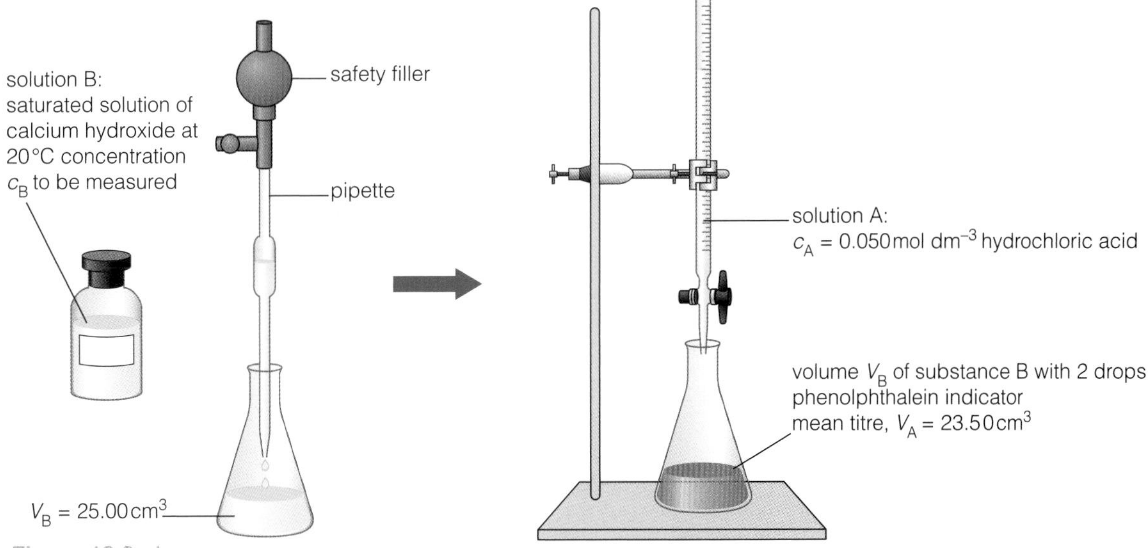

Figure 12.6 ▲
A titration to determine the solubility of calcium hydroxide

The equation for the titration reaction is:

$$Ca(OH)_2(aq) + 2HCl(aq) \rightarrow CaCl_2(aq) + 2H_2O(l)$$

So 1 mol of the alkali reacts with 2 mol of the acid.

Substituting the values in the titration formula gives:

$$\frac{0.050 \times 23.50}{c_B \times 25.00} = \frac{2}{1}$$

Rearranging:

$$c_B = \frac{0.050 \times 23.50}{2 \times 25.00} = 0.0235\ \text{mol dm}^{-3}$$

So the concentration of saturated calcium hydroxide solution is 0.0235 mol dm^{-3}.
The molar mass of calcium hydroxide is 74.1 g mol^{-1}
So the concentration of saturated calcium hydroxide solution

$$= 0.0235\ \text{mol dm}^{-3} \times 74.1\ \text{g mol}^{-1} = 1.74\ \text{g dm}^{-3}$$

Test yourself

9 A 25.0 cm^3 sample of nitric acid was neutralised by 18.0 cm^3 of 0.15 mol dm^{-3} sodium hydroxide solution. Calculate the concentration of the nitric acid.

10 A 2.65 g sample of anhydrous sodium carbonate was dissolved in water and the solution made up to 250 cm^3. In a titration, 25.0 cm^3 of this solution was added to a flask and the end-point was reached after adding 22.5 cm^3 of hydrochloric acid. Calculate the concentration of the hydrochloric acid.

11 A 41.0 g sample of the acid H_3PO_3 was dissolved in water and the volume of solution was made up to 1 dm^3. 20.0 cm^3 of this solution was required to react with 25.0 cm^3 of 0.80 mol dm^{-3} sodium hydroxide solution. What is the equation for the reaction?

Activity

Finding the percentage of calcium carbonate in egg shell

Egg shells contain calcium carbonate. It is possible to determine the percentage of calcium carbonate in an egg shell by titration.

Calcium carbonate reacts with acids but it is not possible to titrate this directly with an acid from a burette. Instead it has to be determined by a procedure called 'back titration'.

The analyst adds an excess of a standard solution of hydrochloric acid to a weighed sample of the crushed shell. When the reaction with calcium carbonate in the egg shell is complete, the next step is to transfer all the solution to a graduated flask. Water is then added to the mark and the diluted solution is well mixed. Finally, the analyst titrates measured volumes of the diluted solution with a standard solution of sodium hydroxide to determine the amount of unreacted acid left over.

The results of such a titration allow the analyst to measure the amount of excess hydrochloric acid, and hence the amount that must have reacted with calcium carbonate in the egg shell.

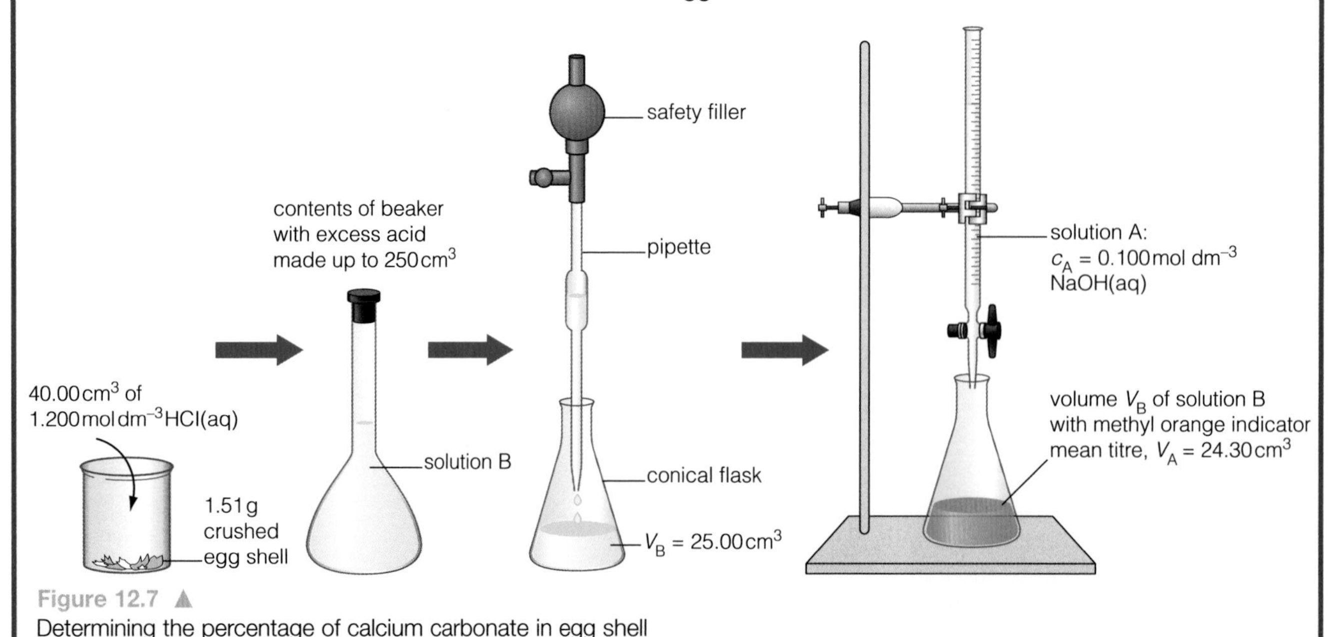

Figure 12.7 ▲
Determining the percentage of calcium carbonate in egg shell

1 Give two reasons why titrating the calcium carbonate in an egg shell with hydrochloric acid from a burette is not possible.

2 Write a balanced equation for the reaction taking place during the titration.

3 Use the data about the titration (from Figure 12.7) to calculate the concentration of the excess hydrochloric acid in the graduated flask (solution B).

4 Use your answer to question **3** to calculate the amount of excess hydrochloric acid, in moles, left over after reaction with the egg shell.

5 Calculate how much hydrochloric acid, in moles, was added to the sample of egg shell in the beaker.

6 Use your answers to questions **4** and **5** to calculate how much hydrochloric acid, in moles, reacted with calcium carbonate in the sample of egg shell.

7 Write the balanced equation for the reaction of calcium carbonate with hydrochloric acid.

8 Use the equation and your answer to question **6** to work out the amount, in moles, of calcium carbonate in the sample of egg shell.

9 Calculate the molar mass of calcium carbonate, and hence the mass of it in the 1.51 g of egg shell.

10 Calculate the percentage of calcium carbonate in the egg shell.

12.8 Iodine–thiosulfate titrations

Many oxidising agents rapidly convert iodide ions to iodine by taking away electrons.

$$2I^-(aq) \rightarrow I_2(aq) + 2e^-$$

The iodine which forms can be titrated with thiosulfate ions and be reduced back to iodide ions.

$$2S_2O_3^{2-}(aq) \rightarrow S_4O_6^{2-}(aq) + 2e^-$$

$$I_2(aq) + 2e^- \rightarrow 2I^-(aq)$$

These two half-equations show that 2 mol of thiosulfate ions reduce 1 mol of iodine molecules.

This system can be used to investigate quantitatively any oxidising agent that can oxidise iodide ions to iodine. The procedure is:

- add excess potassium iodide to a measured quantity of the oxidising agent, which then converts iodide ions to iodine
- titrate the iodine formed with a standard solution of sodium thiosulfate.

Iodine is not soluble in water. It dissolves in potassium iodide solution to form a solution which is dark brown when concentrated but pale yellow when dilute.

At the end-point, the pale yellow iodine colour disappears to give a colourless solution. Adding a few drops of starch solution just before the end-point makes the colour change much sharper. Starch gives a deep blue colour with iodine which disappears at the end-point.

Finding the concentration of domestic bleach

Domestic bleach is a solution of sodium chlorate(I). If the solution is acidified it produces chlorine. The amount of chlorine produced from a measured quantity of bleach is sometimes described as the 'available' chlorine.

The available chlorine from bleach can be estimated by adding acid and excess potassium iodide, then titrating with sodium thiosulfate (Figure 12.8). The bleach has to be diluted quantitatively to give a reasonable titre with 0.100 mol dm^{-3} sodium thiosulfate solution.

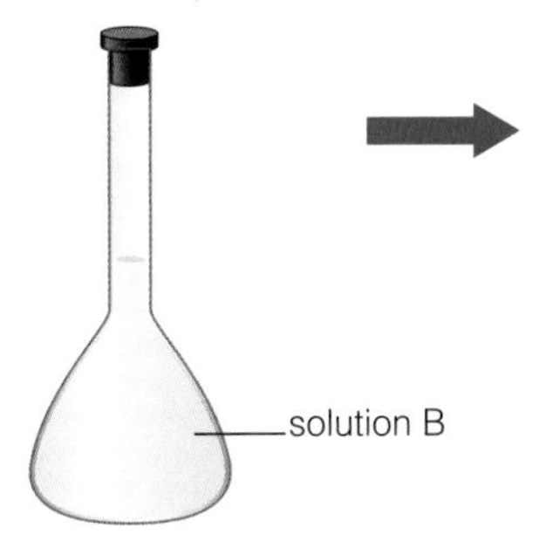

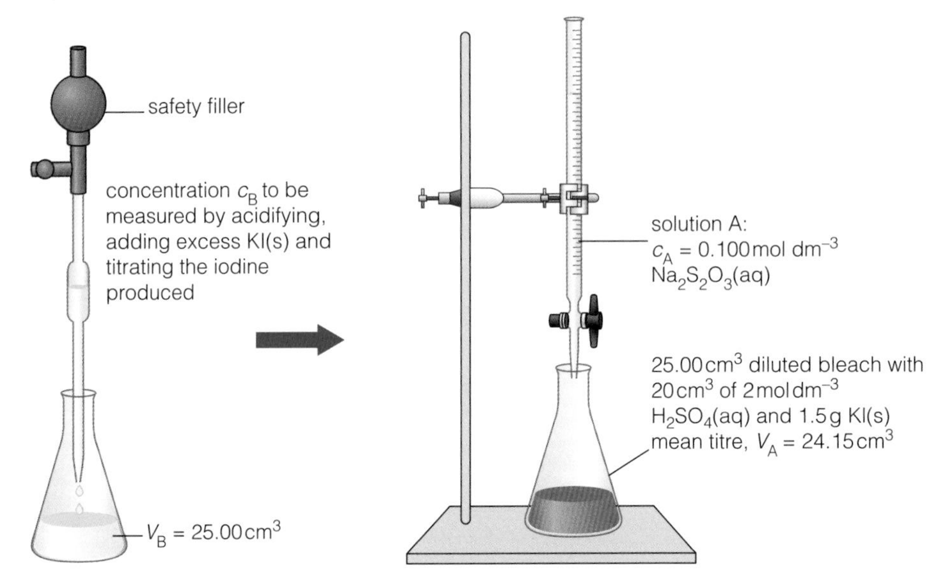

Figure 12.8 ▲
Measuring the concentration of chlorine bleach

Note

When using thiosulfate to titrate iodine formed by an oxidising agent, iodide ions are first oxidised to iodine and then reduced back to iodide. So you do not need to include the iodine in the calculation. It is present as a 'go between'. Relate the thiosulfate directly to the oxidising agent with the help of the equations.

The equation for the titration reaction is:

$$2S_2O_3^{2-}(aq) + I_2(aq) \rightarrow S_4O_6^{2-}(aq) + 2I^-(aq)$$

The iodine was formed by oxidation of iodide ions by bleach:

$$ClO^-(aq) + 2H^+(aq) + 2I^-(aq) \rightarrow I_2(aq) + H_2O(l) + Cl^-(aq)$$

So 2 mol of thiosulfate ions react with 1 mol of iodine, which was produced by reaction with 1 mol of chlorate(I) ions. The ratio of moles of thiosulfate to moles of chlorate(I) is 2 : 1.

Substituting the values from Figure 12.8 in the titration formula gives:

$$\frac{0.100 \times 24.15}{c_B \times 25.00} = \frac{2}{1}$$

Rearranging:

$$c_B = \frac{0.100 \times 24.15}{2 \times 25.00} = 0.0483\ \text{mol dm}^{-3}$$

So the concentration of chlorate(I)ions in diluted bleach = 0.0483 mol dm^{-3}

20.0 cm^3 of the original bleach was diluted to 250 cm^3, so the concentration of chlorate(I) ions in the concentrated bleach is given by:

$$= \frac{250}{20.0} \times 0.0483\ \text{mol dm}^{-3} = 0.604\ \text{mol dm}^{-3}$$

The 'available chlorine' in chlorate bleach is set free on adding acid according to the equation:

$$ClO^-(aq) + 2H^+(aq) + Cl^-(aq) \rightarrow Cl_2(aq) + H_2O(l)$$

This shows that 1 mol of chlorate(I), ClO^-, produces 1 mol of chlorine, Cl_2.

So the 'available chlorine' in concentrated bleach = 0.604 mol dm^{-3}

The molar mass of chlorine, Cl_2 = 71.0 g mol^{-1}

So the 'available chlorine' in the concentrated bleach = 42.9 g dm^{-3}

Test yourself

12 a) Write a balanced equation for the reaction of iodate(V) ions with iodide ions in the presence of acid to produce iodine.

b) A solution of potassium iodate(V) is made up by dissolving 0.89 g in water and making this up to 250 cm^3 in a graduated flask. What is the concentration of the solution in mol dm^{-3}?

c) Excess, acidified potassium iodide solution is added to 25.0 cm^3 of the potassium iodate(V) solution from part **b)**. The iodine formed is titrated with sodium thiosulfate solution. At the end-point the volume of thiosulfate solution added is 20.0 cm^3. What is the concentration of the sodium thiosulfate solution?

12.9 Evaluating results

Every time an analyst carries out a titration, there is some uncertainty in the result. It is important to be able to assess measurement uncertainty because this shows how reliable the result is. Important decisions are based on the results of chemical analysis in healthcare, in the food industry, in law enforcement and in many other areas of life. It is important that the people making these decisions understand the extent to which they can rely on the data from analysis.

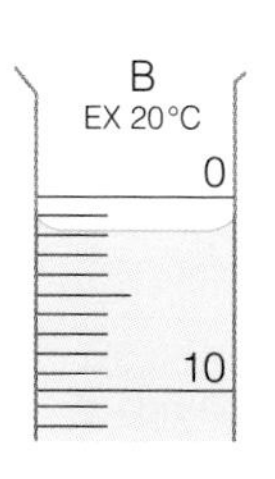

It is difficult to determine accurately the volume of liquid in a burette if the meniscus lies between two graduation marks.

The material used to prepare a standard solution may not be 100% pure.

A 250 cm^3 volumetric flask may actually contain 250.3 cm^3 when filled to the calibration mark due to permitted variation in the manufacture of the flask.

A burette is calibrated by the manufacturer for use at 20 °C. When it is used in the laboratory the temperature may be 23 °C. This difference in temperature will cause a small difference in the actual volume of liquid in the burette when it is filled to a calibration mark.

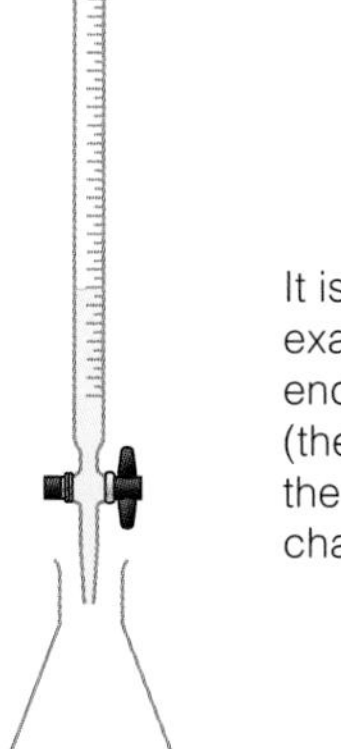

It is difficult to make an exact judgement of the end-point of a titration (the exact point at which the colour of the indicator changes).

The display on a laboratory balance will only show the mass to a certain number of decimal places.

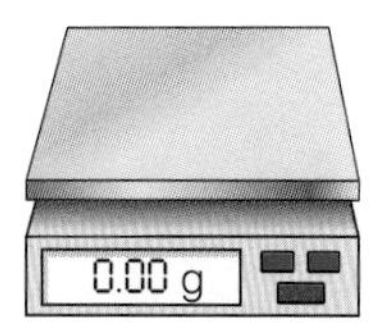

Figure 12.9 ◀ Sources of uncertainty in volumetric analysis

Random errors in titrations

Every time an analyst carries out a titration, there are small differences in the results. This is not because the analyst has made mistakes but because there are factors that are impossible to control.

Unavoidable random errors arise in judging when the bottom of the meniscus is level with the graduation on a pipette, in judging the colour change at the end-point and when taking the reading from a burette scale.

If these random errors are small then the results are all close together – in other words, they are precise. The precision of a set of results can be judged from the range in a number of repeated titrations.

Systematic errors in titrations

Systematic errors mean that the results differ from the true value by the same amount each time. The measurement is always too high or too low.

One source of systematic error is the tolerance allowed in the manufacture of graduated glassware. The tolerance for grade B $250\,cm^3$ graduated flasks is $\pm 0.3\,cm^3$. So when an analyst uses a particular flask, filled correctly to the graduation mark, the volume of solution may be as little as $249.7\,cm^3$ or as much as $250.3\,cm^3$. The error is the same every time that the particular flask is used. Similar tolerances are allowed for pipettes and burettes – for a class B $25\,cm^3$ pipette the tolerance is $\pm 0.06\,cm^3$, while for a class B $50\,cm^3$ burette the tolerance is $\pm 0.1\,cm^3$.

Systematic errors can be allowed for by calibrating the measuring instruments. It is possible to calibrate a pipette, for example, by using it to measure out pure water and then weighing the water with an accurate balance.

Definitions

Results are **precise** if repeat measurements have values that are close together. Precise measurements have a small random error.

Bias arises from systematic errors which affect all the measurements in the same way, making them all higher or lower than the true value. Systematic errors do not average out.

Results are **accurate** if they are precise and free from bias.

Figure 12.10 ▶ Throwing darts at the bullseye of a dartboard illustrates the notions of precision and bias. Reliable players throw precisely and without bias so that their darts hit the centre of the board accurately.

A darts player is practising throwing darts at a board. The aim is to get all the darts close together near the centre of the board. The results of some of the attempts are shown below.

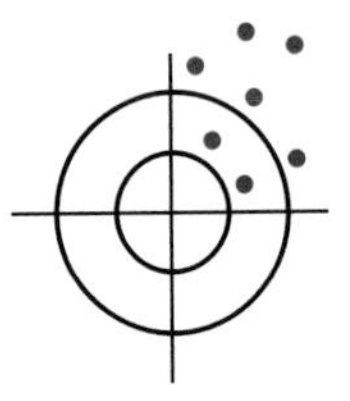

1st attempt: The shots are quite widely scattered and some have not even hit the board. The shots show *poor precision* as they are quite widely scattered. There is also a *bias* in where the shots have landed – they are grouped in the top right-hand corner, not near the centre of the board.

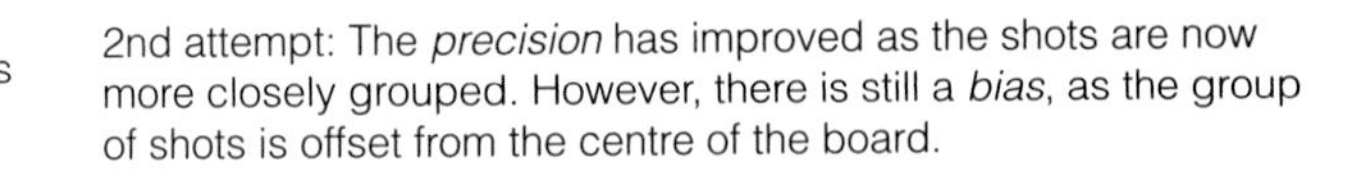

2nd attempt: The *precision* has improved as the shots are now more closely grouped. However, there is still a *bias*, as the group of shots is offset from the centre of the board.

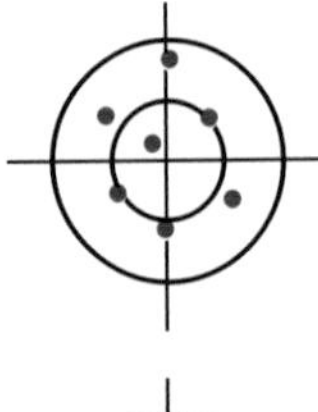

3rd attempt: The player has improved to *reduce* the *bias* – all the shots are now on the board and scattered round the centre. Unfortunately the *precision* is *poor* as the shots are quite widely scattered.

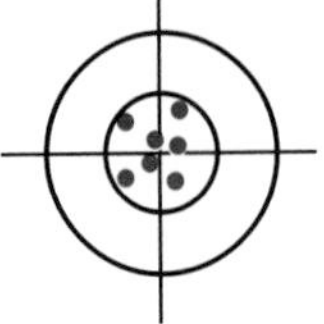

Some time later: The shots are *precise* and *unbiased* – they are all grouped close together in the centre of the board.

Practical guidance

Test yourself

13 Identify examples of random and systematic error when:
a) using a pipette
b) using a burette
c) making up a standard solution in a graduated flask.

REVIEW QUESTIONS

Extension questions

1 Identify the following salts and account for the observations.

a) X is a white solid which colours a flame bright yellow. No precipitate forms on mixing a solution of X with sodium hydroxide solution. On heating, X gives off a colourless gas that relights a glowing splint. **(4)**

b) Y is a white, crystalline solid which colours a flame green. Adding dilute nitric acid followed by silver nitrate solution to a solution of Y produces a white precipitate. **(2)**

c) Z is a white crystalline solid which colours a flame lilac. Z is soluble in water – the solution does not change the colour of indicators. Mixing the solution with a solution of silver nitrate produces a cream precipitate that is insoluble in dilute ammonia but soluble in concentrated ammonia. A solution of Z turns orange on adding aqueous chlorine. **(4)**

2 Describe the observations and write equations to explain how each of the following reagents can be used to distinguish between sodium bromide and sodium iodide:

a) aqueous chlorine **(6)**

b) aqueous silver nitrate **(6)**

c) concentrated sulfuric acid. **(6)**

3 A carbonate of metal M has the formula M_2CO_3. In a titration, a 0.245 g sample of M_2CO_3 was found to neutralise 23.6 cm^3 of 0.150 mol dm^{-3} hydrochloric acid. Follow these steps to identify the metal M.

a) Write the equation for the reaction of M_2CO_3 with hydrochloric acid. **(1)**

b) Calculate the amount, in moles, of hydrochloric acid needed to react with the sample of the metal carbonate. **(1)**

c) Use the equation to calculate the amount, in moles, of M_2CO_3 in the sample. **(1)**

d) Use your answer to part **c)** and the mass of the sample to calculate the relative formula mass of M_2CO_3. **(2)**

e) Calculate the relative atomic mass of metal M. **(1)**

f) Identify the metal M. **(1)**

4 A 20.0 cm^3 sample of water from a swimming pool was added to excess potassium iodide solution. The iodine formed was titrated with 0.0050 mol dm^{-3} sodium thiosulfate solution. The volume of sodium thiosulfate solution needed to reach the end-point was 19.4 cm^3.

a) Write ionic equations:

i) for the reaction of chlorine with iodide ions

ii) for the reaction of iodine molecules with thiosulfate ions. **(3)**

b) Name the indicator used to detect the end-point of the titration. **(1)**

c) Describe the colour changes observed at each stage of the analysis. **(3)**

d) Calculate the concentration of chlorine in the swimming pool water. **(4)**

13 Kinetics

The study of rates of reaction is important because it helps chemists to control reactions both in the laboratory and on a large scale in industry.

Definition

Chemical kinetics is the study of the rates of chemical reactions.

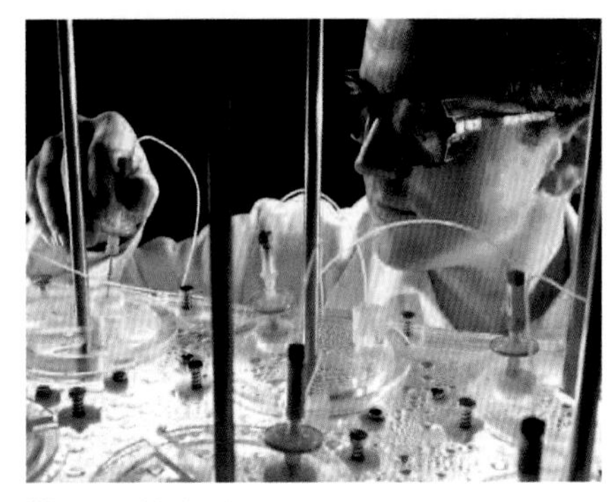

Figure 13.1 ▲
A pharmacy technician monitoring the rate at which a drug is released from a medicine tablet in conditions similar to those in the stomach

13.1 Reaction rates

In the chemical industry, manufacturers aim to get the best possible yield in the shortest time. The development of new catalysts to speed up reactions is therefore one of the frontier aspects of modern chemistry. The aim is to make manufacturing processes more efficient so that they use less energy and produce little or no harmful waste. The need for greater efficiency in chemical processes is now more pressing than ever as people become more aware of the harm that waste chemicals can do to our health and to the environment.

Kinetics is important in many other fields too. The study of rates of reaction helped environmental scientists to explain why CFCs and other chemicals are destroying the ozone layer in the upper atmosphere. Pharmacologists who study the chemistry of drugs must study the speed at which they change to other chemicals or break down in the human body. Then the pharmacists who formulate and supply medicines need to know about the rate at which the chemicals slowly degrade in the bottle or pack. For many medicines, the shelf life is the time for which they can be stored before the concentration of the active ingredient has dropped by 10%.

Chemical reactions happen at a variety of speeds. Ionic precipitation reactions are very fast and explosions are even faster. However, the rusting of iron and other corrosion processes are slow and may continue for years.

Figure 13.2 ▲
Firefighters have to know how to slow down and stop burning. Water cools the burning materials as it evaporates and the steam produced can help to keep out the air.

Test yourself

1 How would you slow down or stop these reactions:
  a) iron corroding
  b) toast burning
  c) milk turning sour?
2 How would you speed up these reactions:
  a) fermentation in dough to make the bread rise
  b) solid fuel burning in a stove
  c) epoxy glues (adhesives) setting
  d) the conversion of chemicals in engine exhausts to harmless gases?

13.2 Measuring reaction rates

Balanced chemical equations tell us nothing about how quickly the reactions occur. In order to get this information, chemists have to do experiments to measure the rates of reactions under various conditions.

The amounts of the reactants and products change during any chemical reaction – products form as reactants disappear. The rates at which these changes happen give a measure of the rate of reaction.

The rate of the reaction between magnesium and hydrochloric acid

$$Mg(s) + 2HCl(aq) \rightarrow MgCl_2(aq) + H_2(g)$$

can be measured by:

- the rate of loss of magnesium
- the rate of loss of hydrochloric acid
- the rate of formation of magnesium chloride
- the rate of formation of hydrogen.

In this example, it is probably easiest to measure the rate of formation of hydrogen by collecting the gas and recording its volume with time (Figure 13.3).

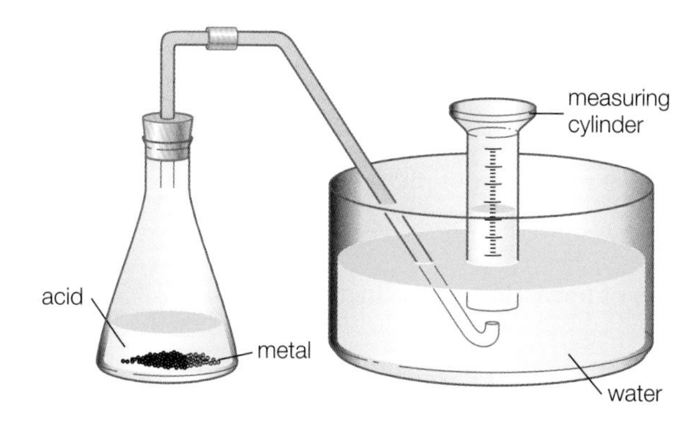

Figure 13.3 ◄
Collecting and measuring the gas produced when magnesium reacts with acid. A gas syringe can be used instead of a measuring cylinder full of water.

Chemists design their rate experiments to measure a property which changes with the amount or concentration of a reactant or product. Then:

$$\text{Rate of reaction} = \frac{\text{change recorded in the property}}{\text{time for the change}}$$

In most chemical reactions the rate changes with time. The graph in Figure 13.4 is a plot of the results from a study of the reaction of magnesium with dilute hydrochloric acid. The graph is steepest at the start, when the reaction

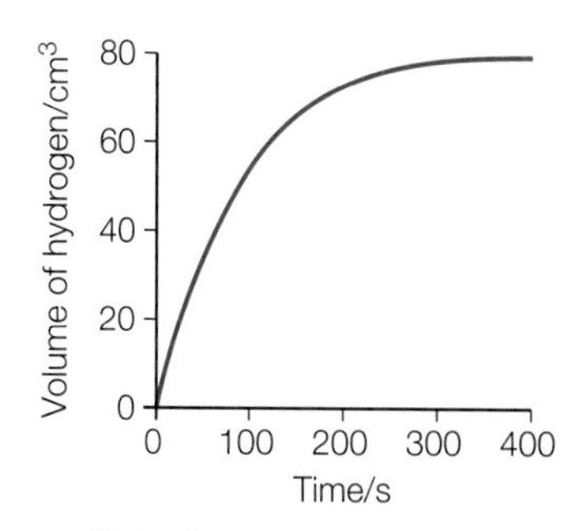

Figure 13.4 ▲
Volume of hydrogen plotted against time for the reaction of magnesium with hydrochloric acid

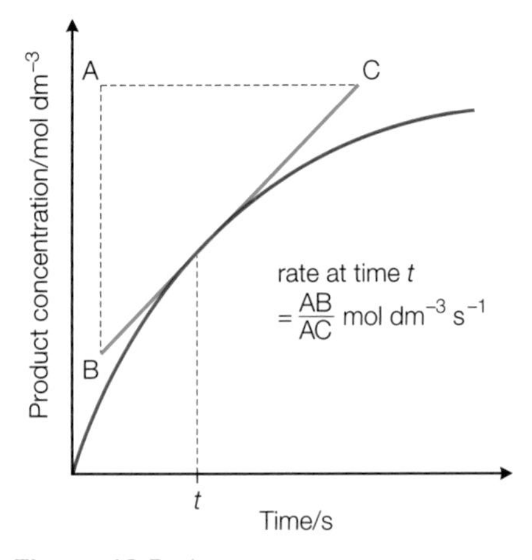

Figure 13.5 ▲
Graph showing the concentration of a product plotted against time. The gradient at any point measures the rate of reaction at that time.

is at its fastest. As the reaction continues it slows down, until it finally stops. This happens because one of the reactants is being used up until none of it is left.

The gradient at any point on a graph showing amount or concentration plotted against time measures the rate of reaction (Figure 13.5).

Test yourself

3 In an experiment to study the reaction of magnesium with dilute hydrochloric acid, 48 cm^3 of hydrogen forms in 10 s at room temperature. Calculate the average rate of formation of:
a) hydrogen in $cm^3 s^{-1}$
b) hydrogen in mol s^{-1} (Section 1.7)
c) the rates of appearance or disappearance of the other product and the reactants in mol s^{-1}.

A useful way of studying the effect of changing the conditions on the rate of a reaction is to find a way of measuring the rate just after mixing the reactants. Figure 13.6 is a graph for two different sets of conditions. One of the reactants was more concentrated to produce line A. Near the start, it took t_A seconds to produce x mol of product. When the same reactant was less concentrated, the results gave line B. This time near the start it took t_B seconds to produce x mol of product. The reaction was slower when the concentration was lower, so it took longer to produce x mol of product.

Figure 13.6 ▶
Formation of the same amount (x mol) of product starting with different concentrations of one of the reactants

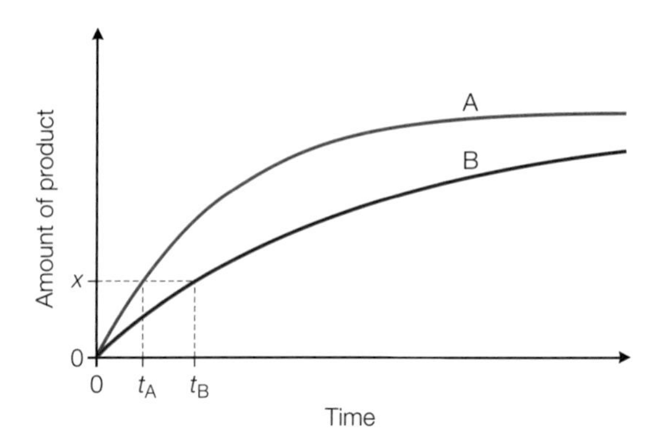

$$\text{The average rate of formation of product on line A} = \frac{x}{t_A}$$

$$\text{The average rate of formation of product on line B} = \frac{x}{t_B}$$

If x is kept the same, it follows that the average rate near the start $\propto \frac{1}{t}$

This means that it is possible to arrive at a measure of the initial rate of a reaction by measuring how long the reaction takes to produce a small fixed amount of product, or use up a small fixed amount of reactant.

13.3 What factors affect reaction rates?

Concentration

In general, the higher the concentration of the reactants, the faster the reaction. For gas reactions, a change in pressure has the same effect as changing the concentration – a higher pressure compresses a mixture of gases and increases their concentration. So, in a mixture of reacting gases, the higher the pressure the faster the reaction.

Activity

Investigation of the effect of concentration on the rate of a reaction

Figure 13.7 illustrates an investigation of the effect of concentration on the rate at which thiosulfate ions in solution react with hydrogen ions to form a precipitate of sulfur.

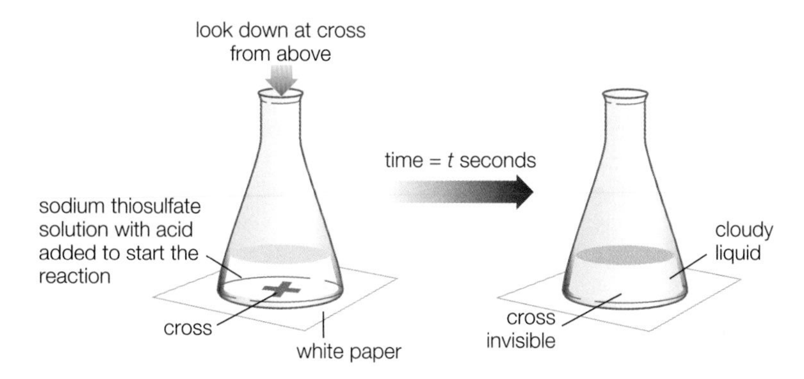

Figure 13.7 ▲
Investigating the effect of the concentration of thiosulfate ions on the rate of reaction in acid solution. The hydrogen ion concentration is the same in each experiment.

$$S_2O_3^{2-}(aq) + 2H^+(aq) \rightarrow S(s) + SO_2(aq) + H_2O(l)$$

The observer records the time taken for the sulfur precipitate to obscure the cross on the paper under the flask. In this example, the quantity *x* in Figure 13.6 is the amount of sulfur needed to hide the cross on the paper. This is the same each time – so, the rate of reaction is proportional to $1/t$. The results of the investigation are shown in Table 13.1.

Experiment	Concentration of thiosulfate ions/mol dm^{-3}	Time, *t*, for the cross to be obscured/s	Rate of reaction, $\frac{1}{t}$/s^{-1}
1	0.15	43	0.023
2	0.12	55	
3	0.09	66	0.015
4	0.06	105	0.0095
5	0.03	243	0.0041

Table 13.1 ▲
Results of the investigation in Figure 13.7

1. How would you prepare 50 cm^3 of a solution of sodium thiosulfate solution with a concentration of 0.12 mol dm^{-3} from a solution with a concentration of 0.15 mol dm^{-3}?
2. Calculate the value for the rate of reaction when the concentration of thiosulfate ions is 0.12 mol dm^{-3}.
3. Plot a graph to show how the rate of reaction varies with the concentration of thiosulfate ions.
4. What is the relationship between reaction rate and concentration of thiosulfate for this reaction according to your graph?

Surface area of solids

Breaking a solid into smaller pieces increases the surface area in contact with a liquid or gas. This speeds up any reaction happening at the surface of the solid. This effect also applies to reactions between liquids which do not mix. Shaking breaks up one liquid into droplets which are then dispersed in the other liquid, thereby increasing the surface area for reaction.

Activity

Investigating the effect of surface area on the rate of a reaction

Figure 13.8 illustrates an investigation of the rate of reaction of lumps of calcium carbonate (marble) with dilute nitric acid. The results are given in Table 13.2. Both sets of results were obtained using 20 g of marble chips and 40 cm^3 of 2.0 mol dm^{-3} nitric acid. The marble was in excess.

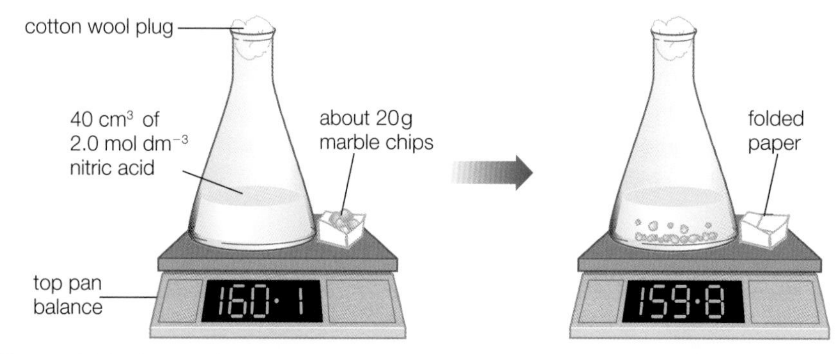

Figure 13.8 ▲
Apparatus for comparing the reaction rate of calcium carbonate with nitric acid

Time/s	Mass of carbon dioxide formed/g	
	Small marble chips	Large marble chips
30	0.45	0.18
60	0.85	0.38
90	1.13	0.47
120	1.31	0.75
180	1.48	1.05
240	1.54	1.25
300	1.56	1.38
360	1.58	1.47
420	1.59	1.53
480	1.60	1.57
540	1.60	1.59
600	1.60	1.60

Table 13.2 ▲
Results of experiments to compare the reaction rate of calcium carbonate with nitric acid using the same mass of larger and smaller marble chips

1 Plot the two sets of results on the same axes.

2 Which reaction had the greater initial rate?

3 a) After what time did the reaction stop for each set of results?
  b) Why did the reaction stop?

4 Why was the same mass of carbon dioxide formed in both sets of results?

5 For a given mass of marble, how is surface area related to particle size?

6 What is the effect on this reaction of changing the surface area of the solid?

7 Sketch on your graph the results you would expect if you repeated the experiment with 20 g small marble chips and 40 cm^3 of 1.0 mol dm^{-3} nitric acid.

Temperature

Raising the temperature is a very effective way of increasing the rate of a reaction. In general, a 10 °C rise in temperature roughly doubles the rate of reaction (see Figure 13.9).

Bunsen burners, hot-plates and heating mantles are common items of equipment in laboratories because it is often convenient to speed up reactions by heating the reactants. For the same reason, many industrial processes are carried out at high temperatures.

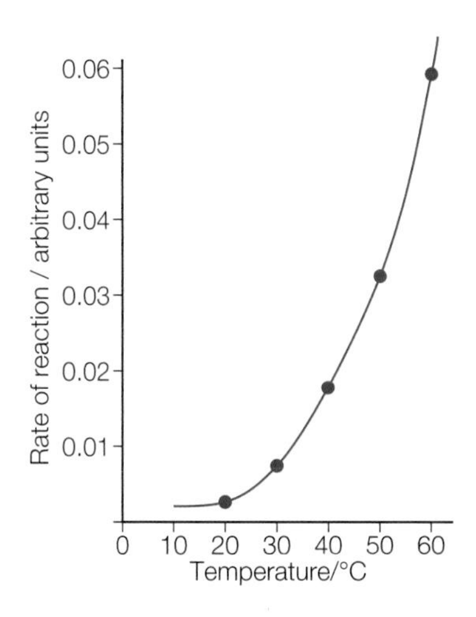

Figure 13.9 ▲
The effect of temperature on the rate of decomposition of thiosulfate ions to form sulfur

Catalysts

Catalysts have an astonishing ability to speed up the rates of some chemical reactions without themselves changing permanently. Very small quantities of active catalysts can speed reactions to produce many times their own weight of chemicals.

Catalysts work by removing or lowering the barriers preventing reaction – they bring reactants together in a way that makes a reaction more likely.

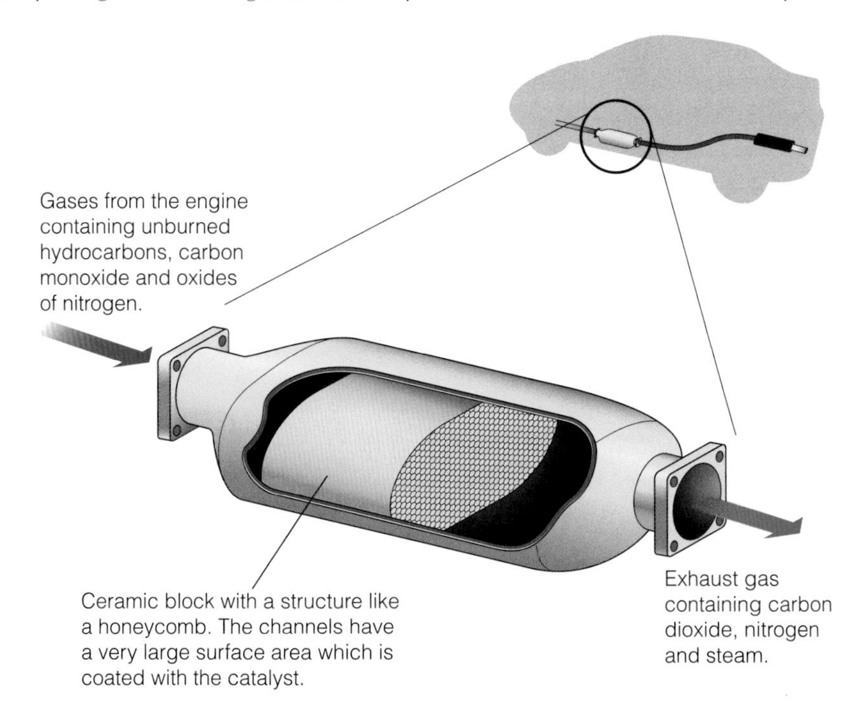

Figure 13.10 ◀
The catalyst in a catalytic converter is made from a combinations of platinum, palladium and rhodium. The catalyst speeds up reactions which remove pollutants from motor vehicle exhausts. The reactions convert oxides of nitrogen to nitrogen and oxygen. They also convert carbon monoxide and unburned hydrocarbons to carbon dioxide and steam.

Catalysts can also be extraordinarily selective – a catalyst may increase the rate of only one very specific reaction. Enzymes, the catalysts in living cells, are especially selective.

Most industrial processes involve catalysts. One of the targets of chemists in industry is to develop catalysts that make manufacturing processes more efficient, so that they use less energy and produce less waste (see Topic 18).

Catalysts get involved in reactions, but they are not reactants and they do not appear in the overall chemical equation. In theory, catalysts can be used over and over again, but in practice there is some loss of catalyst. Sometimes catalysts become contaminated, sometimes they are hard to recover completely from the products and sometimes the catalyst changes its state, such as from lumps to a fine powder, which means that it is no longer useable.

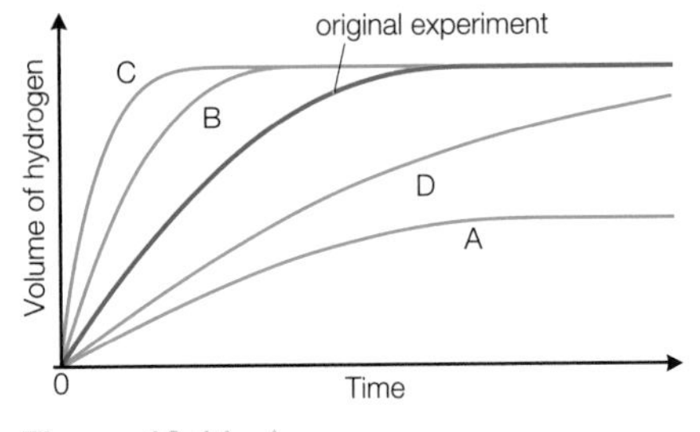

Figure 13.11 ▲

Test yourself

4 Why do you think the catalyst in a catalytic converter is present as a very thin layer on the surface of many fine holes running through a block of inert ceramic?

5 Figure 13.11 illustrates the effect of changing the conditions on the reaction of zinc metal with sulfuric acid. The red line shows the volume of hydrogen plotted against time using an excess of zinc turnings and 50 cm^3 of 2.0 mol dm^{-3} sulfuric acid at 20 °C.

a) Write a balanced equation for the reaction.

b) Draw the apparatus which could be used to obtain the results to plot the graph.

c) Identify which line of the graph shows the effect of carrying out the same reaction under the same conditions with the following changes:

i) adding a few drops of copper(II) sulfate solution to act as a catalyst

ii) raising the temperature to 30 °C

iii) using the same mass of zinc but in larger pieces

iv) using 50 cm^3 of 1.0 mol dm^{-3} sulfuric acid.

Figure 13.12 ▲
The hot air balloon flights of the Montgolfier brothers in the later eighteenth century inspired scientists to study the behaviour of gases.

13.4 Collision theory

Scientists can explain the factors that affect reaction rates. They use a model which gives a picture of what happens to atoms, molecules and ions as they react. This model works best for gases but it can also be applied to reactions in solution.

Gas molecules in motion

The model that scientists use to explain the behaviour of gases assumes that the molecules in a gas are in rapid random motion and colliding with each other. They call this particles-in-motion model the 'kinetic theory'. The name comes from a Greek word for movement.

The kinetic theory makes a number of assumptions about the molecules of a gas. Applying Newton's laws of motion to the collection of particles leads to equations that can describe the properties of gases very accurately.

The assumptions of the kinetic theory model are that:

- gas pressure results from the collisions of the molecules with the walls of the container
- there is no loss of energy when the molecules collide with the walls of any container
- the molecules are so far apart that the volume of the molecules can be neglected in comparison with the total volume of gas
- the molecules do not attract each other
- the average kinetic energy of molecules is proportional to their temperature on the Kelvin scale.

A gas that behaves exactly as this model predicts is called an 'ideal gas'. Real gases do not behave exactly like this. Under laboratory conditions, however, there are gases which are close to behaving like an ideal gas. These are the gases which, at room temperature, are well above their boiling temperature, such as helium, nitrogen, oxygen and hydrogen.

The assumptions built into the model help to explain why real gases approach ideal behaviour at high temperatures and low pressures:

- at high temperatures, the molecules are moving so fast that any small attractive forces between them can be ignored
- at low pressures the volumes are so big that the space taken up by the molecules is insignificant.

The theory also helps to explain why real gases deviate from ideal gas behaviour as they get nearer to becoming a liquid. As a gas liquefies, the molecules get very close together and the volume of the molecules cannot be ignored. Also, gases could not liquefy unless there were some attractive (intermolecular) forces between the molecules to hold them together.

The Dutch physicist Johannes van der Waals (1837–1923) developed his theory of intermolecular forces (Section 9.1) by studying the behaviour of real gases and their deviations from the ideal gas behaviour.

The Maxwell–Boltzmann distribution

Two physicists used the kinetic theory to explore the distribution of energies among the molecules in gases. They worked out the proportion of molecules with a given energy at a particular temperature. The two physicists were James Maxwell (1831–1879) in Britain and Ludwig Boltzmann (1844–1906) in Austria.

Figure 13.13 shows the distribution of energies for the molecules of a gas under two sets of conditions. This Maxwell–Boltzmann distribution helps to explain the effects of temperature changes and catalysts on the rates of reactions.

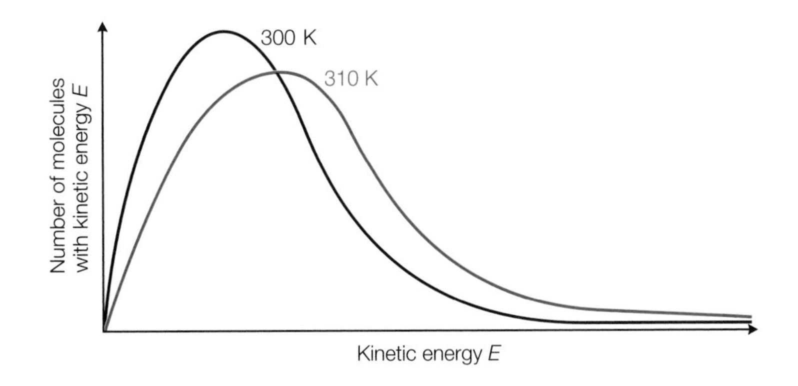

Figure 13.13 ◀ The Maxwell–Boltzmann distribution of kinetic energies of the molecules of a gas at two temperatures. The area under the curve gives the total number of molecules. This area does not change as the temperature rises, so the peak height falls as the temperature rises and the curve spreads to the right.

Explaining the effects of concentration, pressure and surface area on reaction rates

In any reaction mixture the atoms, molecules or ions are forever bumping into each other. When they collide there is a chance that they will react.

Raising the pressure of a gas means that the reacting particles are closer together. There are more collisions and therefore the reaction goes faster. Increasing the concentration of reactants in solution has a similar effect.

In a reaction of a solid with either a liquid or a gas, the reaction is faster if the solid is broken up into smaller pieces. Crushing the solid increases its surface area – collisions can be more frequent and the rate of reaction is faster.

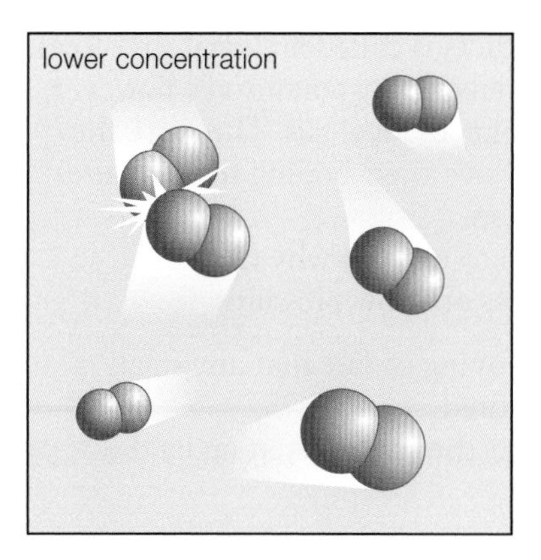

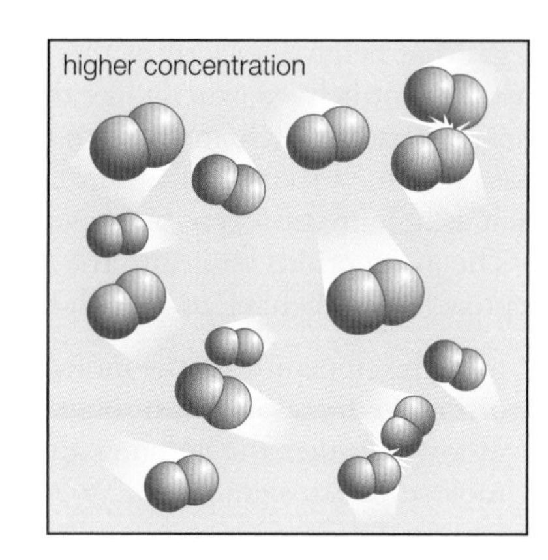

Figure 13.14 ▶
Raising the pressure, or concentration, means that the reacting atoms, molecules or ions are closer together. There are more collisions and the reaction is faster.

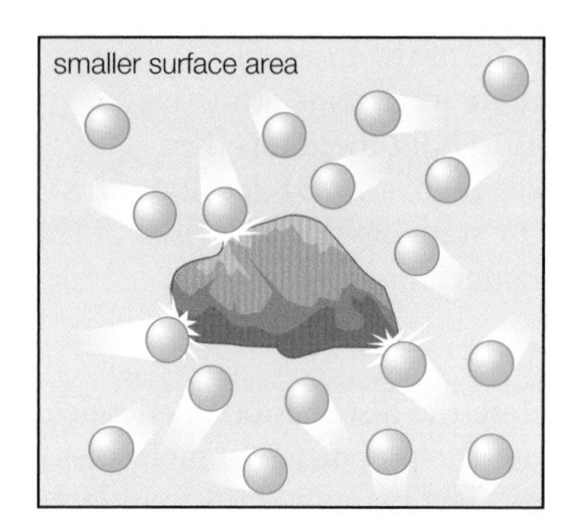

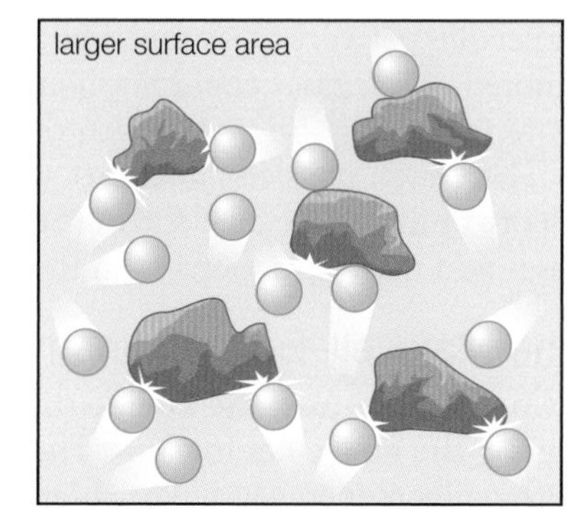

Figure 13.15 ▶
Breaking a solid into smaller pieces increases the surface area exposed to reacting chemicals in a gas or in solution.

Explaining the effects of temperature on reaction rates

It is not enough simply for the molecules to collide. Some collisions do not result in a reaction – in soft collisions the molecules simply bounce off each other. Molecules are in such rapid motion that if every collision led to a reaction, most reactions would be explosive. Only those molecules which collide with enough energy to stretch and break chemical bonds can lead to new products.

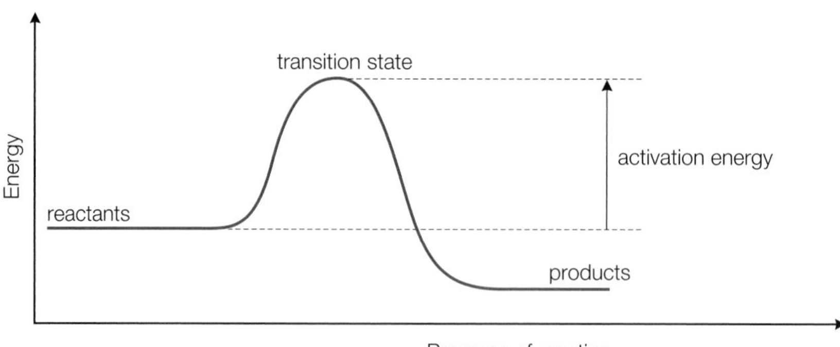

Figure 13.16 ▶
Reaction profile showing the activation energy for a reaction

Chemists use the term 'activation energy' to describe the minimum energy needed in a collision between molecules if they are to react. Activation energies account for the fact that reactions go much more slowly than would be expected if every collision between atoms and molecules led to a reaction. Only a very small proportion of collisions bring about chemical change. Molecules can only react if they collide with enough energy for bonds to stretch and then break so that new bonds can form. At around room temperature, only a small proportion of molecules have enough energy to react.

Definitions

A **reaction profile** is a graph which shows how the total enthalpy (energy) of the atoms, molecules or ions changes during the progress of a reaction from reactants to products.

The **activation energy** is the height of the energy barrier separating reactants and products during a chemical reaction. It is the minimum energy needed for a reaction between the amounts, in moles, shown in the equation for the reaction.

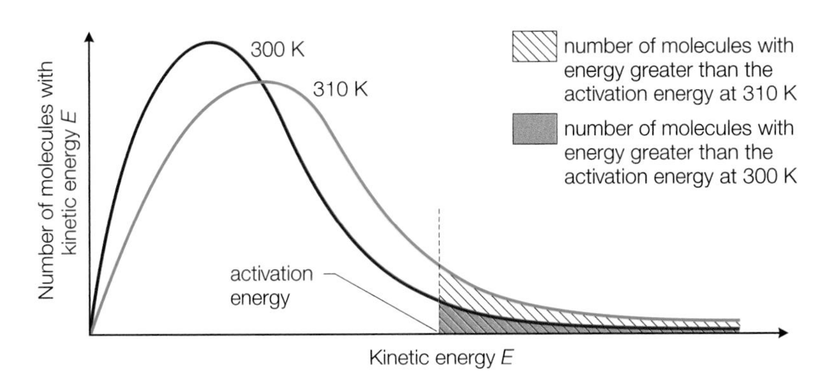

Figure 13.17 ◀
The Maxwell–Boltzmann distribution of kinetic energies in the molecules of a gas at 300 K and 310 K. The area under each curve is a measure of the number of molecules. At 310 K, more molecules have enough energy to react when they collide with other molecules.

The Maxwell–Boltzmann distribution in Figure 13.17 shows that the proportion of molecules with energies greater than the activation energy is small around 300 K.

The shaded areas in Figure 13.17 show the proportions of molecules having at least the activation energy for a reaction at 300 K and 310 K. This area is larger at the higher temperature. So at a higher temperature, more molecules have enough energy to react when they collide, and the reaction goes faster. Also, when the molecules are moving faster they collide more often.

Test yourself

6 Two factors explain why reactions go faster when the temperature rises. Identify these two factors in terms of the speed and energy of molecules, atoms and ions.

Explaining the effects of catalysts on reaction rates

A catalyst works by providing an alternative pathway for the reaction with a lower activation energy. Lowering the activation energy increases the proportion of molecules with enough energy to react.

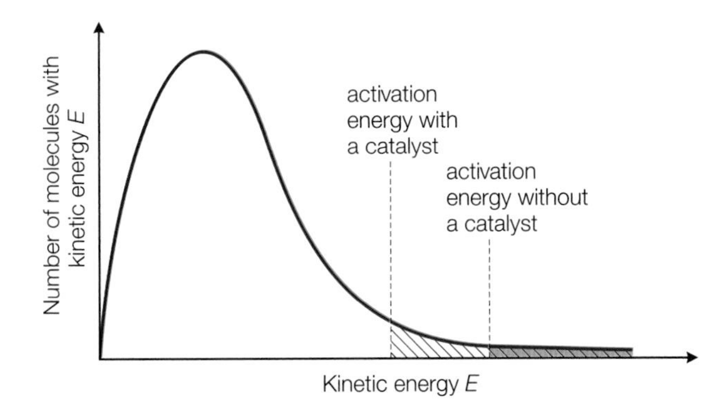

Figure 13.18 ◀
Distribution of molecular energies in a gas showing how the proportion of molecules able to react increases when a catalyst lowers the activation energy

A catalyst often changes the mechanism of a reaction and makes a reaction more productive by increasing the yield of the desired product and reducing waste.

One of the ways in which a catalyst can change the mechanism of a reaction is to combine with the reactants to form an intermediate. The intermediate is a stage in the transition from reactants to products – it breaks down to give the products and the catalyst is released. This frees the catalyst to interact with further reactant molecules and the reaction continues.

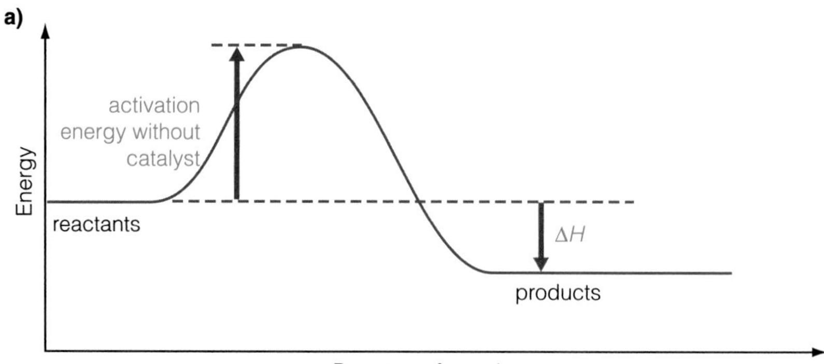

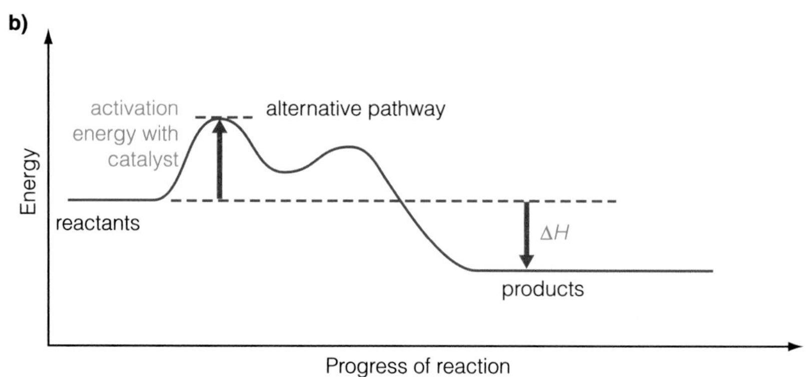

Figure 13.19 ▶ Reaction profiles for a reaction a) without a catalyst and b) with a catalyst. The dip in the curve of the pathway with a catalyst shows where an unstable intermediate forms.

Test yourself

7 Which part of Figure 13.19b shows the formation of an intermediate?

8 a) Why is a match or spark needed to light a Bunsen burner?
  b) Why does the gas keep burning once it has been lit?

9 Suggest a reason why catalysts are often specific for a particular reaction.

REVIEW QUESTIONS

1 Give one example to illustrate the effect of each of these factors on the rate of a chemical reaction:

a) concentration (1)

b) pressure (1)

c) temperature (1)

d) surface area (1)

e) catalyst. (1)

2 Hydrogen peroxide solution, H_2O_2(aq), decomposes slowly releasing oxygen. The reaction is catalysed by manganese(IV) oxide. The table shows the volume of oxygen collected at regular intervals when one measure of MnO_2 powder is added to 50 cm³ of a hydrogen peroxide solution at 20 °C.

Time/s	Volume of oxygen/cm³
0	0
20	10
40	20
60	26
80	32
100	35
120	38
140	39
160	40
180	40

a) Write an equation for the reaction. (1)

b) Draw a graph of the results on axes with a vertical scale showing the volume of oxygen up to 100 cm³. (2)

c) Explain the shape of your graph. (2)

d) On the same axes, sketch the graphs you would expect if, in separate experiments, all the conditions are the same except that:

i) the temperature is raised to 40 °C

ii) the volume of hydrogen peroxide solution is $100\,cm^3$

iii) manganese(IV) oxide granules are used in place of powder

iv) the concentration of the hydrogen peroxide solution is halved. (8)

3 a) Explain why most collisions in a reaction mixture do not result in a reaction. (2)

b) How can the collision frequency between molecules in a gas be increased without changing the temperature? (1)

c) Why can a small increase in temperature lead to a large increase in the rate of a reaction? (3)

4 Sketch the reaction profile for a reaction taking place with a catalyst given that:

- the reaction is endothermic
- the activation energy for the formation of an intermediate is higher than the activation energy for the conversion of the intermediate to the products. (3)

5 a) Sketch a graph, with labelled axes, to show the Maxwell–Boltzmann distribution of the energies of the reactant molecules. (3)

b) On the energy axis mark plausible values for the activation energy of a reaction without a catalyst and the activation energy with a catalyst. (2)

c) Use your graph to explain why a catalyst speeds up the rate of a reaction. (3)

14 Chemical equilibrium

The study of reversible reactions helps chemists to answer the questions 'How far?' and 'In which direction?' – questions they need to answer when trying to make new chemicals in laboratories and in industry.

Figure 14.1 ▲
Frying an egg is not reversible – once cooked, the egg cannot be uncooked.

14.1 Reversible changes

Some changes go in only one direction – like baking bread. Once baked in an oven, there is no way to reverse the process and split a loaf back into flour, water and yeast. Some chemical reactions are like this, but many other changes are reversible. Haemoglobin, for example, combines with oxygen as red blood cells flow through the lungs, but then releases the oxygen for respiration as blood flows in the capillaries throughout the rest of the body.

Burning a fuel, such as natural gas or petrol, is an example of a one-way process. Once these fuels have burned in air to make carbon dioxide and water, it is impossible to turn the products back to natural gas and petrol. The combustion of the fuels is an irreversible process.

Figure 14.2 ▲
Stalactites and stalagmites form in limestone caves because the reaction of calcium carbonate with carbon dioxide and water is reversible.

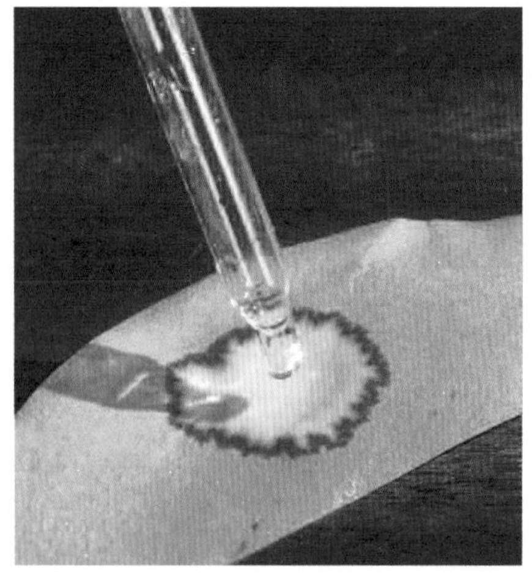

Figure 14.3 ▲
Using cobalt chloride paper to test for water

Many other chemical reactions are reversible. One example is the basis of a simple laboratory test for water. Hydrated cobalt(II) chloride is pink and so is a solution of the salt in water. Heating filter paper soaked in the solution in an oven makes the paper turn blue because water is driven off from the solution leaving anhydrous cobalt(II) chloride on the paper.

$$CoCl_2.6H_2O(s) \rightarrow CoCl_2(s) + 6H_2O(g)$$

pink blue

The blue paper provides a sensitive test for water as it turns pink again if exposed to water or water vapour. At room temperature water rehydrates the blue salt.

$$CoCl_2(s) + 6H_2O(l) \rightarrow CoCl_2.6H_2O(s)$$

blue pink

The reaction of ammonia with hydrogen chloride is another reaction in which the direction of change depends on the temperature. At room temperature, the two gases combine to make a white smoke of ammonium chloride.

$NH_3(g) + HCl(g) \rightarrow NH_4Cl(s)$

Heating reverses the reaction and ammonium chloride decomposes at high temperatures to give hydrogen chloride and ammonia.

$NH_4Cl(s) \rightarrow NH_3(g) + HCl(g)$

The apparatus in Figure 14.4 can be used to show that ammonium chloride decomposes into two gases on heating. Ammonia gas diffuses through the glass wool faster than hydrogen chloride. After a short time the alkaline ammonia rises above the plug of glass wool and turns the red litmus blue. A while later both strips of litmus paper turn red as the acid hydrogen chloride arrives. A smoke of ammonium chloride appears above the tube when both gases meet and cool.

Changing the temperature is not the only way to alter the direction of change. Hot iron, for example, reacts with steam to make iron(III) oxide and hydrogen. Supplying plenty of steam and 'sweeping away' the hydrogen means that the reaction continues until all the iron changes to its oxide.

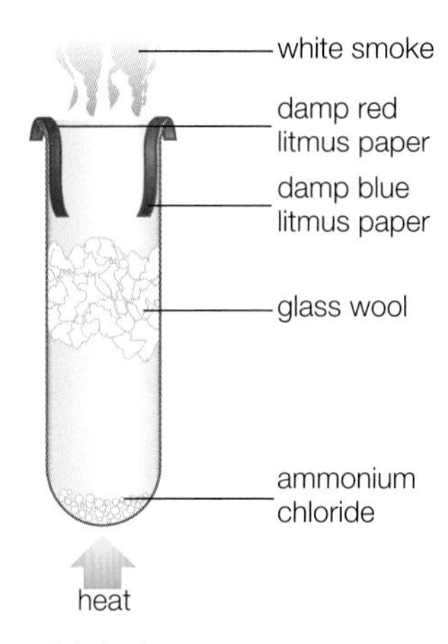

Figure 14.4 ▲ Investigating the thermal decomposition of ammonium chloride

$3Fe(s) + 4H_2O(g) \rightarrow Fe_3O_4(s) + 4H_2(g)$

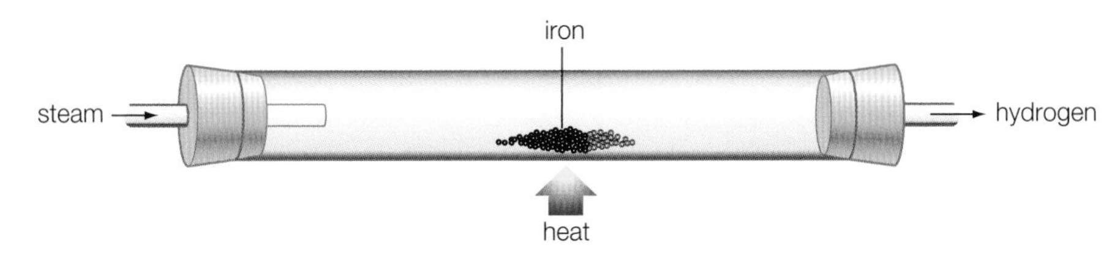

Figure 14.5 ◄ The forward reaction goes when the concentration of steam is high and the hydrogen is swept away, keeping its concentration low

Altering the conditions brings about the reverse reaction. A stream of hydrogen reduces all the iron(III) oxide to iron so long as the flow of hydrogen sweeps away the steam that has formed.

$3Fe(s) + 4H_2O(g) \leftarrow Fe_3O_4(s) + 4H_2(g)$

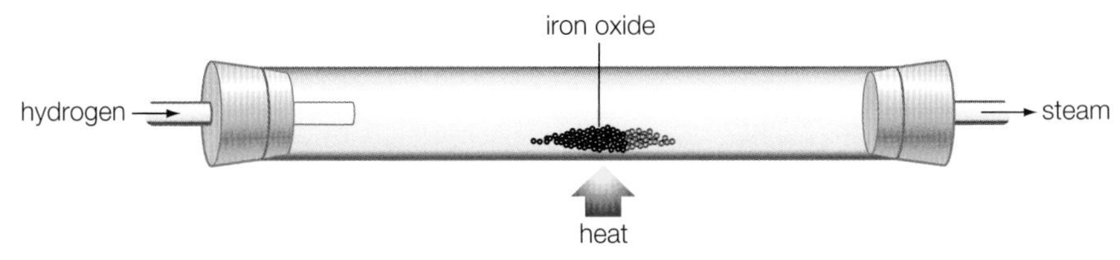

Figure 14.6 ▲ The backward reaction goes when the concentration of hydrogen is high and the steam is swept away, keeping its concentration low.

Test yourself

1 How can the following changes be reversed, either by changing the temperature or by changing the concentration of a reactant or product:
  a) freezing water to ice
  b) changing blue litmus to its red form
  c) converting blue copper(II) sulfate to its white form?

2 Explain why wet washing does not dry if kept in a plastic laundry bag, but does dry if hung out on a line.

Note

In an equation the chemicals on the left-hand side are the reactants – those on the right are the products. The 'left-to-right' reaction is the 'forward' reaction and the 'right-to-left' reaction is the backward reaction.

14.2 Reaching an equilibrium state

Reversible changes often reach a state of balance, or equilibrium. What is special about chemical equilibria is that nothing appears to be happening, but at a molecular level there is ceaseless change.

When chemists ask the question 'How far?' they want to know what the state of a reaction will be when it reaches equilibrium. At equilibrium, the reaction shown by an equation may be well to the right (mostly new products), well to the left (mostly unchanged reactants) or somewhere in between.

Balance points exist in most reversible reactions when neither the forward nor the reverse reaction is complete. Reactants and products are present together and the reactions appear to have stopped – this is the state of chemical equilibrium.

One way to study the approach to equilibrium is to watch what happens on shaking a small crystal of iodine in a test tube with hexane and a solution of potassium iodide, KI(aq). The liquid hexane and the aqueous solution do not mix.

Iodine freely dissolves in hexane, which is a non-polar solvent (Section 9.4). The non-polar iodine molecules mix with the hexane molecules – there is no reaction. The solution is a purple–violet colour, the same colour as iodine vapour. Iodine hardly dissolves in water but it does dissolve in a solution of potassium iodide. The solution is yellow, orange or brown depending on the concentration. In the solution, iodine molecules, I_2, react with iodide ions, I^-, to form triiodide ions, I_3^-.

Figure 14.7 is a study of changes which can be summed up by this equilibrium:

$$I_2(\text{in hexane}) + I^-(aq) \rightleftharpoons I_3^-(aq)$$

Note

The symbol $\rightleftharpoons$ represents a reversible reaction at equilibrium. In theory it is only possible to achieve a state of equilibrium in a closed system (Section 2.1).

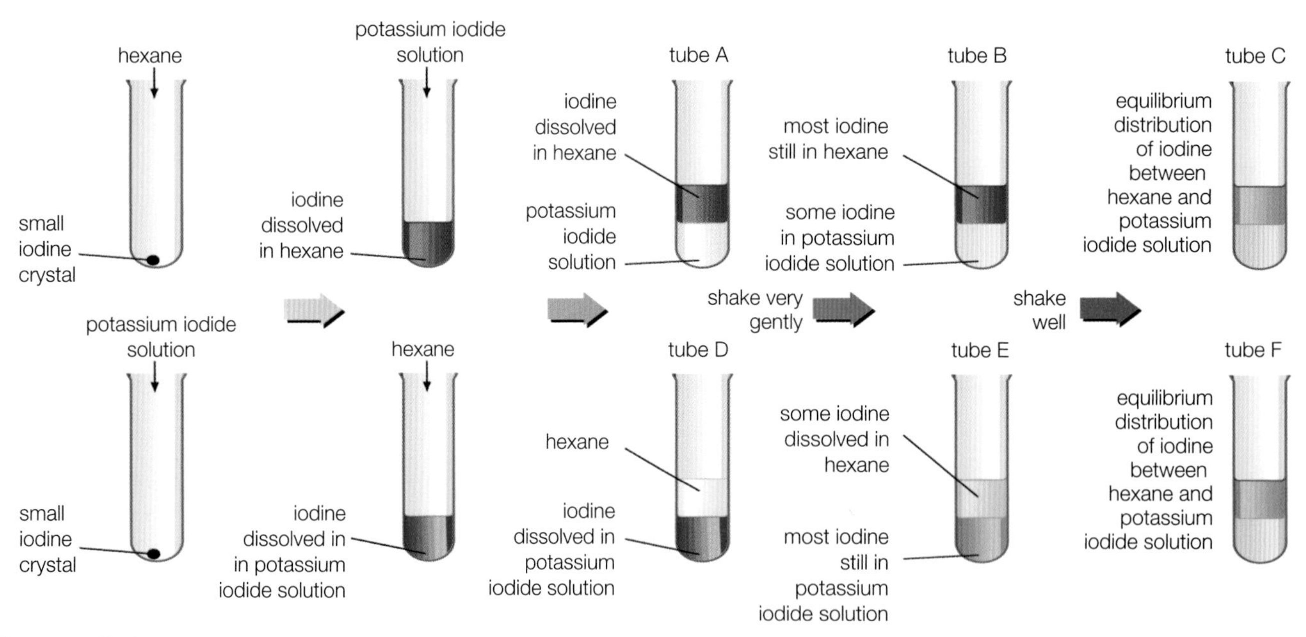

Figure 14.7 ▲
Two approaches to the same equilibrium state

The graphs in Figure 14.8 show how the iodine concentration in the two layers changes with shaking. After a little while, no further change seems to take place – tubes C and F look just the same. Both contain the same equilibrium system.

This demonstration shows two important features of equilibrium processes:

- at equilibrium the concentration of reactants and products does not change
- the same equilibrium state can be reached from either the 'reactant side' or the 'product side' of the equation.

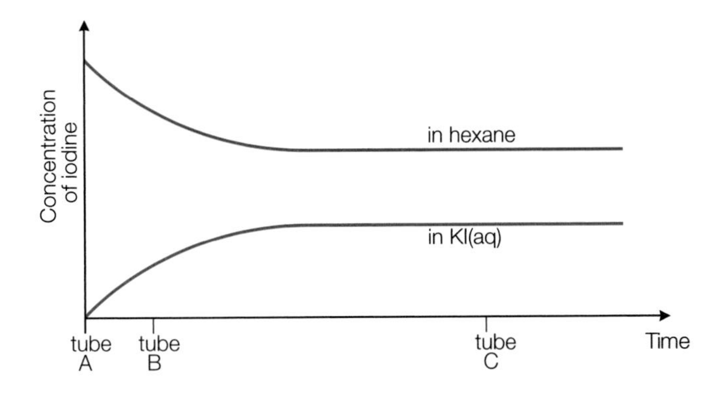

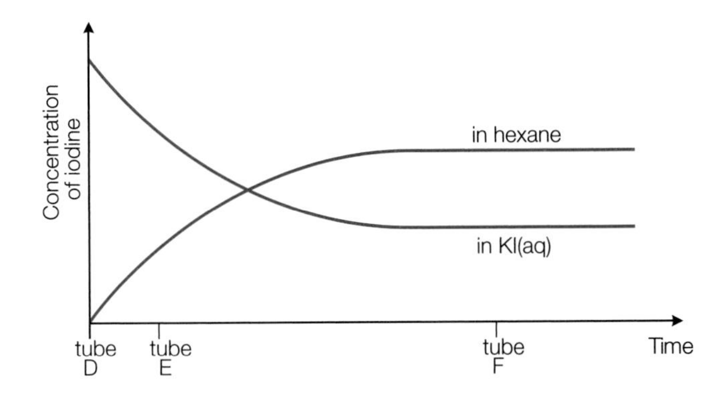

Figure 14.8 ▲
Change in concentration of iodine with time in the mixtures shown in Figure 14.7

14.3 Dynamic equilibrium

Figure 14.9 shows what is happening at a molecular level – not what your eye can see – in the equilibrium involving iodine moving between hexane and a solution of potassium iodide. Consider tube A in Figure 14.7. All the iodine molecules start in the upper hexane layer. On shaking, some move into the aqueous layer. At first, molecules can move in only one direction (the forward reaction). The forward reaction begins to slow down as the concentration in the upper layer falls.

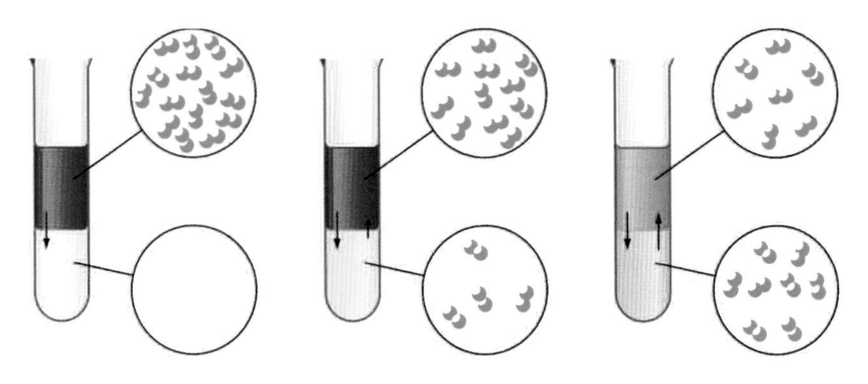

Figure 14.9 ◀
Iodine molecules reaching dynamic equilibrium between hexane and a solution of potassium iodide. The formation of I_3^- ions in the aqueous layer is ignored in this diagram.

Once there is some iodine in the aqueous layer, the reverse process can begin with iodine returning to the hexane layer. This backward reaction starts slowly but speeds up as the concentration of iodine in the aqueous layer increases.

In time, both the forward and backward reactions will happen at the same rate. Movement of iodine between the two layers continues but overall there is no change. In tube C in Figure 14.7 each layer is gaining and losing iodine molecules at the same rate. This is an example of dynamic equilibrium.

Definition

In **dynamic equilibrium**, the forward and backward reactions continue, but at equal rates, so that the overall effect is no change. At the molecular level, there is continuous movement. At the macroscopic level, nothing appears to be happening.

Test yourself

3 Under what conditions are these in equilibrium:
   a) water and ice
   b) water and steam
   c) copper(II) sulfate crystals and copper(II) sulfate solution?

4 Draw a diagram to represent the movement of particles between a crystal and a saturated solution of the solid in a solvent.

14.4 Factors affecting equilibria

Changing the conditions can disturb a system at equilibrium. At equilibrium the rate of the forward and backward reactions is the same. Anything which changes the rates can shift the balance.

Predicting the direction of change

Le Chatelier's principle is a qualitative guide to the effect of changes in concentration, pressure or temperature on a system at equilibrium. The principle was suggested as a general rule by the French physical chemist Henri le Chatelier (1850–1936).

The principle states that when the conditions of a system at equilibrium change, the system responds by trying to counteract the change.

Changing the concentration

Table 14.1 shows the effects of changing the concentration in the generalised equilibrium system

$$A + B \rightleftharpoons C + D$$

Disturbance	How does the equilibrium mixture respond?	The result
Concentration of A increases	System moves to the right – some A is removed by reaction with B	More C and D form
Concentration of D increases	System moves to the left – some of the added D is removed by reaction with C	More A and B form
Concentration of D decreases	System moves to the right to make up for the loss of D	There is more C and less A and B in the new equilibrium

Table 14.1 ▲

The reaction of bromine with water provides examples of predictions based on le Chatelier's principle. A solution of bromine in water is a yellow–orange colour because it contains bromine molecules in the equilibrium:

$$\underbrace{Br_2(aq)}_{\text{orange}} + H_2O(l) \rightleftharpoons \underbrace{OBr^-(aq) + Br^-(aq) + 2H^+(aq)}_{\text{colourless}}$$

Adding alkali turns the solution almost colourless. Hydroxide ions in the alkali react with hydrogen ions, removing them from the equilibrium. As the hydrogen ion concentration falls, the equilibrium shifts to the right, converting orange bromine molecules to colourless ions. Lowering the hydrogen ion concentration slows down the backward reaction, while the forward reaction goes on as before. The position of equilibrium shifts until once again the rates of the forward and backward reactions are the same.

Adding acid increases the concentration of hydrogen ions – this speeds up the backward reaction and makes the solution turn orange–yellow again. The equilibrium shifts to the left reducing the hydrogen ion concentration and increasing the bromine concentration until, once again, the forward and backward reactions are in balance.

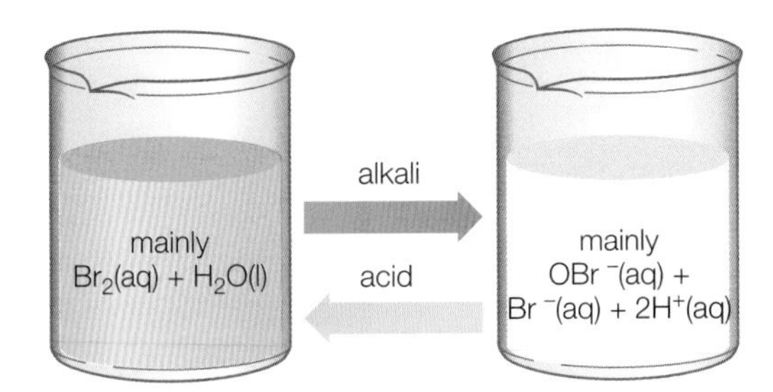

Figure 14.10 ▲
The visible effects of adding alkali and acid to a solution of bromine in water

Test yourself

5 Write an ionic equation for the reversible reaction of silver(I) ions with iron(II) ions to form silver atoms and iron(III) ions. Make a table similar to Table 14.1 to show how le Chatelier's principle applies to this equilibrium.

6 Yellow chromate(VI) ions, $CrO_4^{2-}(aq)$, react with aqueous hydrogen ions, $H^+(aq)$, to form orange dichromate(VI) ions, $Cr_2O_7^{2-}(aq)$, and water molecules. The reaction is reversible. Write an equation for the system at equilibrium. Predict how the colour of a solution of chromate(VI) ions changes:
   a) on adding acid
   b) followed by adding hydroxide ions (OH^-), which neutralise hydrogen ions (Section 1.4).

7 Heating limestone, $CaCO_3$, in a closed furnace produces an equilibrium mixture of calcium carbonate with calcium oxide, CaO, and carbon dioxide gas. Heating the solid in an open furnace decomposes the solid completely into the oxide. How do you account for this difference?

Changing the pressure and temperature

Many industrial processes happen in the gas phase. High pressures and high temperatures are often needed even when there is a catalyst. One important example of this is the Haber process used to make ammonia. The equilibrium system involved is:

$$N_2(g) + 3H_2(g) \rightleftharpoons 2NH_3(g) \quad \Delta H = -92.4\,\text{kJ mol}^{-1}$$

The reaction takes place in a reactor packed with an iron catalyst. Adding a catalyst does not affect the position of equilibrium. A catalyst simply speeds up both the forward and the back reactions, so it shortens the time taken to reach equilibrium.

Le Chatelier's principle helps to explain the conditions chosen for the Haber process. There are 4 mol of gases on the left-hand side of the equation but only 2 mol on the right. Increasing the pressure makes the equilibrium shift from left to right, because this reduces the number of molecules and tends to reduce the pressure. So, increasing the pressure increases the proportion of ammonia at equilibrium.

The reaction is exothermic from left to right, and therefore endothermic from right to left. Le Chatelier's principle predicts that raising the temperature makes the system shift in the direction which takes in energy (tending to lower the temperature). So, raising the temperature lowers the proportion of ammonia at equilibrium.

Tutorial

Activity

Ice that burns

An unexpected catch

From time to time, people who go deep-sea fishing get a surprise when they haul in their catch from deep oceans. In among the fish, they find chunks of white stuff that looks like snowballs. This strange material turns out to be a combination of ice and methane. As the icy material melts, it releases the trapped methane.

Figure 14.11 ▲
Methane hydrate. The methane can catch fire as the hydrate melts and releases the gas.

Cage compounds

Ice has a very open structure – so much so that it can trap molecules within the spaces between the water molecules. One example of this is the methane hydrate mentioned above.

Methane hydrate is unstable at room temperature and pressure. For a time, scientists thought that it could only exist in the outer regions of the Solar System, where temperatures are very low and ice is common. However, oceanographers have discovered that there are huge quantities of methane hydrate in deep sediments below the sea-bed where the water is cool and the pressure is high. Methane hydrate is also found under the ground in regions of permafrost, where there are layers of soil below the surface which are permanently frozen – as they have been for thousands of years.

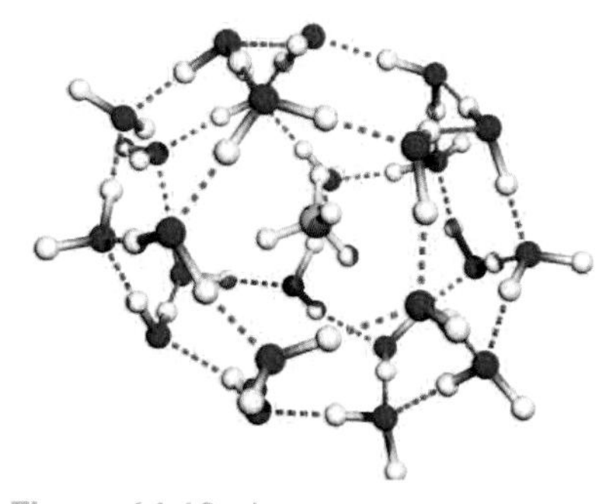

Figure 14.12 ▲
Computer model of the structure of methane hydrate. The methane molecule is trapped in a cage of water molecules. This whole structure is an example of a clathrate. Note that in this computer model, the colour code for atoms differs from that generally used in molecular models.

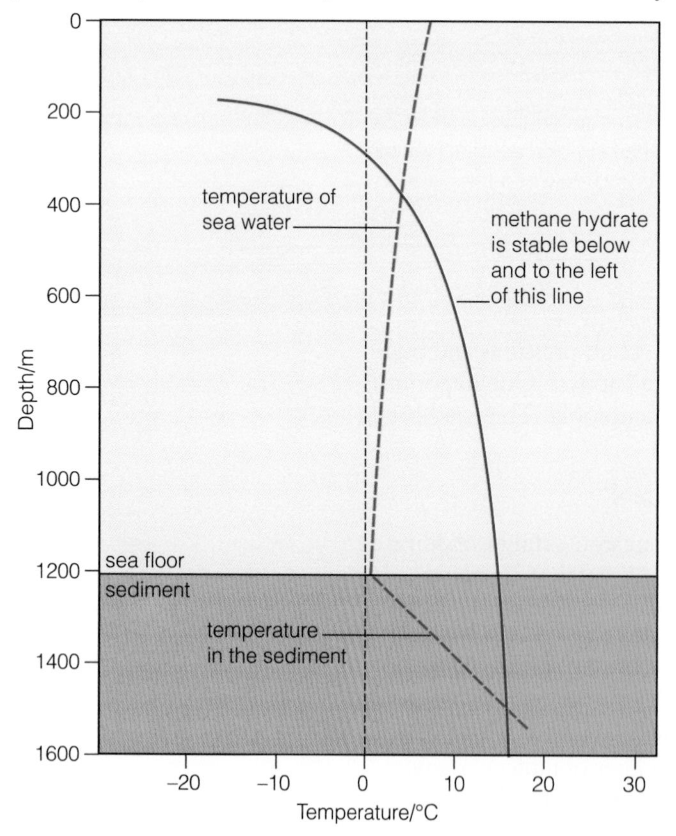

Figure 14.13 ▲
Conditions for the formation of methane hydrate in ocean sediments. The blue line shows how temperature varies with depth. The red line shows the temperatures and pressures at which methane hydrate is in equilibrium with water and free methane. Below and to the left of this line, methane hydrate is stable.

Although its composition varies, methane hydrate typically contains 1 mole of methane for every 5.75 moles of water. Its density is about $0.9\,g\,cm^{-3}$.

A vast resource and a threat

On the margins of the continents, below the sea floor, methane hydrate exists in vast quantities. Some estimates suggest that the amount of methane is between 2 and 10 times the known reserves of natural gas.

There are also large quantities of methane hydrate in regions where the soil remains frozen throughout the year. This permafrost is found in alpine, Arctic and Antarctic regions where the summer sun is not strong enough to melt the soil below the surface.

Any large-scale release of the trapped methane could have catastrophic consequences because methane is a more efficient greenhouse gas than carbon dioxide.

1 Explain the name ‘ice that burns’ for methane hydrate.

2 The reversible decomposition of methane hydrate can be represented by the word equation:

$$\text{methane hydrate(s)} \rightleftharpoons \text{methane(g)} + \text{water(l)}$$

a) Predict whether the decomposition of methane hydrate is exothermic or endothermic. Give reasons for your answer.
b) Explain why methane hydrate is more stable at high pressures.
c) Use le Chatelier’s principle and your answers to parts **a)** and **b)** to account for the general shape of the red line in Figure 14.13.

3 Estimate, from Figure 14.13, the depths of ocean sediment in which methane hydrate might be found. Explain your answer.

4 What might be the source of methane for the formation of methane hydrate in ocean sediments?

5 Estimate the volume of methane, measured at room temperature and pressure, that is released when $1\,dm^3$ of methane hydrate decomposes.

6 The density of liquid methane is $0.46\,g\,cm^{-3}$. The boiling temperature of methane is $-162\,°C$. Suggest reasons why some energy suppliers have considered converting methane to its hydrate, instead of liquefying it for transport by seagoing vessels.

7 Suggest reasons why it may be difficult to obtain fuel economically from methane hydrate.

8 Explain why the effect of global warming on permafrost regions could lead to a catastrophic effect on the Earth’s climate.

REVIEW QUESTIONS

1 Carbon dioxide is dissolved in water under pressure to make sparkling mineral water. In a bottle of mineral water, there is an equilibrium between carbon dioxide dissolved in the drink, $CO_2(aq)$, and carbon dioxide in the gas above the drink, $CO_2(g)$.

a) Write an equation to represent the equilibrium between carbon dioxide gas and carbon dioxide in solution. (1)

b) Use this example to explain the term 'dynamic equilibrium'. (2)

c) Explain why lots of bubbles of gas form when a bottle of sparkling mineral water is opened. (2)

d) Less than 1% of the dissolved carbon dioxide reacts with water. It forms hydrogencarbonate ions:

$$CO_2(g) + H_2O(l) \rightleftharpoons HCO_3^-(aq) + H^+(aq)$$

Use this equation to explain why carbon dioxide is much more soluble in sodium hydroxide solution than in water. (2)

2 Haemoglobin is a large molecule in red blood cells that can be represented by the symbol Hb. Haemoglobin carries oxygen in the blood from our lungs to the cells in our bodies:

$$Hb(aq) + 4O_2(g) \rightleftharpoons HbO_8(aq)$$

a) Suggest a reason why haemoglobin takes up oxygen as blood passes through the blood vessels in the lungs. (2)

b) Suggest a reason why haemoglobin releases oxygen as blood passes through the blood vessels between cells in muscles. (2)

c) Haemoglobin molecules are affected by the presence of carbon dioxide. The molecules hold onto oxygen less strongly if the carbon dioxide concentration is higher. Why does this help the blood deliver oxygen to cells in muscles? (2)

3 Methanol is manufactured from carbon monoxide and hydrogen by the reaction

$$CO(g) + 2H_2(g) \rightleftharpoons CH_3OH(g) \qquad \Delta H = -91\,\text{kJ mol}^{-1}$$

The reaction is carried out in the presence of aluminium oxide pellets coated with a mixture of copper and zinc oxides.

The process runs at a pressure that is about 100 times atmospheric pressure and at a temperature of 575 K

a) Explain why the process operates at a high pressure. (2)

b) What is the purpose of the pellets coated with metal oxides? (1)

c) i) What is the effect of raising the temperature on the yield of methanol at equilibrium? Explain your answer. (2)

ii) Suggest why a temperature as high as 575 K is used. (2)

15 Alcohols and halogenoalkanes

This topic covers two families of organic compounds with carbon atoms bonded to electronegative atoms – oxygen atoms in alcohols and halogen atoms in the halogenoalkanes. The presence of these electronegative atoms means that there are polar bonds in the molecules. This makes them more reactive than alkanes, especially with ionic reagents.

15.1 Alcohol names and structures

Ethanol is the best known member of the family of alcohols – it is the alcohol in beer, wine and spirits. Alcohols are useful solvents in the home, in laboratories and in industry. Understanding the properties of the –OH functional group in alcohols helps to make sense of the reactions of some important biological chemicals, especially carbohydrates such as sugars and starch.

Figure 15.1 ▲
Checking pipes in a distillery in Uttar Pradesh, India. The plant produces ethanol by fermenting sugar from sugar cane. There is worldwide rapid growth in the use of ethanol as a fuel.

Alcohols are compounds with the formula R–OH, where R represents an alkyl group. The hydroxy group, –OH, is the functional group which gives alcohols their characteristic reactions.

The IUPAC rules name alcohols by changing the ending of the corresponding alkane to '-ol' – so ethane becomes ethanol.

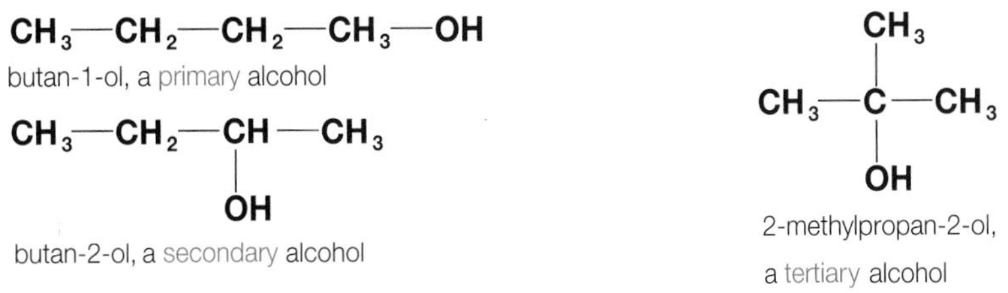

$CH_3—CH_2—CH_2—CH_2—OH$
butan-1-ol, a primary alcohol

$CH_3—CH_2—CH(OH)—CH_3$
butan-2-ol, a secondary alcohol

$(CH_3)_3C—OH$
2-methylpropan-2-ol, a tertiary alcohol

Figure 15.2 ▲
The names and structures of alcohols

Definitions

The terms **primary**, **secondary** and **tertiary** are used with organic compounds to show the positions of functional groups. A primary alcohol has the –OH group at the end of the chain. A secondary alcohol has the –OH group somewhere along the chain, but not at a branch in the chain or at the ends. A tertiary alcohol has the –OH group at a branch in the chain.

Data

Test yourself

1 Draw the structures of these alcohols, and state whether they are primary, secondary or tertiary compounds:
a) ethanol
b) propan-1-ol
c) propan-2-ol
d) 2-methylbutan-2-ol.

2 Draw the skeletal formula and name one isomer with the formula $C_5H_{11}OH$ that is:
a) a primary alcohol
b) a secondary alcohol
c) a tertiary alcohol.

15.2 Physical properties of alcohols

The simplest alcohols, methanol and ethanol, are liquids at room temperature because of hydrogen bonding between –OH groups (Section 9.3). For the same reason, alcohols are much less volatile than hydrocarbons that have roughly the same molar mass. Alcohols with relatively short hydrocarbon chains also mix freely with water.

Test yourself

Data

3 Use the data from the data sheets on the Dynamic Learning Student website to show that alcohols are less volatile than alkanes with similar molar masses.

4 a) Draw a diagram to show the hydrogen bonding between a methanol molecule and a water molecule.

b) Explain why hydrogen bonding accounts for the fact that methanol, at room temperature, is a liquid that mixes freely with water, while ethane is a gas.

15.3 Chemical properties of alcohols

Alcohols are much more reactive than alkanes because the C–O and O–H bonds in the molecules are polar (Section 5.2).

Combustion

Alcohols burn in air with a clean, colourless flame. Methanol and ethanol are both common fuels (Section 19.4) and fuel additives.

$$2CH_3OH(l) + 3O_2(g) \rightarrow 2CO_2(g) + 4H_2O(l)$$

Reaction with sodium

When a small piece of sodium is added to ethanol, it produces a steady stream of hydrogen gas. The metal does not melt. The reaction is similar to the reaction of sodium with water but is much less violent.

Alcohols and water react in chemically similar ways with sodium because both contain the –OH group. With water, the products are sodium hydroxide and hydrogen; with ethanol, the products are sodium ethoxide and hydrogen.

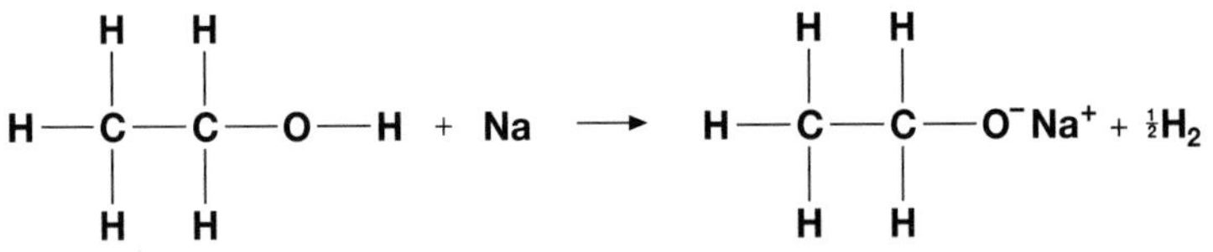

$$CH_3CH_2\text{–O–H} + Na \longrightarrow CH_3CH_2\text{–}O^-Na^+ + \tfrac{1}{2}H_2$$

sodium ethoxide

Figure 15.3 ▶ Reaction of ethanol with sodium. Note the ionic bond in sodium ethoxide.

Test yourself

5 a) Describe what happens on adding a small piece of sodium to water.

b) Show the reaction of sodium with water in a similar style to Figure 15.3.

6 Write equations to summarise the reactions of propan-1-ol with:

a) oxygen when it burns

b) sodium.

Substitution of a halogen atom for an –OH group

Alcohols react with hydrogen halides. In the reaction the –OH group is replaced by a halogen atom. This is an example of a substitution reaction (Section 16.1).

The reaction is more rapid with tertiary alcohols, and slower with primary alcohols. A tertiary alcohol, such as 2-methylpropan-2-ol, reacts with concentrated hydrochloric acid at room temperature.

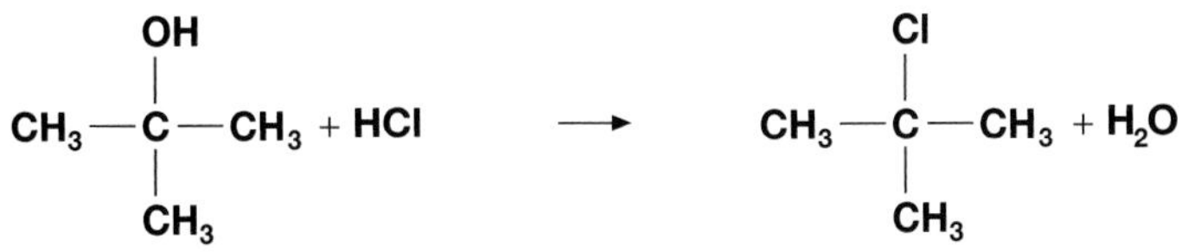

$$CH_3-\underset{CH_3}{\overset{OH}{C}}-CH_3 + HCl \longrightarrow CH_3-\underset{CH_3}{\overset{Cl}{C}}-CH_3 + H_2O$$

Figure 15.4 ◄
Substitution reaction to replace an –OH group by a chlorine atom

Using more vigorous conditions, it is possible to convert a primary alcohol to a halogenoalkane in a similar way. For example, butan-1-ol reacts with hydrogen bromide on heating in the presence of sulfuric acid to form 1-bromobutane. It is often convenient to make the hydrogen bromide in the reaction flask by mixing 50% sulfuric acid with sodium bromide.

All alcohols react rapidly at room temperature with phosphorus pentachloride to give chloroalkanes. The reaction also produces the gas hydrogen chloride. This makes it a useful test for the presence of –OH groups in molecules.

$$C_3H_7OH(l) + PCl_5(s) \rightarrow C_3H_7Cl(l) + POCl_3(l) + HCl(g)$$

A similar method converts alcohols to iodoalkanes. Phosphorus iodide is unstable so it is made in the reaction flask by using a mixture of red phosphorus and iodine, which combine to make phosphorus triiodide.

The reactions described above are used to prepare halogenoalkanes (Section 15.7).

Test yourself

7 Name the product of the reaction shown in Figure 15.4.

8 Write an equation for the formation of 1-bromobutane from butan-1-ol and hydrogen bromide.

9 What tests would you use to detect the hydrogen chloride formed when PCl_5 reacts with an alcohol?

10 Write an equation for the reaction of PI_3 with propan-1-ol, given that the inorganic product is the acid H_3PO_3.

11 Why is it not possible to convert an alcohol to an iodoalkane using a mixture of potassium iodide and concentrated sulfuric acid?

12 Does the C–O or the O–H bond break when an alcohol reacts with:
a) sodium
b) PCl_5?

Oxidation by potassium dichromate(VI)

An acidified solution of potassium dichromate(VI) is orange. It turns green on warming with an alcohol such as ethanol. During the reaction, potassium dichromate(VI) oxidises primary and secondary alcohols, but there is no reaction with tertiary alcohols.

Oxidation of a primary alcohol

This takes place in two steps to produce an aldehyde and then a carboxylic acid. During the reaction the potassium dichromate(VI) turns from the orange colour of $Cr_2O_7^{2-}$(aq) to the green colour of Cr^{3+}(aq).

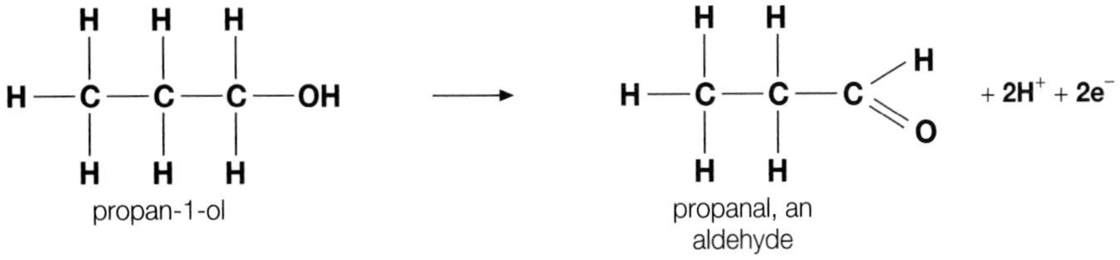

$$H-\underset{H}{\overset{H}{C}}-\underset{H}{\overset{H}{C}}-\underset{H}{\overset{H}{C}}-OH \longrightarrow H-\underset{H}{\overset{H}{C}}-\underset{H}{\overset{H}{C}}-C\begin{matrix}H\\O\end{matrix} + 2H^+ + 2e^-$$

propan-1-ol → propanal, an aldehyde

Figure 15.5 ◄
Oxidation of propan-1-ol to propanal by acidified $Cr_2O_7^{2-}$(aq). The oxidising agent takes away the electrons (Section 10.1).

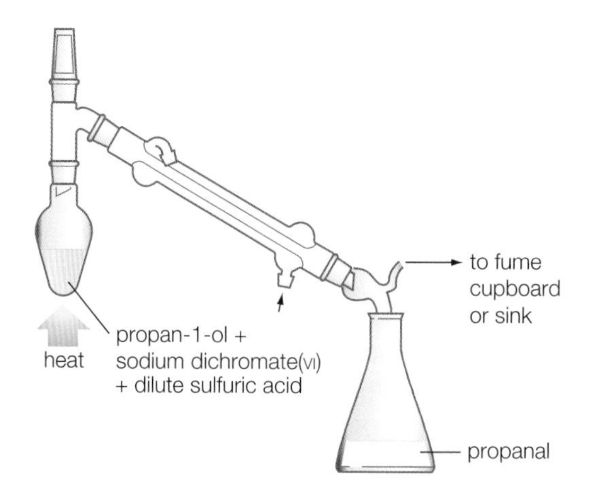

Figure 15.6 ▶
Apparatus used to oxidise a primary alcohol to an aldehyde. The aldehyde distils off as it forms. This prevents further oxidation of the aldehyde.

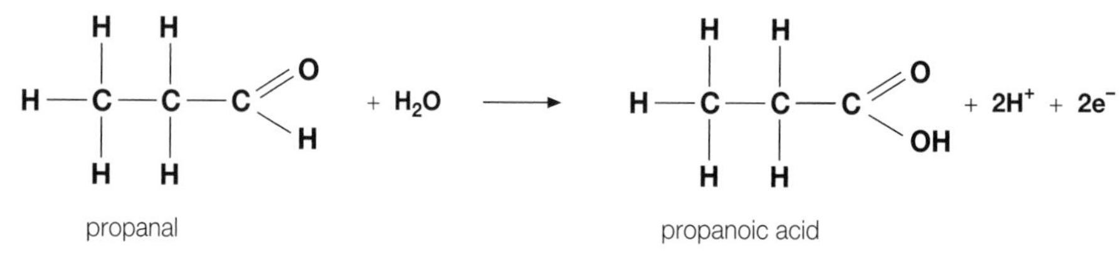

Figure 15.7 ▲
Oxidation of propanal to propanoic acid. Oxidation is completed when a primary alcohol is heated with acidic potassium dichromate(VI) in the apparatus shown in Figure 15.8. This converts the alcohol first to an aldehyde, then to a carboxylic acid.

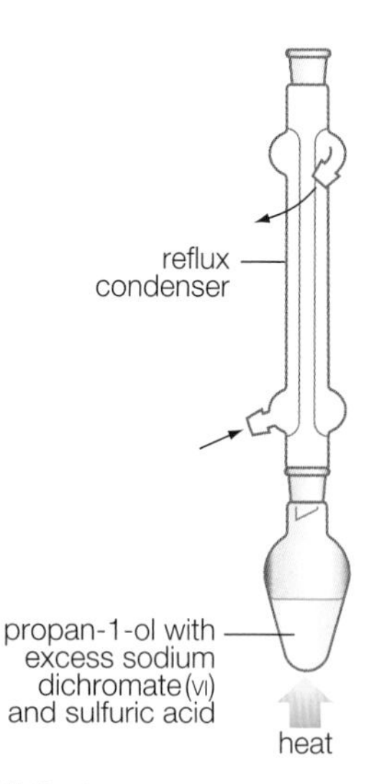

Figure 15.8 ▲
Apparatus used to oxidise a primary alcohol to a carboxylic acid. The reflux condenser ensures that any volatile aldehyde condenses and flows back into the flask, where excess oxidising agent ensures complete conversion.

Oxidation of a secondary alcohol

This produces a ketone.

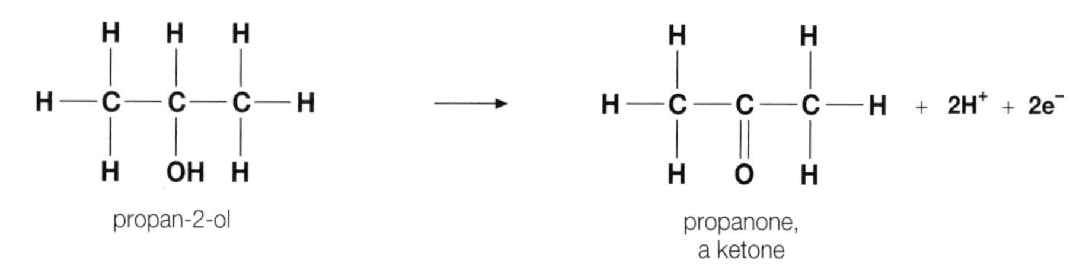

Figure 15.9 ▲
Oxidation of propan-2-ol produces propanone, a ketone

Definition

A **reflux condenser** is fitted to a flask to prevent vapour escaping while a liquid is being heated. Vapour from the boiling reaction mixture condenses and flows back into the flask.

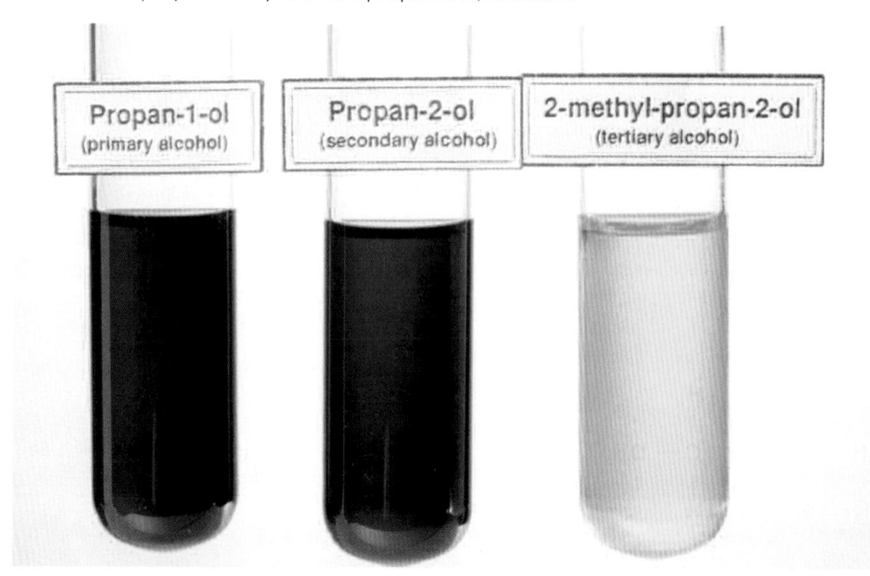

Figure 15.10 ▶
The result of warming three alcohols with an acidic solution of potassium dichromate(VI). Dichromate(VI) ions are reduced to green chromium(III) ions if there is a reaction. Tertiary alcohols do not react with potassium dichromate(VI).

These oxidation reactions help to distinguish between primary, secondary and tertiary alcohols.

Infrared spectroscopy can be used to detect the functional groups in organic molecules. It is an analytical tool that can be used to show the change in functional groups when alcohols are oxidised (Section 17.2).

Test yourself

13 Use oxidation numbers to show that it is the chromium that is reduced when $Cr_2O_7^{2-}$(aq) ions turn into Cr^{3+}(aq) ions.

14 Predict the products, if any, of oxidising the following alcohols with acidified dichromate(VI):
a) butan-1-ol when the product is distilled off immediately
b) butan-1-ol when the reagents heated under reflux for some time
b) butan-2-ol
c) 2-methylbutan-2-ol.

15.4 Halogenoalkanes

Halogenoalkanes are important to organic chemists both in research and in industry. The reason for this is that many halogenoalkanes are reactive compounds which can be converted into other more valuable products. This makes them useful as intermediates for converting one chemical to another.

One particular class of unreactive halogenoalkanes has become notorious in recent years because of the damaging effect they have on the ozone layer. These are called chlorofluorocarbons (CFCs) and were developed in the early twentieth century, supposedly as safe alternatives to toxic refrigerants such as ammonia and sulfur dioxide. As CFCs are also powerful greenhouse gases, their use is doubly damaging.

In the structure of a halogenoalkane, one or more of the hydrogen atoms in an alkane molecule is replaced with a halogen atom.

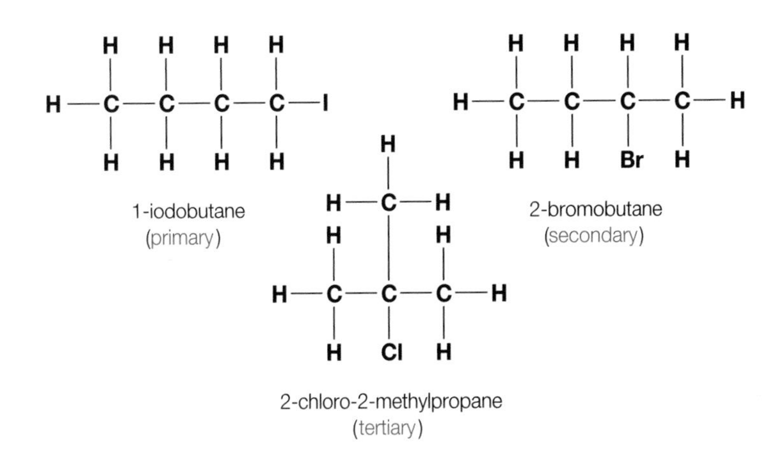

Figure 15.11 ◀ Names and structures of three halogenoalkanes

Test yourself

15 Draw the structures of the following compounds, and identify them as primary, secondary or tertiary compounds:
a) 1-iodopropane
b) 2-chloro-2-methylbutane
c) 3-bromopentane.

15.5 Physical properties of halogenoalkanes

Chloromethane, bromomethane and chloroethane are gases at room temperature. Most other halogenoalkanes are colourless liquids which do not mix with water.

Figure 15.12 ▶ Bromomethane is still used in the USA to sterilise soils and protect strawberry crops from pests.

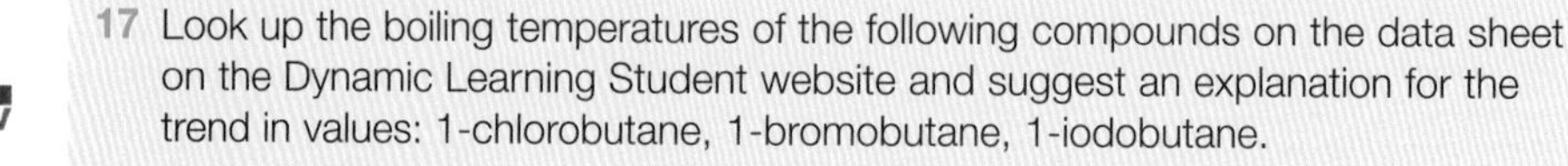

16 Which of the following molecules are polar and which are non-polar: $CHCl_3$, CH_2Cl_2, $CHCl_3$ and CCl_4?

17 Look up the boiling temperatures of the following compounds on the data sheet on the Dynamic Learning Student website and suggest an explanation for the trend in values: 1-chlorobutane, 1-bromobutane, 1-iodobutane.

Data

18 Compare the boiling temperatures of the isomers 1-bromobutane, 2-bromobutane and 2-bromo-2-methylpropane. Suggest an explanation for the differences in boiling temperatures of the primary, secondary and tertiary compounds.

15.6 Chemical reactions of halogenoalkanes

The two important reactions of halogenoalkanes are substitution and elimination (Section 16.1).

Definition

Hydrolysis is a reaction in which a compound splits apart in a reaction involving water.

Substitution by reaction with water

Cold water slowly hydrolyses halogenoalkanes, replacing the halogen atoms with an –OH group to form alcohols.

$$CH_3CH_2CH_2I(l) + H_2O(l) \rightarrow CH_3CH_2CH_2OH(l) + H^+(aq) + I^-(aq)$$

The rate of hydrolysis of different halogenoalkanes can be compared by carrying out the reaction in the presence of silver ions. The halogen atoms in halogenoalkanes are covalently bonded to carbon and give no precipitate of a silver halide. Hydrolysis releases halide ions, which immediately precipitate as the silver halide (Section 11.8).

Halogenoalkanes do not mix with water or aqueous solutions. For this reason the reaction is carried out in the presence of ethanol, which can dissolve the halogenoalkane and mix with the aqueous silver nitrate.

Test yourself

19 Explain the use of the terms 'substitution' and 'hydrolysis' to describe the reaction of halogenoalkanes with water.

20 Write equations for the two reactions which take place when 2-bromobutane reacts with water in the presence of silver ions, Ag^+.

Substitution by reaction with hydroxide ions

Replacement of the halogen atom of a halogenoalkane by an –OH group is much quicker with an aqueous solution of an alkali, such as sodium or potassium hydroxide. Heating increases the rate of reaction even further.

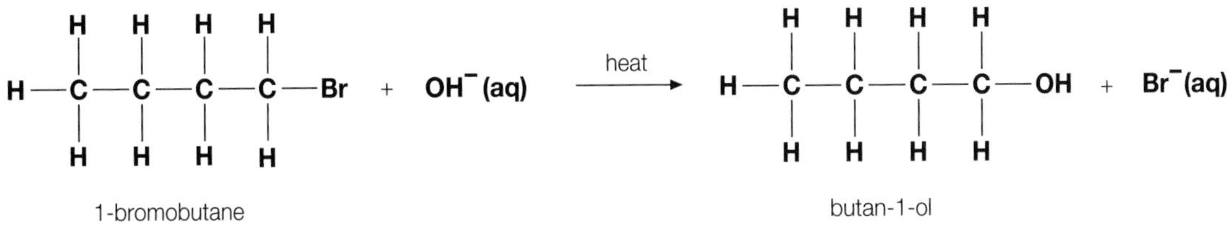

$$CH_3CH_2CH_2CH_2Br + OH^-(aq) \xrightarrow{\text{heat}} CH_3CH_2CH_2CH_2OH + Br^-(aq)$$

1-bromobutane → butan-1-ol

Figure 15.13 ▲
Reaction of a halogenoalkane with an alkali on heating

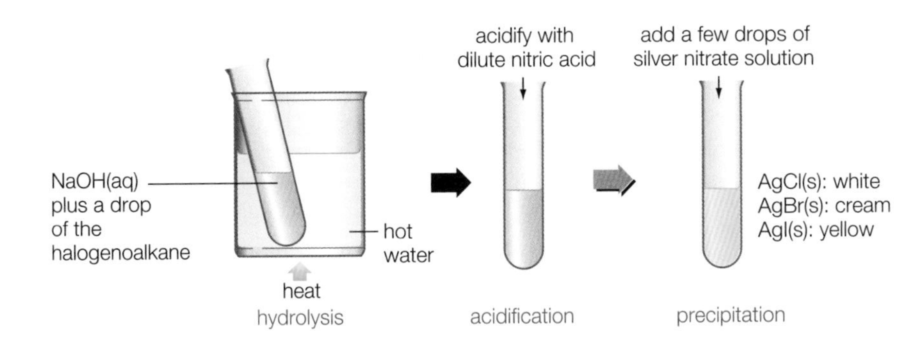

Figure 15.14 ◄
Hydrolysing halogenoalkanes makes it possible to distinguish between chloro-, bromo- and iodo- compounds. Heating the compound with an alkali releases halide ions. Acidifying with nitric acid and then adding silver nitrate produces a precipitate of the silver halide.

The rates of reaction of halogenoalkanes with water and alkali are in the order RI > RBr > RCl, where R represents an alkyl group.

The reaction rates do not correlate with bond polarity in the halogenoalkane. Chlorine is the most electronegative and iodine the least electronegative, so the C–Cl bond is the most polar and the C–I bond the least polar. So bond polarity is not the key factor which determines the rates of reaction.

The reaction rates do, however, correlate with the strength of the bonds. The C–I bond is the longest and the weakest (as measured by the mean bond enthalpy). The C–Cl bond is the shortest and the strongest. This suggests that the C–I bond breaks more readily than the C–Cl bond.

Test yourself

21 Refer to Figure 15.14 in answering this question.
 a) Why is hydrolysis necessary before testing with silver nitrate?
 b) Why must nitric acid be added before the silver nitrate solution?
 c) Write the equations for the three reactions that take place when detecting bromide ions in 1-bromobutane by this method.

Substitution by reaction with ammonia

Warming a halogenoalkane with a solution of ammonia in ethanol produces an amine (Section 5.2). The other product is a hydrogen halide, which reacts with excess ammonia to form an ammonium salt.

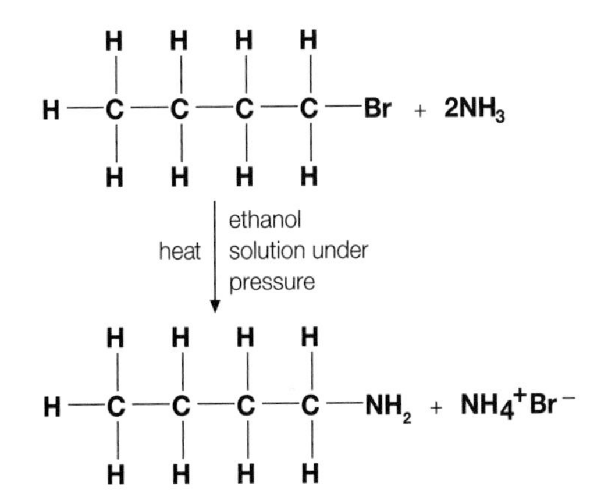

Figure 15.15 ◄
Reaction of 1-bromobutane with ammonia to make 1-aminobutane (butylamine)

The reactivity of ammonia depends on the lone pair of electrons on its nitrogen atom (Section 4.9). The problem in this case is that there is also a lone pair on the nitrogen atom of the product that is even more reactive. So the product can also react with the halogenoalkane. Fortunately, it is possible to limit this reaction by using an excess of concentrated ammonia solution so that there is a much greater chance of halogenoalkane molecules reacting with ammonia molecules.

Elimination reactions

Heating a halogenoalkane with a base can bring about elimination of a hydrogen halide instead of substitution.

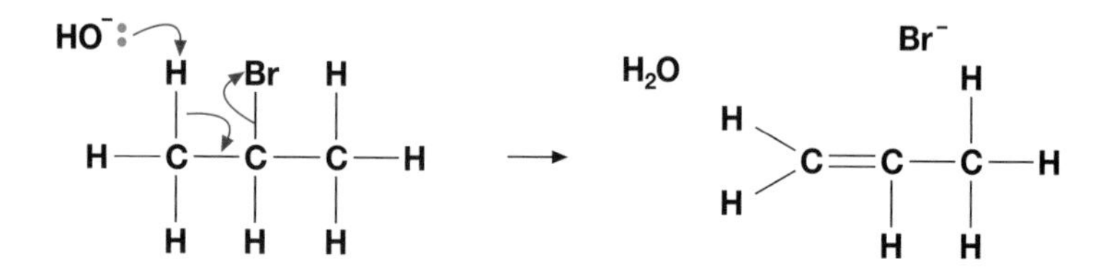

Figure 15.16 ▶ Elimination of hydrogen bromide from 2-bromopropane on heating with a solution of potassium hydroxide in ethanol. The red curly arrows show the transfer of electron pairs.

The hydroxide ion provides both the electrons needed to form a new bond to the hydrogen atom. The C–H bond breaks and the pair of electrons from that bond forms a second bond between the two carbon atoms. At the same time, the C–Br bond breaks heterolytically. In this case, both electrons in the bond leave with the bromine atom, which is set free as a bromide ion.

Hydrolysis leading to substitution is much more likely using potassium or sodium hydroxide dissolved in water. Elimination is more likely if there is no water and the alkali is dissolved in ethanol.

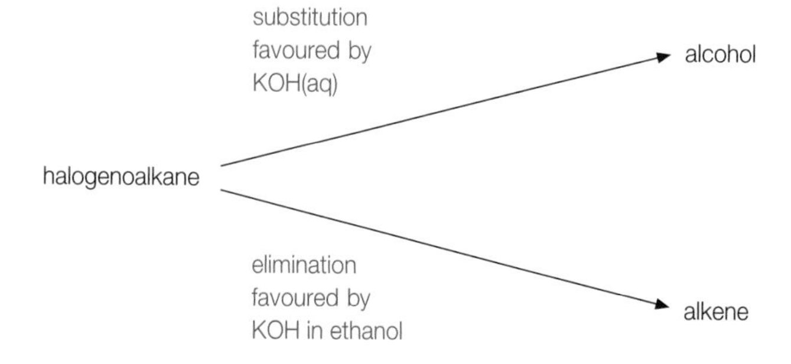

Figure 15.17 ▶ Alternative reactions of a halogenoalkane with solutions of hydroxide ions

The result of these reactions is often a mixture of products. Generally, elimination happens more readily with secondary or tertiary halogenoalkanes.

Test yourself

22 Give the name and structure of the main organic product when 2-bromo-2-methylpropane reacts on heating with:
a) a solution of ammonia in ethanol
b) a solution of potassium hydroxide in ethanol.

23 Suggest the structure of the organic product when 1-aminobutane reacts with 1-bromobutane. Can this product also react with 1-bromobutane? If so, what is formed?

15.7 The preparation of halogenoalkanes

Three types of reaction can be used to produce organic molecules containing halogen atoms.

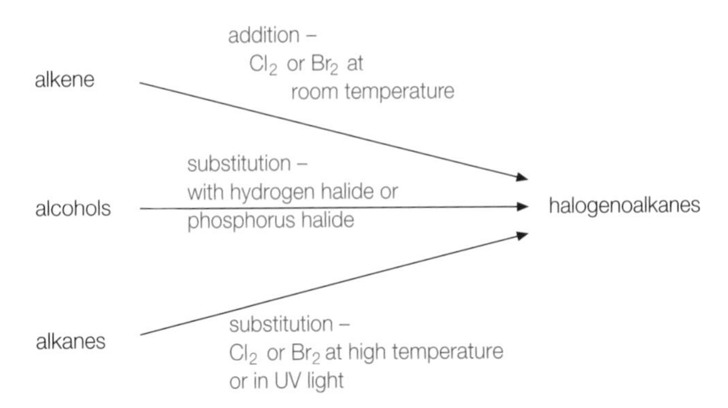

Figure 15.18 ◀ Reactions used to make halogenoalkanes

On a laboratory scale, the usual preparative methods are based on the reactions of alcohols with a hydrogen halide, or with a phosphorus halide (Section 15.3).

Practical guidance

Test yourself

24 Write an equation to illustrate each of the three methods for making halogenoalkanes shown in Figure 15.18. Name the reactants and products, and state the conditions required for each reaction.

25 The products of the reaction of 2-bromobutane with potassium hydroxide solution depend on the conditions.
   a) What conditions favour the formation of an alcohol?
   b) What conditions favour the formation of an alkene?

15.8 The uses of halogenoalkanes and their impacts

The discovery, production and use of halogenoalkanes during the last century have led to dramatic examples of the way in which science-based technology can provide us with products we value, and which makes our lives more comfortable and safer. Yet, at the same time these new technologies can turn out to have unintended and undesirable consequences.

There are now increasing restrictions on the uses of halogenoalkanes because of concerns about their hazards to health, their persistence in the environment and their effect on the ozone layer.

Figure 15.19 ◀ A recycling plant in South Wales which recovers CFCs from the coolant system of old refrigerators and freezers. The CFCs are removed from the canisters as gases, liquefied and then chemically destroyed.

Activity

The preparation of 1-bromobutane from butan-1-ol

Figure 15.20 shows the steps in synthesising a pure sample of 1-bromobutane from butan-1-ol. The alcohol reacts with hydrogen bromide formed from sodium bromide and 50% sulfuric acid.

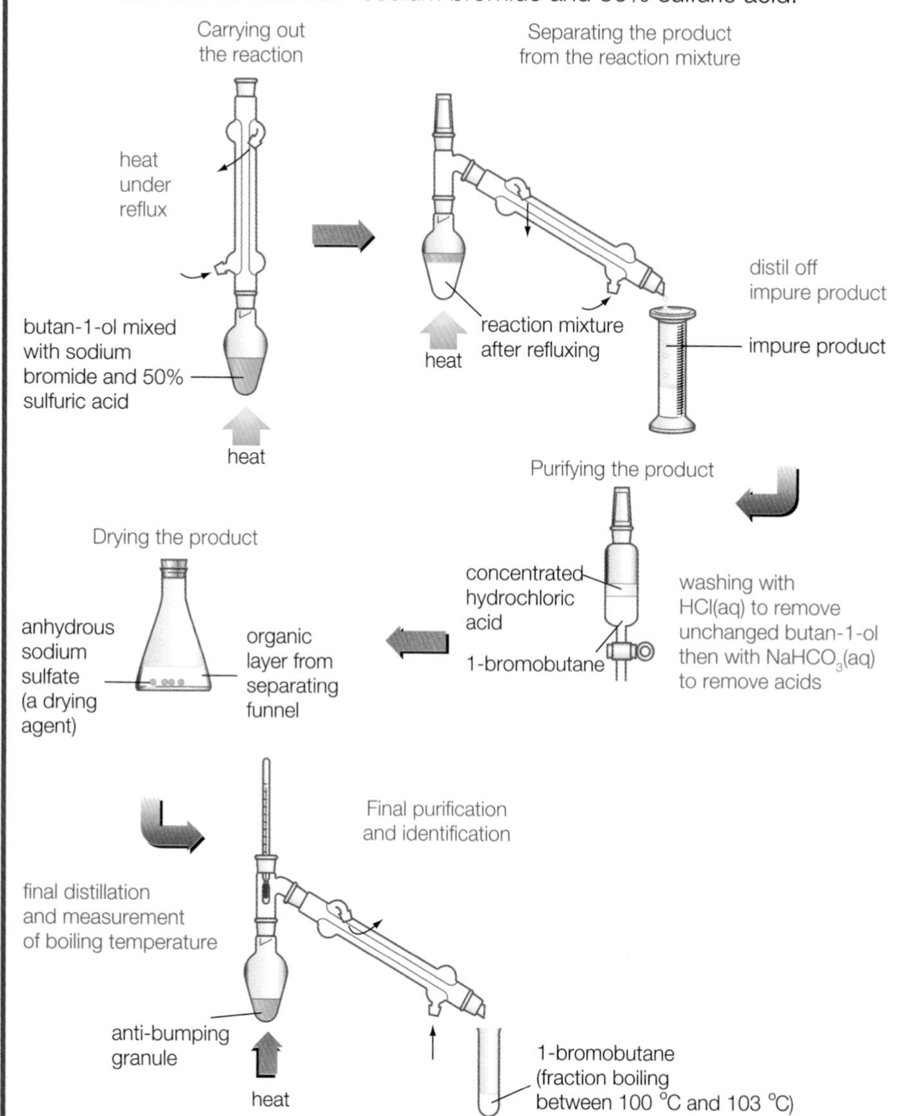

Figure 15.20 ▲
Steps in the synthesis of 1-bromobutane from butan-1-ol

1 a) Identify three aspects of this preparation that might be hazardous.

b) What steps would you take to reduce the risks from these hazards?

2 Write an equation for the reaction of sodium bromide with 50% sulfuric acid to form hydrogen bromide.

3 Explain why 50% sulfuric acid is used, and not concentrated sulfuric acid.

4 The reaction mixture is heated for about 40 minutes but even after this time some of the alcohol does not turn into the product – suggest a reason why.

5 Explain why the reaction flask is fitted with a reflux condenser.

6 After heating the reaction mixture for some time, the flask contains a mixture of chemicals including 1-bromobutane, unchanged butan-1-ol, hydrogen bromide and unchanged sodium bromide. Which of these chemicals is likely to distil over and collect in the measuring cylinder when separating the impure product?

7 Why are there two layers in the separating funnel when the product is shaken with aqueous reagents?

8 Suggest a reason why shaking the product with hydrochloric acid helps to remove unchanged butan-1-ol from the impure product.

9 Why is aqueous sodium hydrogencarbonate used in the separating funnel, and not aqueous sodium hydroxide, to remove acidic impurities?

10 Explain the term 'fraction' to describe the sample of product collected during the final distillation.

11 When this synthesis was carried out, the yield was 6.8 g of 1-bromobutane from 7.5 cm^3 butan-1-ol. The density of butan-1-ol is 0.81 $g\,cm^{-3}$. Calculate the percentage yield.

Practical guidance

12 Suggest three reasons why the percentage yield is below 100%.

Halogenoalkanes are used as:

- solvents, e.g. dichloromethane
- refrigerants, e.g. CF_3CH_2F, which has replaced the CFC, CF_2Cl_2
- pesticides, e.g. bromomethane, which is being phased out because of its powerful ozone-depleting properties
- fire extinguishers, e.g. halons, such as CBr_2ClF, CF_2BrCl and CF_3Br.

Chlorofluorocarbons (CFCs) are compounds containing just the elements chlorine, fluorine and carbon, such as CCl_3F, CCl_2F_2 and CCl_2FCClF_2. They contain no hydrogen. CFCs have some desirable properties – they are unreactive, do not burn and are not toxic. It is also possible to make CFCs with different boiling temperatures to suit different applications. These properties made CFCs ideal as the working fluid in refrigerators and air-conditioning units. They can also act as the blowing agents to make the bubbles in expanded plastics and insulating foams. CFCs were once valued as good solvents for dry cleaning and for removing grease from electronic equipment.

The manufacture of CFCs and their general use was banned by the Montreal Protocol in 1987 because of the effects on the ozone layer (see Section 19.5). CFCs can also contribute to global warming. In 1987, governments also agreed to ban the production of bromofluoroalkanes, which were then in common use as halon fire extinguishers in homes, on aircraft and on ships, as well as with a wide range of electronic equipment.

These bans set chemists the challenge of finding replacement chemicals with similar properties to CFCs but without the harmful impacts on the environment. The chemists' first step was to try the effect of adding hydrogen to make hydrochlorofluorocarbons (HCFCs). These are much less stable in the lower atmosphere so they break down before reaching the ozone layer.

More recently, new chemicals have been made that contain no chlorine. These are hydrofluorocarbons (HFCs), which survive for an even shorter time in the lower atmosphere. HFCs have no effect on the ozone layer, but they are greenhouse gases.

Some countries still require aircraft to be fitted with firefighting systems based on halons. The halons are now obtained by recycling existing stocks of the chemicals. Alternatives to halons are being developed, tested and brought into service.

Test yourself

26 Suggest reasons why CFCs were so widely adopted by the makers of refrigerators after their development in the 1920s, when the common refrigerants had been ammonia and sulfur dioxide.

27 Suggest reasons why halon fire extinguishers are still installed legally in aircraft and used in military equipment, despite being banned more generally.

Extension questions

REVIEW QUESTIONS

1 Consider the following reaction scheme:

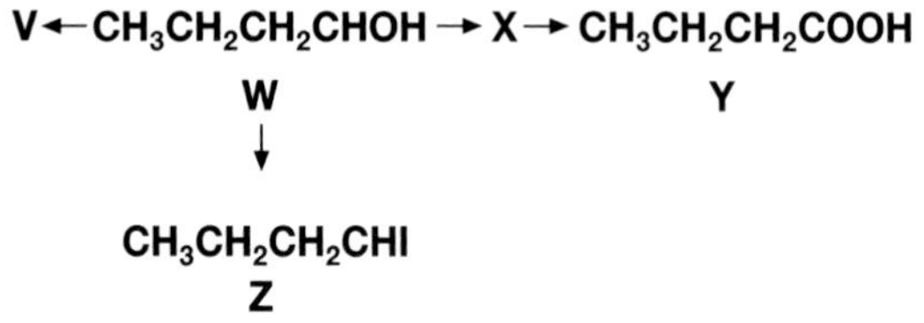

$$V \leftarrow \underset{W}{CH_3CH_2CH_2CHOH} \rightarrow X \rightarrow \underset{Y}{CH_3CH_2CH_2COOH}$$

$$W \downarrow \quad \underset{Z}{CH_3CH_2CH_2CHI}$$

Figure 15.21 ▲

a) State the reagents and conditions needed to convert compound W directly to compound Y. (3)

b) Draw the structure of X and state how the conditions you have given in part **a)** could be modified to produce compound X instead of Y. (2)

c) Compound W reacts with sodium to form compound V.

i) Describe what you would observe during this reaction.

ii) Write an equation for the reaction. (3)

d) Give the reagents and conditions for converting compound W into compound Z. (2)

2 Draw the structure of the organic product on heating:

a) 2-bromo-2-methylpropane under reflux with a solution of potassium hydroxide in ethanol (1)

b) 1-iodopropane with an aqueous solution of potassium hydroxide (1)

c) 1-bromopropane with excess ammonia in ethanol. (1)

3 A few drops of a halogenoalkane were added to 2 cm^3 of ethanol in a test tube and 5 cm^3 of aqueous silver nitrate were then added. Finally, the test tube was placed in a water bath for a few minutes. A cream precipitate formed. This precipitate was soluble in concentrated ammonia.

a) Why was ethanol used in this experiment? (1)

b) Why was the test tube containing the mixture warmed in a water bath? (1)

c) What was the formula of the precipitate? Explain your answer. (2)

d) What can you conclude about the halogenoalkane? (1)

4 'Technologies based on science can provide us with many things that improve our quality of life. Some technologies, however, have unintended and undesirable impacts. These need to be weighed against the benefits.'

Illustrate and discuss this statement with examples based on the uses of halogenoalkanes. (6)

16 Organic reaction mechanisms

An important way of making sense of carbon chemistry is to classify reactions. Chemists classify reactions by describing what happens to the atoms, ions and molecules as a result of oxidation, reduction, hydrolysis, addition, elimination or substitution reactions.

An alternative classification is based on the mechanisms of reactions, showing the types of reagents, the method of bond breaking and the nature of any intermediates.

Figure 16.1 ◀
Using a computer to model the binding of an anti-cancer drug to an enzyme. Understanding the mechanisms of reactions helps researchers to explain how drug molecules work and to design better drugs.

16.1 Types of organic reaction

Addition – adding bits to molecules

During a simple addition reaction, two molecules add together to form a single product. Bromine, for example, adds to ethene to form the addition product 1,2-dibromoethane (Section 6.8).

Addition reactions are characteristic of unsaturated compounds with double bonds, especially alkenes and other compounds with C=C double bonds.

Test yourself

1 Give the structures and names of the products of the addition reactions of pent-1-ene with:
a) hydrogen
b) chlorine
c) hydrogen bromide.

Polymerisation – making long chains

Under the right conditions, small molecules with double bonds can join up in long chains to form polymers. This is addition polymerisation (Section 6.11).

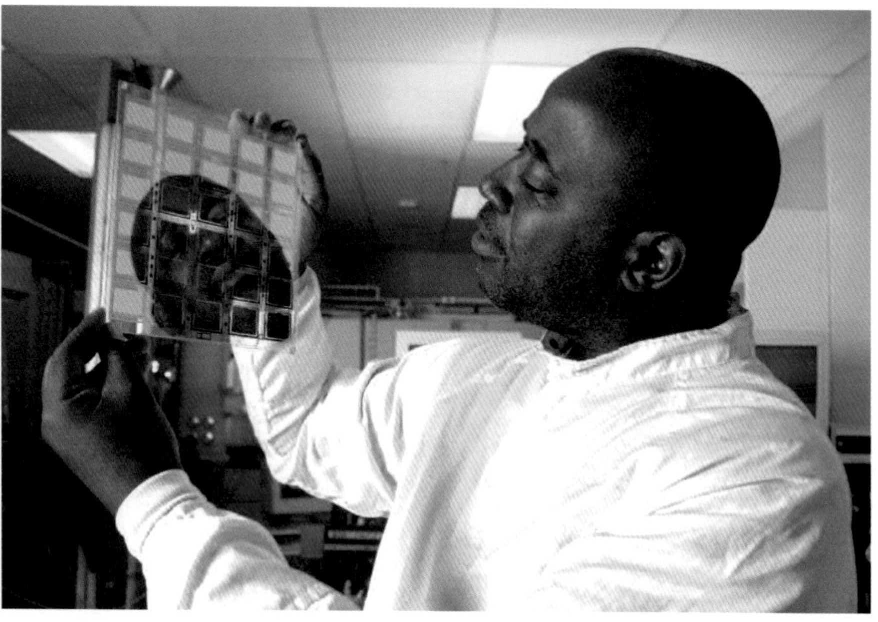

Figure 16.2 ▲
This researcher is holding a panel of organic light-emitting diodes. These ultra-thin polymers are semiconductors that convert an applied voltage into light. Understanding reaction mechanisms can help chemists to produce new polymers with properties that are specially designed and adapted to their uses.

Test yourself

2 Draw a structure to represent the addition polymer ptfe, poly(tetrafluoroethene), formed from tetrafluorethene.

3 Identify the following polymers, and give the name and structural formula of the monomers:

a)
$$\left[\begin{array}{cc} H & H \\ | & | \\ -C & -\ C- \\ | & | \\ H & H \end{array}\right]_n$$

b)
$$\left[\begin{array}{cc} H & H \\ | & | \\ -C & -\ C- \\ | & | \\ H & Cl \end{array}\right]_n$$

Elimination – splitting bits off from molecules

An elimination reaction splits off a simple molecule, such as water or hydrogen chloride, from a larger molecule leaving a double bond.

An example of an elimination reaction is the removal of a hydrogen halide from a halogenoalkane to produce an alkene (Section 15.6).

Test yourself

4 Write an equation for the elimination reaction which happens when 2-bromopropane reacts with a hot solution of potassium hydroxide in ethanol.

Substitution – replacing one or more atoms by others

Substitution reactions replace an atom or a group of atoms by another atom or group of atoms. An example is the reaction of butan-1-ol with hydrogen bromide to make 1-bromobutane (Sections 15.3 and 15.7).

Substitution reactions are characteristic of halogenoalkanes (Section 15.6). Other examples of substitution reactions include the replacement of hydrogen atoms in alkanes by chlorine or bromine atoms (Section 6.3).

Test yourself

5 Identify the reagents and conditions for the substitution reactions which convert:
  a) ethane to chloroethane
  b) pentan-2-ol to 2-iodopentane
  c) 2-bromopropane to 2-aminopropane
  d) 1-bromobutane to butan-1-ol.

Redox

Redox originally meant gaining oxygen or losing hydrogen, but the term now covers all reactions in which atoms, molecules or ions lose and gain electrons. Redox processes have been further extended by defining oxidation as a change in which the oxidation number of an element becomes more positive, or less negative, and reduction as a change in which the oxidation number of an element becomes more negative, or less positive (see Topic 10).

Oxidation number rules apply in both organic and inorganic chemistry, but in organic reactions it is often easier to use the older definitions. As well as describing the oxidation of an alcohol in terms of the loss of electrons (Section 15.3), a common method is to represent oxygen from the oxidising agent as [O].

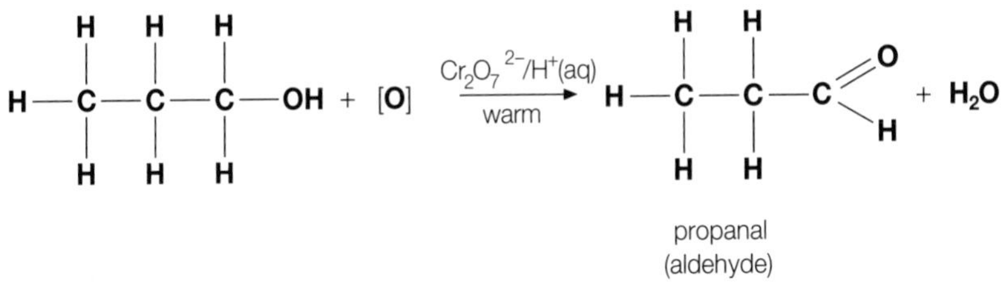

$$CH_3CH_2CH_2OH + [O] \xrightarrow[\text{warm}]{Cr_2O_7^{2-}/H^+(aq)} CH_3CH_2CHO + H_2O$$

propanal (aldehyde)

Figure 16.3 Oxidation of propan-1-ol to propanal by an acidic solution of dichromate(VI) ions. [O] represents oxygen atoms from the oxidising agent.

Oxidation and reduction always occur together in redox reactions. When an acid solution of sodium dichromate(VI) oxidises an alcohol, the orange dichromate(VI) ions are reduced at the same time to green chromium(III) ions. The reaction reduces chromium from +6 in $Cr_2O_7^{2-}$ to +3 in Cr^{3+}.

Reduction is the opposite of oxidation – according to the older definitions, reduction meant the removal of oxygen or the addition of hydrogen.

Test yourself

6 What are the conditions for the oxidation of propan-1-ol to propanal?

7 Write an equation for the oxidation of the following alcohols by an excess of hot, acidified potassium dichromate(VI) solution:
  a) butan-2-ol
  b) pentan-1-ol.

Hydrolysis – splitting apart with water

The word 'hydrolysis' comes from two other words – 'hydro' related to water and 'lysis' meaning splitting. So, the term hydrolysis is used to describe any reaction in which water causes another compound to split apart. Hydrolysis reactions are often catalysed by acids or alkalis. They are also often substitution reactions in which the chemical attack is by water molecules (or hydroxide ions).

An example of hydrolysis is the reaction of halogenoalkanes with water. In this case, bonds in both the water and halogenoalkanes split to form alcohols and hydrogen halides (Section 15.6). The reaction is very slow in the absence of alkali.

Test yourself

8 a) Describe the test you would use to show that iodide ions form during the hydrolysis of 1-iodopropane.
b) What would you see when carrying out the test?

16.2 Bonds and bond breaking

Homolytic fission

A normal covalent bond consists of a shared pair of electrons (Section 4.8). There are two ways in which a bond can break. In one, the covalent bond breaks leaving one electron with each of the atoms that were linked by the bond – this is homolytic bond breaking. This type of bond breaking produces free radicals, which are reactive particles with unpaired electrons (Section 5.6).

Free radicals are often produced as intermediates in reactions taking place:

- in the gas phase at high temperature or in ultraviolet light
- in non-polar solvents, either with an initiator or when irradiated by ultraviolet light.

Examples of free radical processes include the reactions of alkanes with chlorine or bromine (Section 6.3) and the formation and destruction of the ozone layer (Section 19.5).

Note

The symbol for a free radical represents the unpaired electron by a dot. Other paired electrons in the outer shells are not usually shown. A single bromine atom has one unpaired electron in its outer shell and is a free radical.

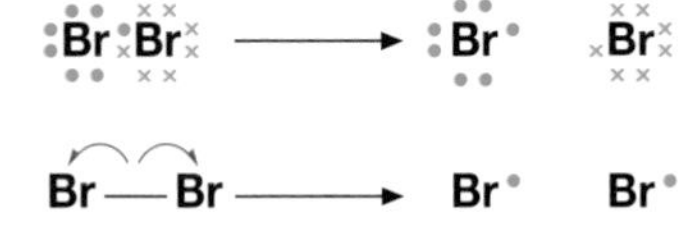

Figure 16.4 ▲
Two ways of showing homolytic bond breaking of the covalent bond in a bromine molecule. A curly arrow with only half an arrow head indicates the movement of a single electron.

Test yourself

9 Why is a high temperature (or ultraviolet light) needed to form free radicals?
10 Draw structures and use curly arrows to show what happens when an ethane molecule splits into two methyl radicals.
11 What is the practical importance of the reaction of alkanes with halogens?
12 a) What are the two main products of the reaction of excess methane with chlorine?
b) Account for the formation of some ethane during the reaction.
c) Explain why the reaction may also produce other products such as dichloromethane, trichloromethane and tetrachloromethane.

Heterolytic fission

Covalent bonds can also break in a way which produces ionic intermediates (Section 5.6). In heterolytic bond breaking, a covalent bond breaks so that one of the two atoms joined by the bond takes both of the electrons in the shared pair. This is the type of bond breaking that usually occurs when the reaction takes place in a polar solvent, such as water.

The bond which breaks to produce ionic intermediates is often polar to start with. So, heterolytic bond breaking is characteristic of compounds with electronegative atoms such as chlorine, oxygen and nitrogen (Section 8.2). Bonds may also break heterolytically if they are easily polarised.

The reagents which attack polar bonds and bring about heterolytic fission are electrophiles and nucleophiles.

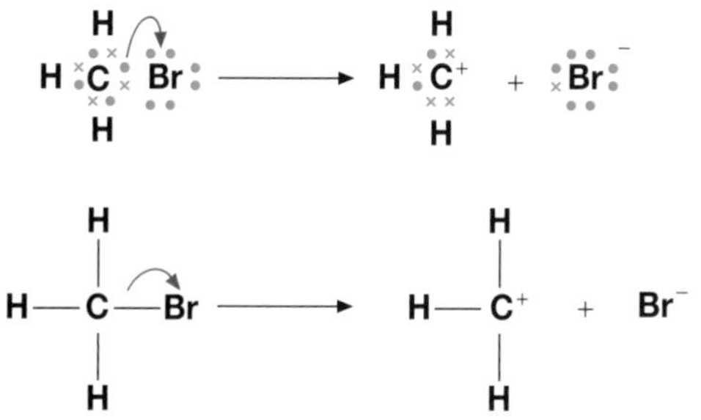

$$H_3C{:}Br \longrightarrow H_3C^+ + {:}Br^-$$

$$H_3C{-}Br \longrightarrow H_3C^+ + Br^-$$

Figure 16.5 ◀ Two ways to show heterolytic bond breaking of a C–Br bond. Note the use of curly arrows with a full arrow head to show what happens to pairs of electrons as the bond breaks. One atom takes both the electrons from the shared pair. The tail of the arrow starts at the electron pair. The head of the arrow points to where the electron pair ends up.

Test yourself

13 Explain why there is a positive charge on the carbon atoms and a negative charge on the chlorine atoms after the C–Cl bonds break heterolytically in chloromethane.

14 Classify these reagents as free radicals, electrophiles or nucleophiles: Cl_2, Cl, H_2O, HBr, Br_2, Br^-, Br, H^+, CH_3, OH^-, NH_3.

15 Draw the displayed formulae of the following molecules and indicate any polar bonds using $\delta+$ and $\delta-$ signs: C_3H_6, C_2H_5Cl, CH_3OH, C_3H_8, CH_3NH_2.

16.3 Investigating reaction mechanisms

The techniques which chemists use to study reactions are becoming more and more sophisticated. With the help of laser beams and spectroscopy, it is now possible to follow extremely fast reactions and to watch molecules breaking apart and rearranging themselves in fractions of a second.

Labelling with isotopes

Chemists can use isotopes as markers to track what happens to particular atoms during a chemical change (Section 3.3). They replace atoms of the normal isotope of an element in a molecule with a different isotope.

Isotopes of an element have identical chemical properties, so it is possible to use them to follow what happens during a change without altering the normal course of a reaction. Radioactive isotopes can usually be tracked fairly easily and conveniently by detecting their radiation. The fate of non-radioactive isotopes can be followed by analysing samples with a mass spectrometer (Section 17.1).

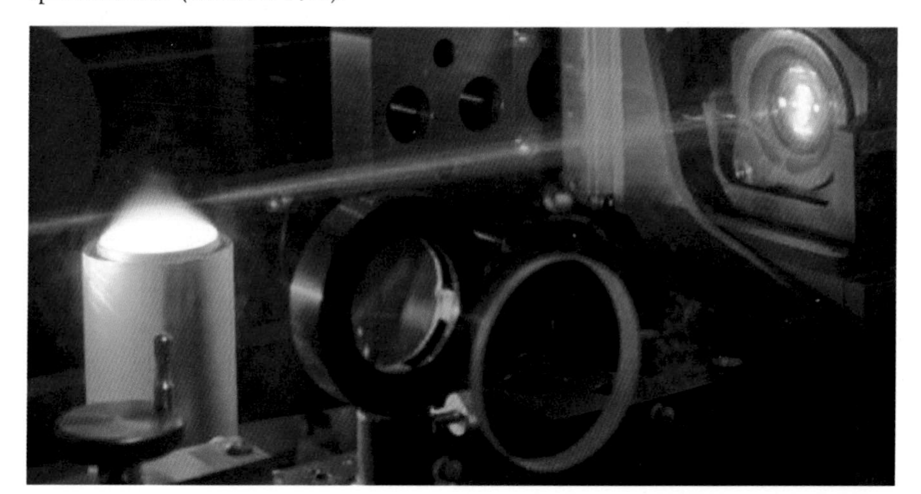

Figure 16.6 ◀ This blue laser is used to investigate how fuels burn. The aim is to find ways to make fuels burn more efficiently and to reduce pollution.

Activity

Investigating the mechanism of a hydrolysis reaction

Alcohols react with carboxylic acids to form esters. Hydrolysis splits esters back to the alcohol and acid. Isotopic labelling has been used to investigate the mechanism of this hydrolysis reaction. The researchers used water labelled with oxygen-18 instead of the normal oxygen-16 isotope. After hydrolysis with $H_2^{18}O$, they found that the heavier oxygen atoms from the water ended up in the acid and not in the alcohol. In this way they were able to identify exactly which bond breaks during hydrolysis of the ester.

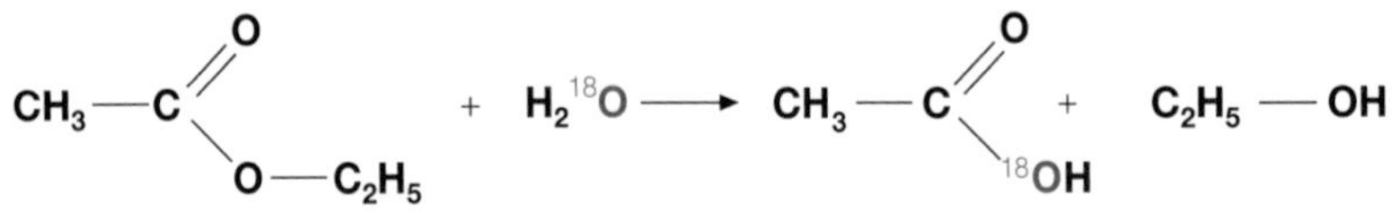

$$CH_3C(=O)O{-}C_2H_5 + H_2{}^{18}O \longrightarrow CH_3C(=O){}^{18}OH + C_2H_5{-}OH$$

Figure 16.7 ▲
Use of labelling to investigate bond breaking during the hydrolysis of an ester

1 Why is the reaction of an ester with water described as 'hydrolysis'?

2 How do atoms of the oxygen-18 and oxygen-16 isotopes differ?

3 Why do oxygen-18 and oxygen-16 isotopes have the same chemical properties?

4 What method of analysis can be used to distinguish ethanoic acid molecules with ^{18}O atoms from those with ^{16}O atoms? (Neither isotope of oxygen is radioactive.)

5 Look closely at Figure 16.7. Which bond in the ester breaks during the reaction?

6 Where would the oxygen-18 atoms have appeared if the mechanism involved breaking the other C–O bond in the ester?

Trapping intermediates

An important key to understanding reaction mechanisms was the realisation that most reactions do not take place in one step, as implied by the balanced equation. Instead, most reactions involve a series of steps (Sections 6.3 and 6.9). In the course of these mechanisms, atoms, molecules and ions, which do not appear in the balanced equation, exist as intermediates as chemicals change from the reactants to the products.

Chemists can often use spectroscopy to detect intermediates which exist for only a short time during a reaction. Intermediates can also be detected chemically. This can be done by adding chemicals to trap the intermediates by reacting with them. This gives rise to products that would never be formed without first producing the intermediate. In this way, chemists showed that the addition of bromine to alkenes must be a two-step reaction (Section 6.9).

Activity

Investigating the mechanism for the addition of bromine to ethene

Ethene molecules react rapidly with bromine molecules, but they do not react with chloride ions. However, when ethene reacts with bromine in the presence of sodium chloride the product is an oily liquid containing roughly equal amounts of 1,2-dibromoethane and 1-bromo-2-chloroethane. This suggests that the mechanism must involve an intermediate that is able to react with chloride ions.

1 Write an equation for the expected reaction of ethene with bromine and name the product.

2 How do the results of the experiment show that the reaction of ethene with bromine does not happen in one step?

3 Why is it reasonable to suggest that the intermediate in the addition reaction is a positive ion?

4 Explain how the mechanism for the addition of bromine to ethene described in Section 6.9 can account for the following results in the presence of sodium chloride:

a) there are two main products, one of which is 1-chloro-2-bromoethane

b) no 1,2–dichloroethane forms.

5 Which bonding electrons between the carbon atoms in ethene (those in the σ bond or those in the π bond) would you expect to take part in the addition reaction?

6 Why is bromine called an electrophile when it reacts with ethene?

Studying the shapes of molecules

Studying the shapes of molecules can also give clues to the details of reaction mechanisms. Part of the evidence that helped to confirm the two-step mechanism for electrophilic addition to alkenes came from studies of the isomers which form when bromine adds to compounds such as cyclohexene.

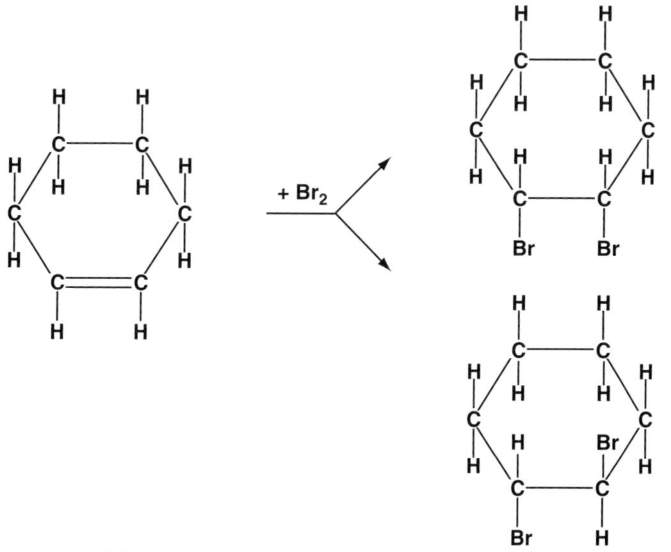

+ Br_2

Figure 16.8 ◀ Two possible products when bromine adds to cyclohexene forming 1,2-dibromocyclohexane

There are two possible isomers when bromine adds to cyclohexene. One has both bromine atoms on the same side of the ring of carbon atoms and the other has the bromine atoms on opposite sides. It turns out that the main product of the reaction is the isomer with the bromine atoms on opposite sides of the ring. This suggests that the bromine does not add directly to alkene molecules as Br_2 molecules.

Test yourself

16 a) What type of isomerism is shown by the two isomers of 1,2-dibromocyclohexane?
   b) What names are used to distinguish the isomers?

17 Why is there a possibility of two isomers forming when bromine adds to cyclohexene, but not when bromine adds to ethene?

18 a) Show how the mechanism of electrophilic addition described in Section 6.9 can account for the formation of the isomer with bromine atoms on opposite sides of the ring of carbon atoms.
   b) Refer to the data sheets on bond lengths and ionic radii on the Dynamic Learning Student website to suggest why the mechanism leads to the formation of the isomer with bromine atoms on the opposite sides of the ring.

Studying reaction rates

Chemists have learned a great deal about reaction mechanisms by studying the rates of chemical reactions. This is illustrated by the mechanisms of substitution reactions of halogenoalkanes (Section 15.6). What these mechanisms show is that the effects of changing the concentrations of the reactants are not the same for primary and for tertiary halogenoalkanes. Taken with other evidence, this suggests that the two types of halogenoalkane react by different mechanisms, as described in the next section.

16.4 Nucleophilic substitution in halogenoalkanes

The carbon–halogen bond in a halogenoalkane is polar, with the carbon atom at the δ+ end of the dipole. The characteristic reactions of these compounds are substitution reactions in which nucleophiles attack the δ+ end of the carbon–halogen bond (Section 5.6).

$$\mathrm{Nu{:}^- + \ C^{\delta+}\!-\!X^{\delta-} \longrightarrow Nu\!-\!C + {:}X^-}$$

nucleophile — halogenoalkane — leaving group

Figure 16.9 ▶ Generalised description of the attack of a nucleophile on a halogenoalkane leading to a substitution reaction

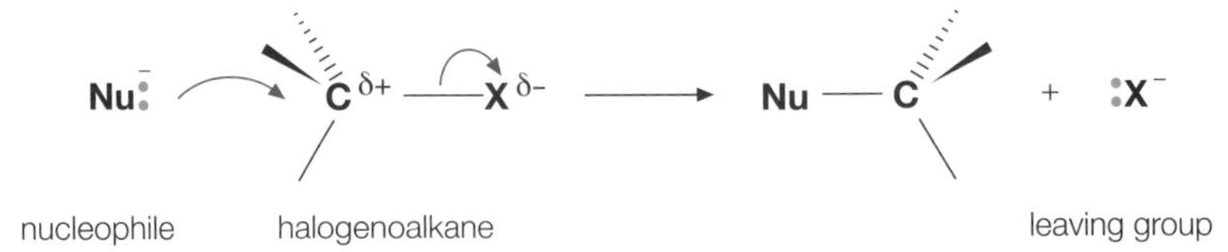

Nucleophiles which react with halogenoalkanes include water molecules, hydroxide ions and ammonia molecules.

The study of the rates of these substitution reactions suggests that there are two possible mechanisms involved. For example, the rate of reaction of a primary halogenoalkane, such as 1-bromobutane, with hydroxide ions depends on the concentrations of both reactants. This suggests that both halogenoalkane molecules and hydroxide ions are involved in the critical step of the reaction.

To account for this, chemists have proposed a mechanism in which the C–Br bond breaks at the same time as the nucleophile, OH^-, forms a new bond with the carbon atom.

$$\mathrm{HO{:}^- + H_2C(C_3H_7)\!-\!Br \longrightarrow \overset{\delta-}{HO}\cdots C(H_2)(C_3H_7)\cdots \overset{\delta-}{Br} \longrightarrow HO\!-\!CH_2C_3H_7 + Br^-}$$

1-bromobutane — transition state — butan-1-ol

Figure 16.10 ▶ Mechanism of nucleophilic substitution of a primary halogenoalkane

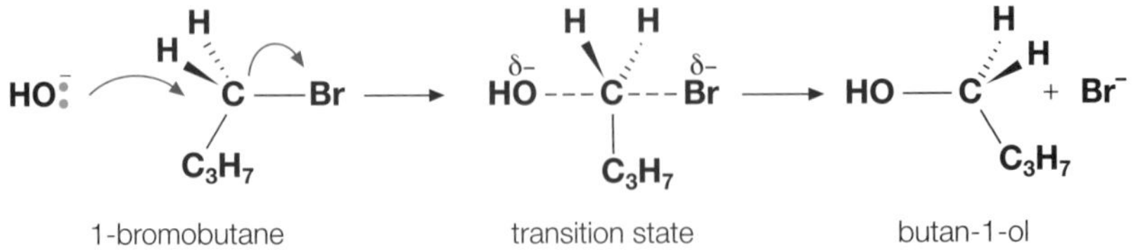

When it comes to the hydrolysis of tertiary halogenoalkanes, such as 2-bromo-2-methylpropane, it turns out that the concentration of the halogenoalkane affects the rate, but that the concentration of hydroxide ions does not affect the rate. So the mechanism has to account for the fact that the hydroxide ions are not involved in the critical step that controls the rate of reaction.

In this case, the suggested mechanism shows that the C–Br bond breaks first to produce a positively charge intermediate. Only after this has happened does the nucleophile, OH^-, form a new bond with carbon.

WWW Tutorial

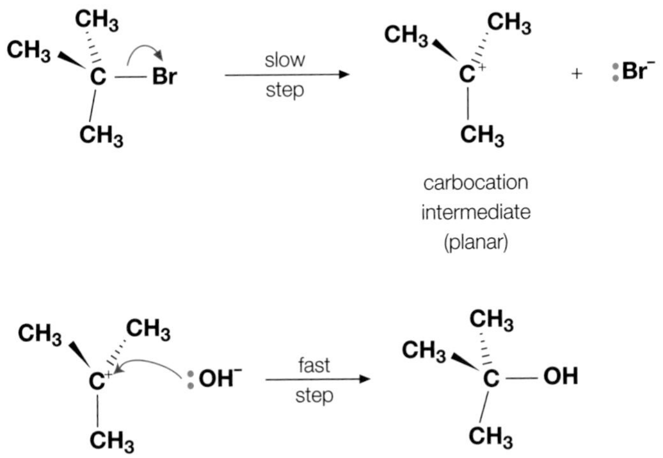

$$(CH_3)_3C{-}Br \xrightarrow[\text{step}]{\text{slow}} (CH_3)_3C^+ + :Br^-$$

carbocation intermediate (planar)

$$(CH_3)_3C^+ + :OH^- \xrightarrow[\text{step}]{\text{fast}} (CH_3)_3C{-}OH$$

Figure 16.11 ▲
Mechanism of nucleophilic substitution of a tertiary halogenoalkane

Test yourself

19 Show the bond forming and bond breaking when a hydroxide ion reacts with 1-iodopropane. Use 'curly arrows' to show the movement of pairs of electrons. Identify any relevant dipoles.

20 In general, the relative rates of hydrolysis of halogenoalkanes are RI > RBr > RCl. Does this trend correlate better with the polarity of the bonds or the strength of the bonds?

REVIEW QUESTIONS

WWW Extension questions

1 Classify the following conversions as *addition*, *elimination*, *substitution*, *oxidation*, *reduction*, *hydrolysis* or *polymerisation* reactions. (Note that a reaction may belong to more than one category.)

a) 2-iodobutane to but-2-ene (1)

b) butane to 1-bromobutane (1)

c) but-1-ene to 1,2-dichlorobutane (1)

d) butanal to butanoic acid (1)

e) 1-bromobutane to butan-1-ol (1)

f) but-1,3-diene to synthetic rubber. (1)

2 a) Use the reaction of chlorine with methane to explain the term free radical. (2)

b) There are three distinct reaction steps in the reaction of chlorine with methane. Name each type of step and illustrate each type with an equation. (6)

c) Why is it sufficient if only a small proportion of the chlorine molecules are split into free radicals by the first step? (2)

d) Why should an excess of methane be used to make chloromethane from methane and chlorine. (1)

3 a) i) Explain the term 'electrophile', giving an example. (2)

ii) Use symbols to describe the mechanism of the electrophilic addition of hydrogen bromide to ethene. Show any relevant dipoles. (5)

b) i) Explain the term 'nucleophile', giving an example. (2)

ii) Use symbols to describe the mechanism of nucleophilic substitution during the reaction of hydroxide ions with bromoethane. Show any relevant dipoles. (5)

4 a) What circumstances favour homolytic bond breaking during organic reactions? (3)

b) What circumstances favour heterolytic bond breaking during organic reactions? (3)

c) Explain why it is helpful for chemists to classify reagents as free radicals, electrophiles or nucleophiles. (4)

17 Instrumental analysis

In modern chemistry, the use of instruments for analysis is more important than chemical analysis (Topic 12). Two of the techniques that chemists use most frequently are mass spectrometry and infrared spectroscopy. Both techniques are very important in the study of organic molecules.

Figure 17.1 ▲
Police use infrared instruments for motorists' breath tests to provide evidence of drink driving.

17.1 Mass spectra of organic compounds

Mass spectrometry is an accurate technique for determining relative atomic masses (Section 3.2). Mass spectrometry can also help to determine the relative molecular masses and molecular structures of organic compounds. In this way it can be used to identify unknown compounds. The technique is extremely sensitive and requires very small samples, which can be as small as one nanogram (10^{-9} g).

Figure 17.2 ▶
A mass spectrometer used at the HFL. The sample is fed into the bottom left of the instrument where it is vaporised and ionised. The ions are accelerated along the U-shaped glass tube and steered by electric and magnetic fields to reach the gold-coloured dectector on the bottom right. This part of the instrument measures about 70 cm × 50 cm.

Inside a mass spectrometer there is a very high vacuum so that it is possible to produce and study ionised molecules and fragments of molecules. The molecular fragments could not exist other than in a high vacuum.

A beam of high-energy electrons bombards the molecules of the sample. This turns them into ions by knocking out one or more electrons.

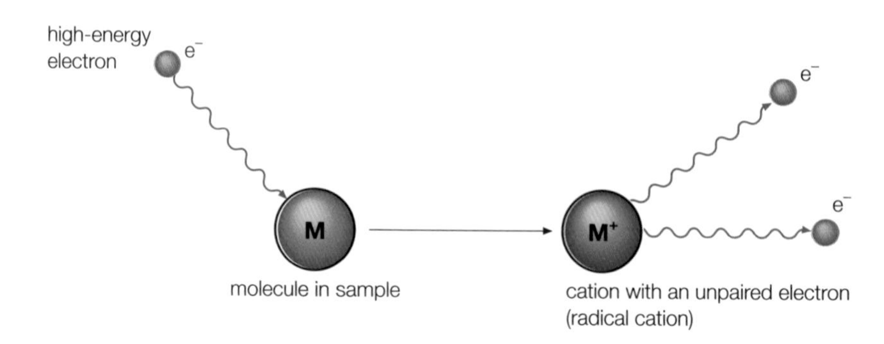

Figure 17.3 ▶
High-energy electrons ionising a molecule in a mass spectrometer. Knocking out one electron leaves a positive ion, shown by the symbol M^+.

Bombarding molecules with high-energy electrons not only ionises them but usually splits them into fragments. As a result, the mass spectrum consists of a 'fragmentation pattern'.

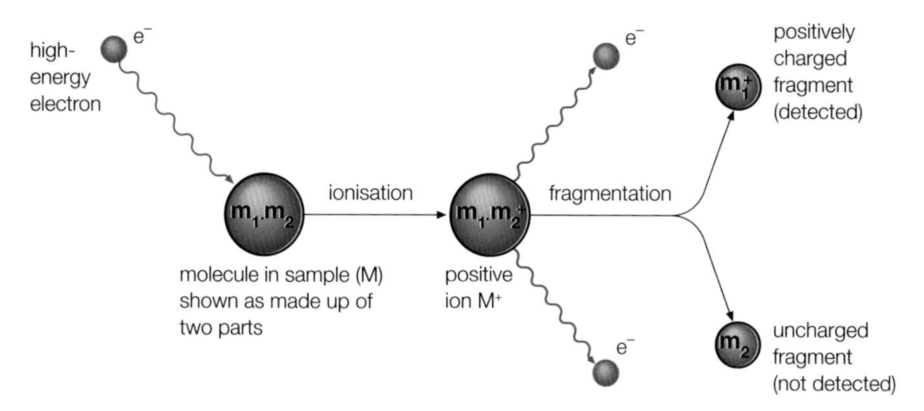

Figure 17.4 ◀
Ionisation and fragmentation of a single molecule $m_1.m_2$ which fragments into two parts m_1^+ and m_2. Only charged species show up in the mass spectrum because electric and magnetic fields have no effect on neutral fragments. So, in this case, the instrument only detects m_1^+.

Molecules break up more readily at weak bonds, or at bonds which give rise to more stable fragments. It turns out that positive ions with the charge on a secondary or tertiary carbon atom are more stable than ions with the charge on a primary carbon atom. Species such as CH_3CO^+ are also more stable because of the presence of the C=O double bond.

After ionisation and fragmentation, the charged species are accelerated and deflected by electric and magnetic fields to produce the mass spectrum (see Figure 3.3). The extent of the deflection depends on the ratio of the mass of the fragment to its charge. When analysing molecular compounds, the peak of the ion with the highest mass is usually the ionised whole molecule. So the mass of this 'parent ion', M, is the relative molecular mass of the compound.

Definition

The **mass to charge ratio** (*m/e*) is the ratio of the relative mass of an ion to its charge, where *e* is the number of charges (1, 2 and so on). In all the examples you will meet *e* = 1.

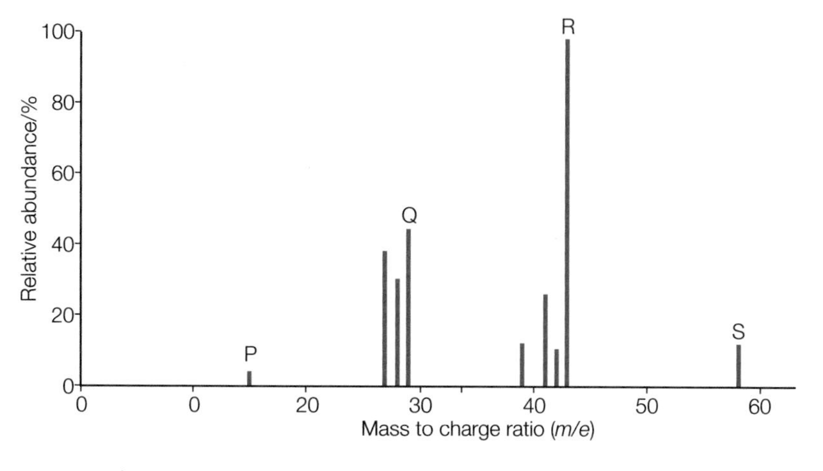

Figure 17.5 ◀
Mass spectrum of butane C_4H_{10}. The pattern of fragments is characteristic of this compound.

Chemists study mass spectra in order to gain insight into the structure of molecules. They identify the fragments from their relative masses, and then piece together likely structures with the help of evidence from other methods of analysis, such as infrared spectroscopy.

Test yourself

1 a) i) Which peak in the mass spectrum of butane in Figure 17.5 corresponds to the parent ion?
ii) What is the relative mass of this ion?
b) Suggest the identity of the fragments labelled P, Q and R.
c) Suggest a reason why the peak at *m/e* = 15 is relatively weak.
d) Use symbols to show one way in which the parent ion of butane could fragment.

Test yourself

2 Put the following stages in order during mass spectrometry: acceleration, fragmentation, ionisation, deflection according to *m/e*, detection, vaporisation.

3 Give two reasons why it is important to have a high vacuum inside a mass spectrometer.

17.2 Infrared spectroscopy

Spectroscopy is a term which covers a range of practical techniques for studying the composition, structure and bonding of compounds. Spectroscopic techniques are now the essential 'eyes' of chemistry.

The range of techniques available covers many parts of the electromagnetic spectrum – including the infrared, visible and ultraviolet regions. The instruments used are called either spectroscopes, which emphasises the use of techniques for making observations, or spectrometers, which emphasises the importance of measurements.

Infrared spectroscopy (or IR spectroscopy) is an analytical technique used to identify functional groups in organic molecules. Infrared radiation from a glowing lamp or fire makes us feel warm. This is because infrared frequencies correspond to the natural frequencies of vibrating atoms in molecules. Our skin warms up as the molecules absorb infrared and vibrate faster.

Most compounds absorb infrared radiation. The wavelengths of the radiation they absorb correspond to the natural frequencies at which vibrating bonds in the molecules bend and stretch. However, it is only molecules that change their polarity as they vibrate which interact with IR.

Definitions

An **absorption spectrum** is a plot showing how strongly a chemical absorbs radiation over a range of frequencies.

Infrared **wavenumbers** range from 400 to 4500 cm^{-1}. Spectroscopists find the numbers more convenient than wavelengths – the wavenumber is the number of waves in 1 cm.

Transmittance on the vertical axis of infrared spectra measures the percentage of radiation which passes through the sample. The troughs appear at those wavenumbers where the compound absorbs strongly.

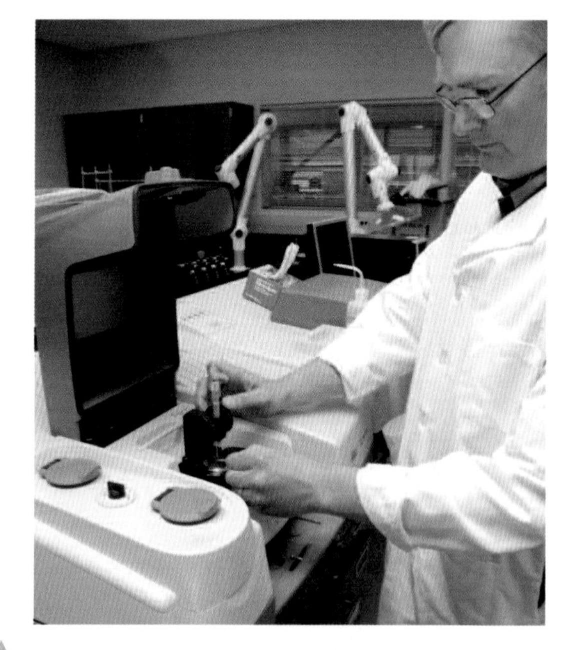

Figure 17.6 ▲
Using an infrared spectrometer. The instrument covers a range of infrared wavelengths and a detector records how strongly the sample absorbs at each wavelength. Wherever the sample absorbs, there is a dip in the intensity of the radiation transmitted which shows up as a dip in the plot of the spectrum.

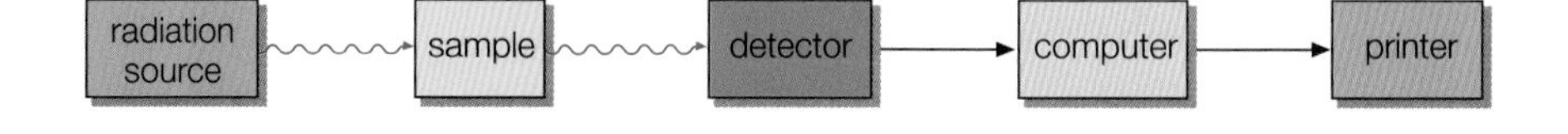

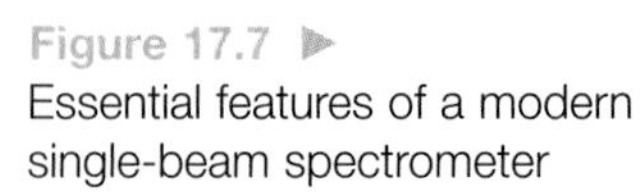

Figure 17.7 ▶
Essential features of a modern single-beam spectrometer

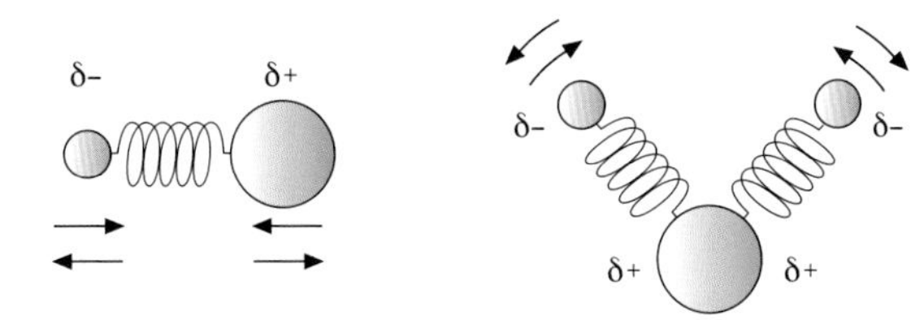

Figure 17.8 ◄
Bond vibrations give rise to absorptions in the infrared region. Vibrations of molecules which cause a fluctuating polarity interact with electromagnetic waves.

Bonds vibrate in particular ways and absorb radiation at specific wavelengths. This means that it is possible to look at an infrared spectrum and identify particular functional groups.

Spectroscopists have found that it is possible to correlate absorptions in the region 4000–1500 cm^{-1} with the stretching or bending vibrations of particular bonds. As a result, infrared spectra give valuable clues about the presence of functional groups in organic molecules. The important correlations between different bonds and observed absorptions are shown in Figure 17.9. Hydrogen bonding broadens the absorption peaks of –OH groups in alcohols, and even more so in carboxylic acids.

Wavenumber ranges

4000 cm^{-1} – 2500 cm^{-1}	2500 cm^{-1} – 1900 cm^{-1}	1900 cm^{-1} – 1500 cm^{-1}	1500 cm^{-1} – 650 cm^{-1}
C–H O–H N–H single bond stretching vibrations	C≡C C≡N triple bond stretching vibrations	C=C C=O double bond stretching vibrations	fingerprint region

Figure 17.9 ▲
The main regions of the infrared spectrum and important correlations between bonds and observed absorptions

Molecules with several atoms can vibrate in many ways because the vibrations of one bond affect others close to it. The complex pattern of vibrations can be used as a 'fingerprint' that can be matched against the recorded infrared spectrum in a database.

Infrared spectroscopy is an analytical tool that can be used to monitor the progress of an organic synthesis. Comparing the spectrum of the final product with the known spectrum in a database can be used to check if the product is pure.

Test yourself

4 Why do the vibrations of O–H, C–O and C=O bonds show up strongly in infrared spectra, while C–C vibrations do not?

5 Figure 17.10 shows the infrared spectra of ethanol, ethanal and ethanoic acid.
  a) Which vibrations give rise to the peaks marked with the letters A–G?
  b) Which spectrum belongs to which compound?
  c) Why do two of the spectra have broad peaks at wavenumbers between 3000 and 3500 cm^{-1}?

6 Suggest reasons why it is better to use infrared spectroscopy to check the purity of a liquid product from a synthesis than to measure its boiling temperature.

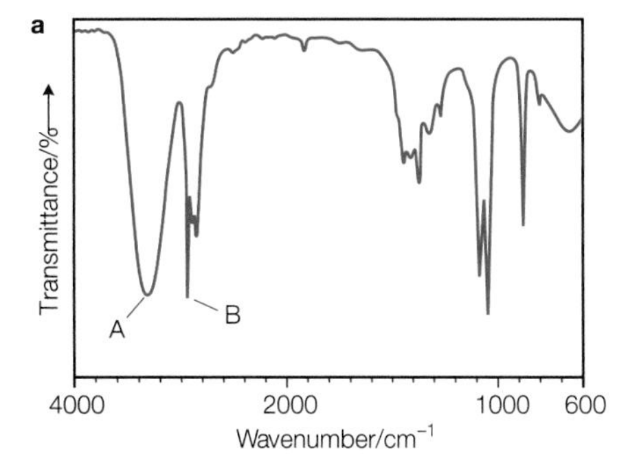

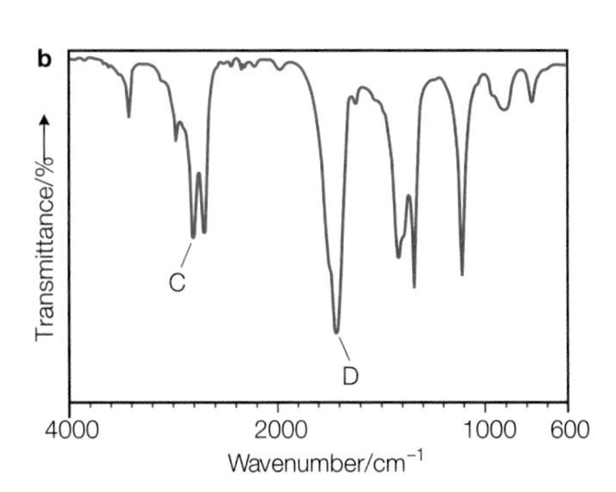

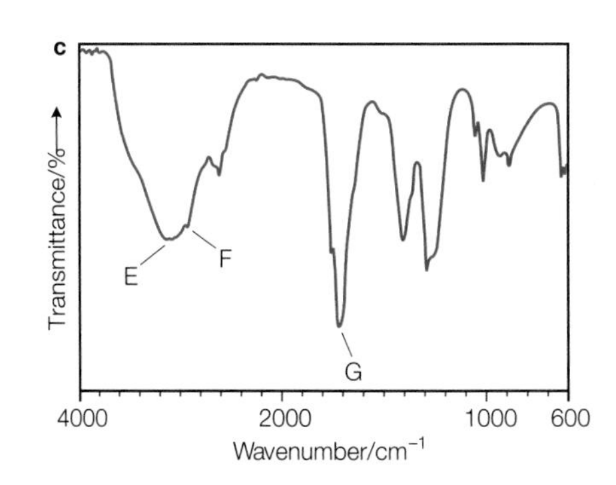

Figure 17.10 ▲
Infrared spectra for three organic compounds. Note that the quantity plotted on the vertical axis is transmittance. This means that the line dips at wavenumbers at which the molecules absorb radiation. Chemists often refer to these dips in the line as 'peaks' because they indicate high levels of absorption.

REVIEW QUESTIONS

1 Figure 17.11 shows the mass spectrum of ethanol.

a) Match the numbered peaks with the formulae of these positive ions from ethanol molecules in a low pressure mass spectrometer: $C_2H_5^+$, CH_2OH^+, $C_2H_5O^+$, $C_2H_5OH^+$, $C_2H_3^+$. (3)

b) Write an equation to represent the formation of the molecular ions. (2)

c) i) Write an equation to show how the molecular ion fragments to give CH_2OH^+.

ii) Why does the other chemical species formed during this fragmentation process not show up in the mass spectrum? (2)

2 With the help of the data sheet on the Dynamic Learning Student website, sketch and label diagrams to predict the pattern of the main peaks in the infrared spectrum of these compounds for wavenumbers between 1500 and 3500 cm^{-1}.

a) cyclohexane (2)

b) 1-bromopropane (2)

c) propanone (2)

3 Oxidation of butan-1-ol gives a product with the infrared spectrum shown in Figure 17.12. Use the data sheets on the Dynamic Learning Student website to interpret the spectrum. State the reagents and conditions used to carry out the oxidation of the alcohol, giving your reasons. (4)

www Extens questic

www Data

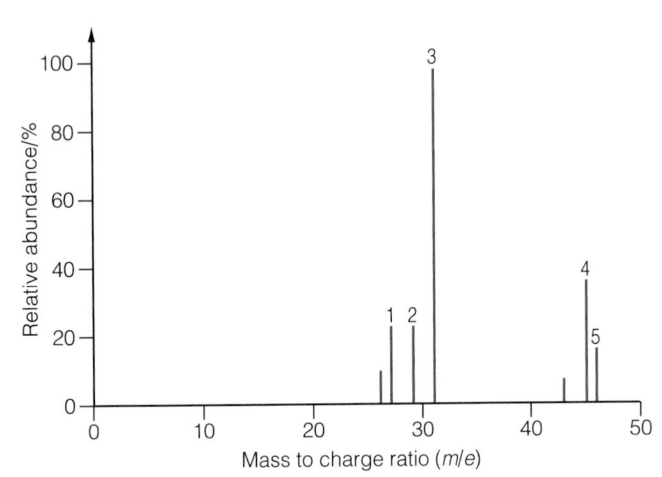

Figure 17.11 ▲

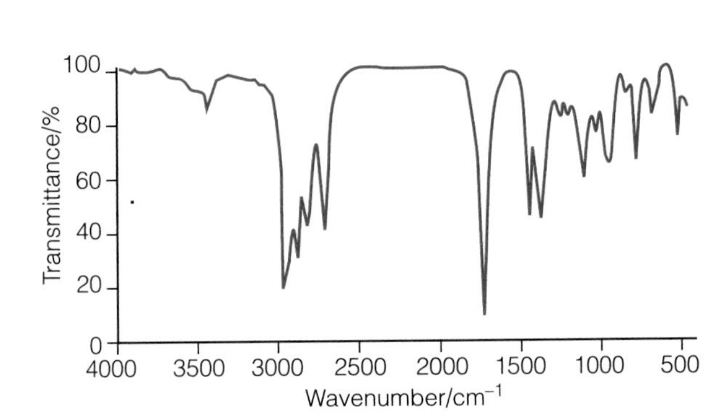

Figure 17.12 ▲

18 Green chemistry

The aim of 'green chemistry' is to meet all our needs for chemicals without damaging our health and the environment. Green chemistry makes the chemical industry more sustainable by using renewable feedstocks, reducing the use of energy resources and cutting down on waste.

18.1 The chemical industry

The chemical industry converts raw materials into useful products. The industry manufactures some chemicals on a scale of thousands, or even millions, of tonnes per year. These are the bulk chemicals such as sodium hydroxide, sulfuric acid, chlorine and ethene.

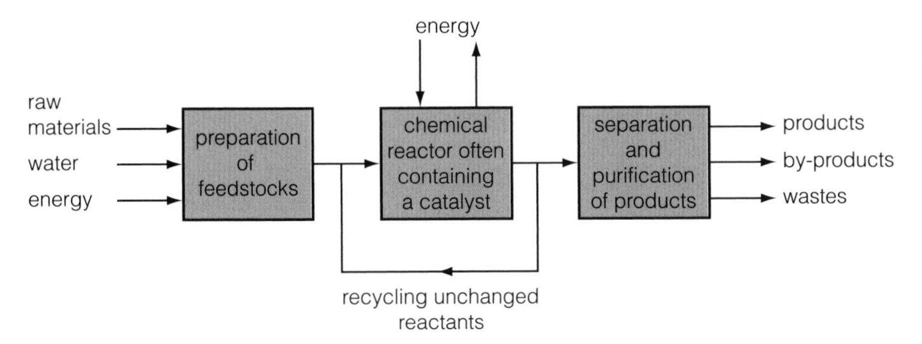

Figure 18.1 ◄
The principles of green chemistry can be applied to all aspects of a chemical process.

Figure 18.2 ▲
The chemical industry helps to provide many of the products we rely on day by day, such as the chemicals used to make clothing fibres, building materials, food preservatives, drugs and medicines. However, some of these synthetic chemicals can be harmful.

The industry makes other chemicals, such as drugs and pesticides, on a much smaller scale. It also makes a wide range of speciality chemicals which are needed by other manufacturers for specific purposes. These include products such as flame retardants, food additives and liquid crystals for flat screen computer displays.

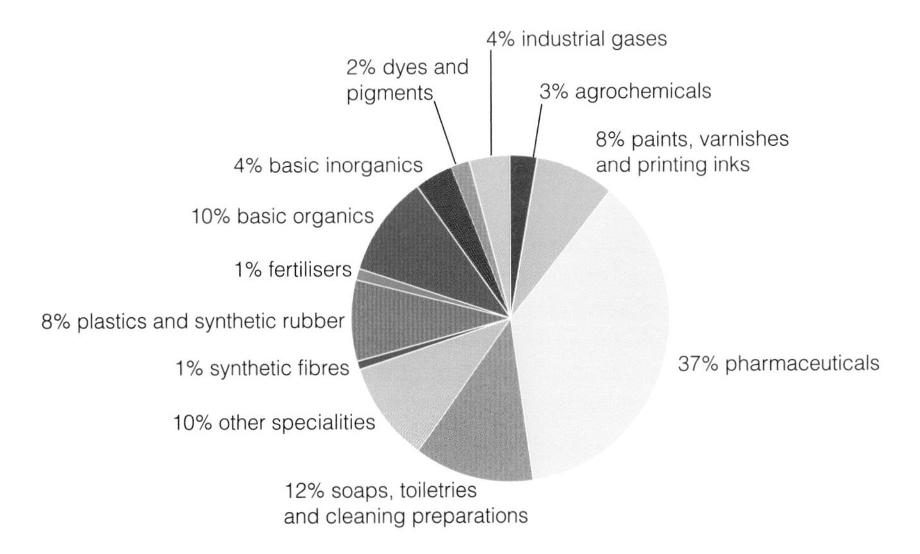

Figure 18.3 ▲

Test yourself

1. Give the name and formula of one bulk chemical.
2. Pharmaceuticals are speciality chemicals, not bulk chemicals. Suggest reasons why the pharmaceutical sector is so large in Figure 18.3.

Definition

Green chemistry is the design of chemical processes and products that reduce or eliminate the production and use of hazardous chemicals.

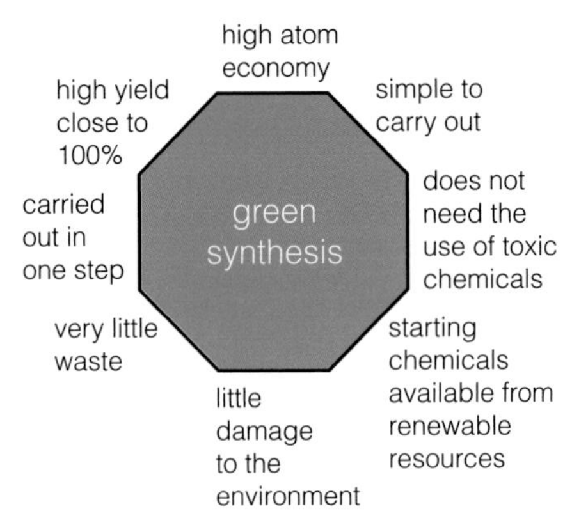

Figure 18.4 ▲
Key features of green synthesis

Definition

Sustainable development meets the needs of the present without compromising the ability of future generations to meet their own needs.

Test yourself

3 Give examples to explain why the traditional chemical industry is not sustainable.

Figure 18.5 ▲
BIOTA spring water is supplied in bottles made of PLA. These bottles degrade within 75–80 days if composted commercially.

18.2 Principles of green chemistry

In 1990, the US Pollution Prevention Act led to new thinking about ways to protect human health and the environment. A major area of interest was the chemical industry. In response to the Act, scientists at the US Environmental Protection Agency were the first to describe the main features of green chemistry.

The key principles of green chemistry cover five broad areas of the production and the use of chemicals.

- ***Changing to renewable resources*** – this means choosing raw materials or chemical feedstocks that can be renewed, instead of relying on crude oil and natural gas.
- ***Finding alternatives to hazardous chemicals*** – this includes producing chemical products that are effective but not harmful, and developing manufacturing processes that avoid toxic intermediates and solvents hazardous to people and the environment.
- ***Developing new catalysts*** – this includes the development of catalysts to make possible processes with high atom economies, and catalysts which are highly selective so that only the desired product is formed, thus cutting down wastes.
- ***Making more efficient use of energy*** – this involves devising processes that run at low temperatures and pressures, and making good use of the energy released by chemical changes.
- ***Reducing waste and preventing pollution*** – this includes minimising waste so that there is less to treat or dispose of; also recycling materials and creating biodegradable products which do not remain in the environment but break down to harmless chemicals. It also includes the increasing use of sensitive methods of analysis to detect pollution.

Many of the changes brought about by applying these principles not only make the chemical industry safer, but also make it more sustainable.

18.3 Greening the chemical industry

Renewable resources

The main raw materials for making organic chemicals are crude oil and natural gas – these are non-renewable resources. The use of renewable resources is still at an early stage in the chemical industry. There are two main approaches:

- ***extract useful chemicals directly from plant materials*** – this is what the perfume industry has done for centuries in obtaining ingredients from rose petals, lavender and other plant materials
- ***break down plant material into simple chemicals for further use*** – for example, this is the way in which ethanol is made by fermentation of starch and sugars worldwide.

A good example of a green plastic made from renewable resources is polylactic acid (PLA). The feedstock is lactic acid made by using bacteria to break down sugars or starch. In the presence of a catalyst, lactic acid molecules join up in pairs to form a cyclic compound. This compound polymerises in the presence of a second catalyst.

PLA can be recycled but it is also biodegradable. It has a range of medical applications, such as stitches for wounds. It is also used for packaging, waste sacks and disposable eating utensils.

Test yourself

4 a) Explain why crude oil is not renewable.
b) Explain why starch and sugars are renewable.

5 Draw the structure of lactic acid, which has the systematic name 2-hydroxypropanoic acid.

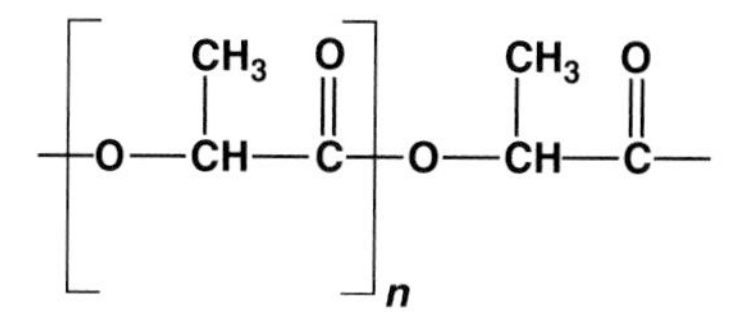

Figure 18.6 ▲
Structure of polylactic acid

Cutting down on hazardous chemicals

The chemical industry uses some highly toxic and corrosive chemicals. An example is hydrogen cyanide, which is acutely toxic and requires special handling to minimise the risks to workers, the community and the environment. The industry has learned to work safely with such chemicals, but accidents do happen and chemicals such as hydrogen cyanide are a threat to people who work on chemical plants and to neighbouring communities.

The chemical company Monsanto, for example, manufactures a herbicide called Roundup®. The original process for making this product required both methanal and hydrogen cyanide – two unpleasant and toxic chemicals.

The key reactions in the original process were exothermic and produced intermediates which became unstable if the temperature was too high. The overall process also created 1 kg of waste for every 7 kg of product, and this waste needed treating before it could be disposed of safely.

Fortunately, Monsanto has been able to develop a new process that relies on a copper catalyst using raw materials that are much less toxic and do not include hydrogen cyanide or methanal. The process is much safer and easier to control because the key reaction is endothermic. There is no waste because the catalyst can be filtered off for reuse, and any chemicals not converted to the product are pure enough to be recycled directly into the reactor. There are fewer steps in the new process and the yield is higher.

Figure 18.7 ◀
This herbicide is now made by a greener process that avoids toxic chemicals such as hydrogen cyanide and methanal.

Most of the organic chemicals used in industry are insoluble in water, so many important reactions must take place in other solvents. Traditionally the industry has used a range of toxic solvents, such as benzene and tetrachloromethane. The most hazardous solvents have largely been phased out but the industry still uses other volatile organic compounds (VOCs) as solvents, and VOCs are pollutants when they evaporate into the air. Research by green chemists aims to replace harmful solvents with safer alternatives.

The pharmaceutical company Pfizer has applied the principles of green chemistry to the manufacture of the drug Zoloft® that is prescribed to people with depression. Zoloft was first made in a multi-stage process which involved the use and recovery of four solvents – dichloromethane, tetrahydrofuran, methylbenzene and hexane. The new process takes place in two key steps, the solvent is ethanol and the use of a more selective palladium catalyst reduces waste.

As well as avoiding the use of hazardous chemicals as reactants and solvents, green chemists aim to reduce the risks from chemicals present in the final products themselves. Chlorofluorocarbons (CFCs), developed as refrigerants, have been replaced by HFCs (hydrofluorocarbons) that do not damage the ozone layer, even though they are powerful greenhouse gases (see Sections 15.8 and 19.3).

Test yourself

6 a) Explain the danger of a runaway reaction in the older method for making Roundup.
   b) Explain why this danger is absent from the newer method.

7 Why is it an advantage to reduce the number of reaction steps involved in the manufacture of a chemical?

8 How has Pfizer improved both worker and environmental safety in redeveloping the process for making Zoloft?

Activity

A fire extinguishing foam

In the 1960s the US Navy devised foam fire-extinguishing systems to deal with burning hydrocarbons. They were effective, but at the temperature of a fire they released hydrofluoric acid and fluorocarbons. The mixture used to make foams included surface-active agents mixed with water to form stable foams.

Figure 18.8 Using foam to put out a fire

Unfortunately the surface-active agents used by the US Navy washed into the ground. They were not biodegradable, so they contaminated water supplies.

In 1993, Pyrocool Technologies developed a fire-extinguishing foam (FEF) which was designed to be much less harmful to the environment. The special foam contains no fluorine compounds and is based on biodegradable surfactants. Very low concentrations are needed and the new foams are very effective.

1 Suggest reasons why a foam of air and water is much more effective than water alone in putting out burning hydrocarbons.

2 Explain why it is hazardous to release hydrogen fluoride into the air.

3 In the presence of electrical equipment, hydrocarbon fires can be put out with halons such as CF_3Br. What environmental damage is caused by halons?

4 Explain why Pyrocool FEF is a 'greener' product than other chemicals used to put out fires.

Developing new catalytic processes

Many of the reactions traditionally used to make drugs and speciality chemicals have involved reagents such as metals, metal hydrides, acids and alkalis. These reagents are used up in the reaction and turn into chemicals that have to be separated from the main product and then be treated as wastes. These processes have low atom economies (see Section 1.8).

One of the challenges for green chemists is to develop catalytic methods to create new chemicals from simple raw materials such as hydrogen, steam, ammonia and carbon dioxide. In an efficient process the catalyst is constantly recycled and not used up. These methods have high atom economies.

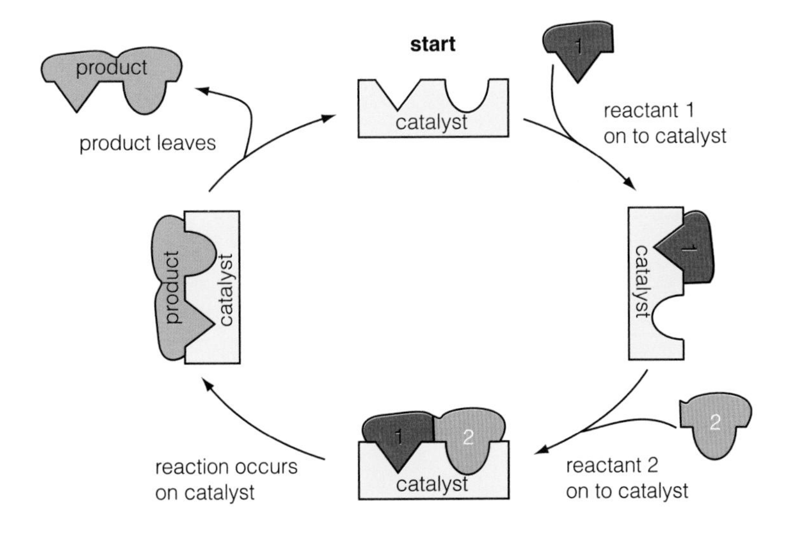

Figure 18.9 ◀ A catalytic cycle. The catalyst is not used up. A long-lasting catalyst can convert a very large amount of reactants into product before it has to be replaced.

Test yourself

9 a) Sketch a graph to show how the energy profile of the system might vary as the reaction proceeds from reactants to products for the catalysed reaction shown in Figure 18.9.

b) Add to the same graph a curve to represent the change in energy for the equivalent uncatalysed reaction.

Over 40 years, developments in catalysts have made the manufacture of ethanoic acid very much more efficient. Over six million tonnes of the acid is made worldwide each year. This is an important bulk chemical. Ethanoic acid is a key intermediate on the way to making a very wide range of other chemicals including polymers, drugs, herbicides, dyes, adhesives and cosmetics.

Activity

Greening the manufacture of ethanoic acid

Until the 1970s, the main method of making ethanoic acid was to oxidise hydrocarbons from crude oil with oxygen in the presence of a cobalt(II) ethanoate catalyst (Figure 18.10). This process ran at 180–200 °C and at 40–50 times atmospheric pressure. The reaction produced a mixture of products which had to be separated by fractional distillation. The economics of the process depended on finding markets for all the products.

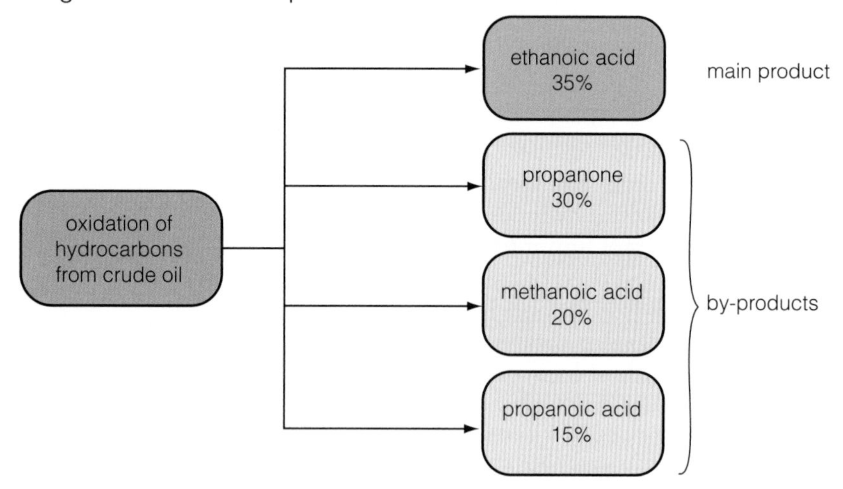

Figure 18.10 ▲
Percentages of the main product and by-products from the manufacture of ethanoic acid by oxidising hydrocarbons. The atom economy of the process is about 35%.

In 1970, Monsanto introduced a new process (Figure 18.11). Methanol and carbon monoxide combine to make ethanoic acid as the only product in the presence of a catalyst made from rhodium and iodide ions. This process runs at 150–200 °C and 30–60 times atmospheric pressure.

This is a very efficient process with high yields and a high atom economy. Much less energy is needed than in the older process based on hydrocarbons. The reaction is fast and the catalyst has a long life.

The Monsanto process does have some disadvantages. Rhodium is very expensive. Also, there has to be a high enough concentration of water present in the reactor to prevent the rhodium and iodide parts of the catalyst combining to form insoluble salts. These salts are inactive and have to be recovered and converted back to the catalyst. Water has to be separated from the product by distillation. Furthermore, rhodium catalyses side reactions such as the reaction of carbon monoxide with water to form carbon dioxide and hydrogen.

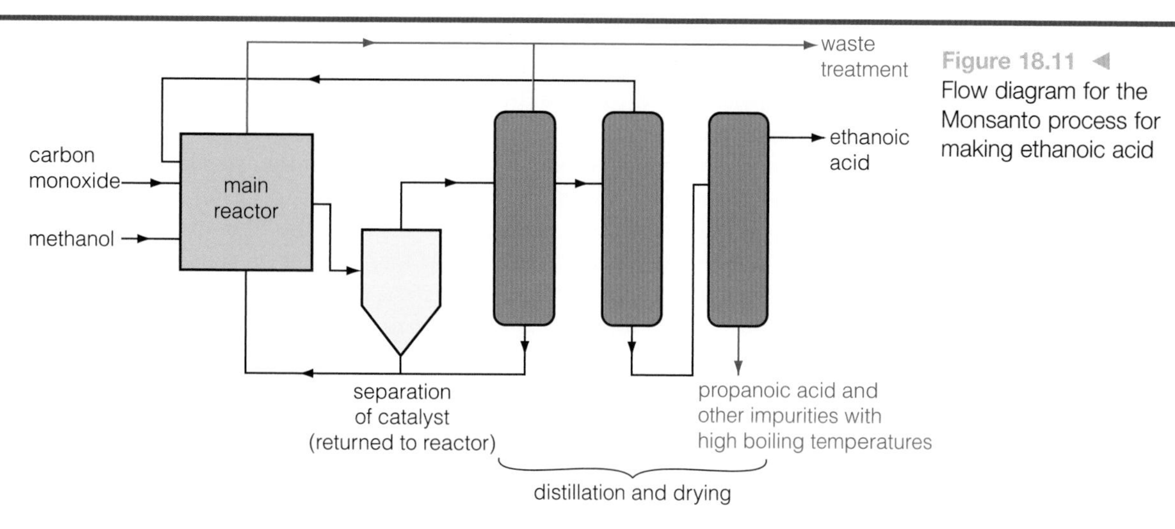

Figure 18.11 ◀ Flow diagram for the Monsanto process for making ethanoic acid

In 1986, the oil company BP bought all the rights to the process from Monsanto. The company now runs a variant of the process called the Cativa process. The reaction of methanol with carbon monoxide is the same but it uses an iridium catalyst instead of a rhodium catalyst. The catalyst is made even more effective by adding ruthenium compounds as catalyst promoters which speed up the reaction by a factor of three.

The Cativa process has several advantages over the Monsanto process:

- iridium is much cheaper than rhodium
- the process is much faster – so more product can be made without increasing the size of the plant
- less carbon monoxide reacts with water – so the yield from CO is over 94%
- the catalyst is more selective in producing ethanoic acid – so less energy is needed to purify the product
- the catalyst is more stable – so it lasts longer
- water concentrations in the plant can be lower – so less energy is needed to dry the product.

1 a) Write an equation for the reaction used to make ethanoic acid in the Monsanto process.

b) What is the atom economy for the Monsanto process?

2 The yield of the Monsanto process is 98% based on methanol, but 90% based on carbon monoxide. Explain why this difference arises.

3 Suggest why the Monsanto process uses much less energy than the older process based on oxidising hydrocarbons.

4 Why is the Monsanto process more economical than the older process in making ethanoic acid?

5 What is the atom economy for the Cativa process?

6 In what ways is the Cativa process greener than the Monsanto process?

7 Metals are present in the catalysts used for all three processes for making ethanoic acid. Where do these metals appear in the periodic table?

Increased energy efficiency

New catalysts can make a big contribution to reducing the energy needed for a particular process. For this, and other reasons, the chemical industry now makes much more efficient use of energy. The average energy needed per tonne of chemical product is less than half of that needed 50 years ago.

A direct way of cutting the use of energy is to install efficient insulation around hot pipes and reaction vessels. This also helps to improve maintenance by preventing loss of steam from leaking valves on steam pipes.

An efficient chemical plant uses the energy from exothermic processes to produce steam for heating and for generating electricity. A notable example of this is the manufacture of sulfuric acid. The first stage of this process involves burning sulfur to make sulfur dioxide. This is so exothermic that a sulfuric acid plant can produce enough steam not only to generate all the electricity needed to run the plant, but also to provide a surplus of electric power that can be sold to the National Grid.

An aim of green chemistry is to make processes more energy efficient. One approach is to devise new methods that run at lower temperatures and pressures. This can be done with the help of biotechnology. DuPont has developed a polymer fibre made from glucose. The polymer, Sorona®, is used to make fibres for clothing and upholstery. A genetically engineered micro-organism converts glucose from starch directly into an organic acid that is the starting chemical for synthesis. The genetically modified (GM) micro-organism makes it possible to replace a chemical from oil as the starting material. This reduces the amount of energy needed and makes the process safer.

Long ago, electric hotplates or heating mantles replaced Bunsen burners in most research and industrial laboratories. Today, microwaves are taking over in many contexts where controlled heating is important. This is especially true in research and development laboratories of pharmaceutical and biotechnology industries. Microwave heating can help chemists to achieve cleaner and more efficient chemical reactions with higher yields, compared to traditional heating methods. Heating with microwaves is very efficient because they heat only the chemicals and not the apparatus or its surroundings.

Test yourself

10 Identify four reasons why the chemical industry needs energy to convert raw materials into useful products.

11 Identify the green features of processes based on micro-organisms.

12 Suggest reasons why microwave heating is used in the pharmaceutical industry but not in the bulk chemical industry.

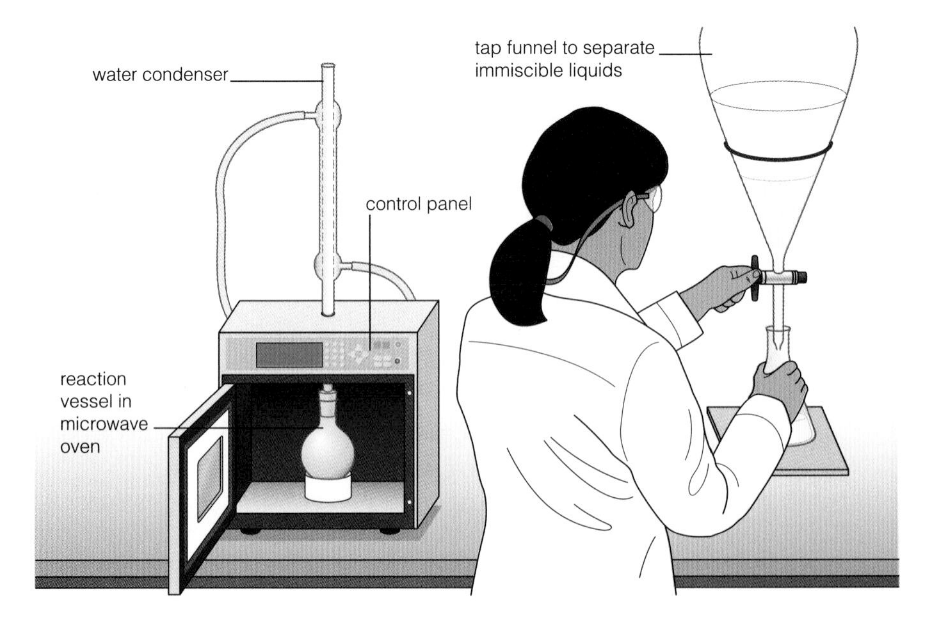

Figure 18.12 ▶ Microwave ovens are used for research. Larger ovens are available for making chemicals on a bigger scale. Ovens can be programmed precisely to vary the rate and timing of the heating.

Polar molecules in a material absorb microwave energy as they rotate backwards and forwards. Internal friction resulting from intermolecular forces spreads the energy and the substance gets hotter as its molecules rotate and vibrate faster. This type of heating can speed up some reactions enormously because the energy is absorbed directly by the molecules. Microwave heating during the synthesis of aspirin completes the process 100 times faster than conventional heating. In catalytic reactions, it is thought that microwaves selectively heat the catalyst surface to a high temperature, greatly increasing the catalytic effect.

Because microwaves heat the chemicals directly, it is often possible to cut down the quantity of solvent needed for a reaction in solution – sometimes no solvent is needed at all. This can be achieved by absorbing the reactants onto a porous form of an inert material, such as alumina. Working like this can help to cut down the materials and energy needed to purify a product.

Figure 18.13 ▲
Toxic waste is hazardous and expensive to dispose of safely.

Reducing waste and cutting pollution

One of the key principles of green chemistry is that it is better to prevent waste than to treat or clean any up after it has formed. This has led to an emphasis on catalytic processes with high atom efficiencies.

The synthesis and manufacture of new drugs often produces relatively large volumes of waste. Syntheses of this kind sometimes involve several steps, each with its own reactants and solvents, as well as chemicals to purify the product. Lilly Research Laboratories has applied the principles of green chemistry to come up with a new process designed to make a drug to stop convulsions. The original process used large volumes of solvent and several toxic chemicals, including chromium oxide which can cause cancer. For every 100 kg of product, the new process cuts out the need for 35 000 litres of solvent and eliminates the production of 300 kg of waste chromium compounds.

Test yourself

13 Suggest three reasons why the chemical industry can become more efficient and more economic by reducing waste from its processes.

REVIEW QUESTIONS

Extension questions

1 a) Calculate the atom economies for the two methods of producing $C_6H_5COCH_3$ in processes A and B below:

A $3C_6H_5CH(OH)CH_3 + 2CrO_3 + 3H_2SO_4 \rightarrow 3C_6H_5COCH_3 + Cr_2(SO_4)_3 + 6H_2O$

B $2C_6H_5CH(OH)CH_3 + O_2 \xrightarrow{\text{catalyst}} 2C_6H_5COCH_3 + 2H_2O$ **(2)**

b) Identify features of the two reactions which make process B greener than process A. **(2)**

2 The main method of making ethanol worldwide is by anaerobic fermentation of sugars with yeast. After fermentation, the solution contains about 14% ethanol. The ethanol is separated from the solution by distillation.

a) Write an equation for the fermentation of glucose ($C_6H_{12}O_6$) to make ethanol and carbon dioxide. **(1)**

b) Work out the atom economy for this process. **(2)**

In the UK, the main method of manufacturing ethanol is by hydrating ethene with steam in the presence of a phosphoric acid catalyst. Some side reactions occur producing other chemicals, such as methanol and ethanal. The process runs at 300 °C and a pressure of 60–70 times atmospheric pressure. The product is purified by distillation.

c) Write an equation for the hydration of ethene to make ethanol. **(1)**

d) Work out the theoretical atom economy for the process shown in the equation. **(1)**

e) Why, in practice, is the efficiency of the process less than your answer to part **d)** would suggest? **(1)**

f) Compare the sustainability of the two processes. **(6)**

3 a) State what you understand by the term 'sustainable development'. **(2)**

b) Explain, with at least two examples, why applying the principles of green chemistry helps to make the chemical industry more sustainable. **(8)**

19 Chemistry in the atmosphere

From the chemist's viewpoint, the atmosphere can be imagined as a giant chemical reactor with chemical inputs from land, sea, from living things and from our human activities. The mixture of chemicals is irradiated with energy from the Sun.

19.1 The atmosphere

The atmosphere is composed of layers. We live in the lowest layer, which is called the troposphere. The air gets cooler with height in the troposphere. There is then an important change where the temperature begins to rise again. This temperature inversion marks the beginning of the second layer, called the stratosphere (Figure 19.1).

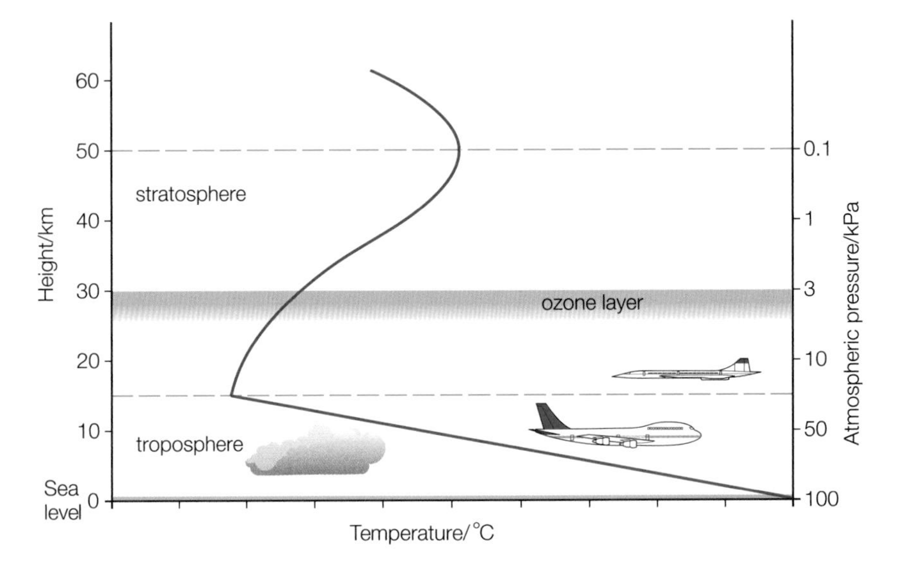

Figure 19.1 ▶ The structure of the Earth's atmosphere

The thin, cold air at the top of the troposphere is trapped between the warmer air of the stratosphere and the warmer, denser air in the troposphere below. This creates a barrier that makes it difficult for molecules to diffuse upwards into the stratosphere.

Human activities change the composition of the atmosphere. These human, or anthropogenic, inputs to the troposphere include pollutant gases from motor vehicles, power stations and industrial processes. Most of these pollutants do not remain unchanged for long enough to reach the stratosphere. An important exception to this is the group of compounds called chlorofluorocarbons, or CFCs. These very stable compounds do not break down in the lower atmosphere. In time, they move up into the stratosphere where they can destroy the ozone that protects us from damaging radiation from the Sun.

Definition

Anthropogenic effects are changes brought about by human activities. They include factors that affect the atmosphere, such as burning fossil fuels, deforestation and intensive agriculture.

19.2 Evidence for atmospheric change

Sensitive techniques of chemical analysis, such as the analysis of glacial ice, have played an important part in obtaining data about changes in the atmosphere that happened hundreds of thousands of years ago.

Scientists have been measuring the carbon dioxide in the air since 1958. The main site for making measurements is at the Mauna Loa observatory in Hawaii. Mauna Loa is the highest mountain in Hawaii and the observatory is 3400 metres above sea level. The air samples are collected through inlet tubes placed several metres above the ground in the barren region of volcanic rock.

The air is analysed using an infrared gas analyser, which continuously records the concentration of carbon dioxide in the stream of gas passing through it.

Test yourself

1 Suggest reasons why the area round the Mauna Loa observatory is a good place for measuring the concentration of carbon dioxide in the air.

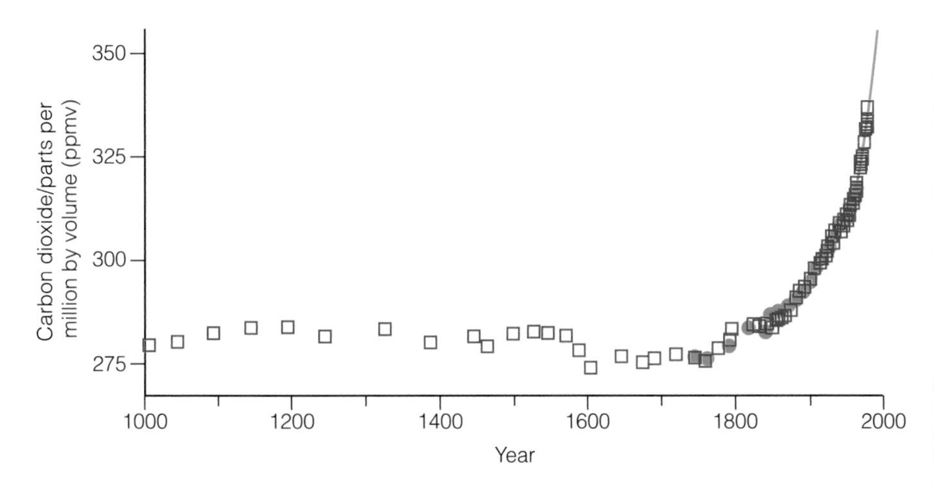

Figure 19.2 ◄
Carbon dioxide concentration in the atmosphere over the last 1000 years. The red squares and green dots are the results from air trapped in glacial ice. The blue line is based on direct measurements of the gas in the atmosphere.

Accurate information about the concentration of gases in the air before 1958 comes from studying ice in the Antarctic and other cold regions of the Earth. Snow falling in the Antarctic never melts. Instead, it is compressed by more snow falling on top of it and turning it to ice. Over hundreds of thousands of years, the Antarctic ice caps have built up to a depth of more than 4 kilometres.

The falling snow traps pollutants and particles. The snow crystals also trap bubbles of air, which remain in the ice forever. Even the water in the snow carries clues to past climates. Most of the water is $H_2{}^{16}O$ but it also contains a small proportion of $H_2{}^{18}O$. The proportions of the two isotopes depend on the Earth's temperature at the time the snow falls. Water with the heavier isotope has a slightly greater tendency to condense out as air moves from warmer regions to the Antarctic. The colder the climate, the more water condenses, as cloud and rain, before reaching the Antarctic where snow falls. So the colder the climate, the higher the ratio of $^{16}O : {}^{18}O$ in the snow.

Figure 19.3 ▲
Dr Eric Wolff, principal investigator at the British Antarctic Survey, studying an ice core in Antarctica

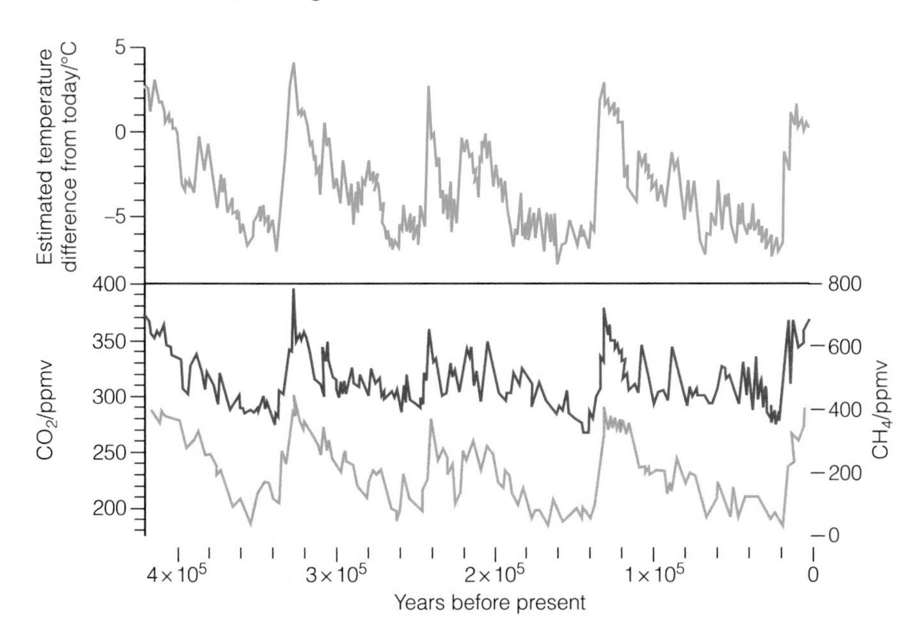

Figure 19.4 ◄
Changes to the temperature and to the concentrations of carbon dioxide and methane in the atmosphere over the last 420 000 years from analysis of ice cores in Antarctica. The red line shows data for methane (CH_4, right axis) and the orange line shows data for carbon dioxide (CO_2, left axis).

Scientists drill down into the ice to extract cylindrical cores about 10 cm in diameter which they store at −20 °C. They cut the cores into sections, some of which are analysed straight away in the field. Other samples are sent to laboratories for analysis by techniques, such as mass spectrometry, which are not possible in the Antarctic.

Test yourself

2 Two of the techniques used to analyse ice samples are mass spectrometry and infrared spectroscopy (Topic 17).
  a) Which technique is suitable for measuring the ratio of oxygen isotopes in ice?
  b) Which technique is suitable for measuring the carbon dioxide concentration in air samples from ice bubbles?

3 Ice samples contain only tiny traces of some pollutants. Suggest two precautions needed to achieve accurate results when samples are sent to laboratories for analysis.

4 What can you conclude about CO_2 changes in the atmosphere from the information in Figure 19.4?

5 Analysis of Greenland ice has showed that the level of lead pollution in 1969 was 200 times the level in 1700. In 1990, the lead level was just 10 times its natural level. Suggest reasons for these changes.

19.3 Greenhouse gases and climate change

Without the atmosphere, the whole Earth would be colder than Antarctica. The greenhouse effect keeps the surface of the Earth about 30 °C warmer that it would be if there was no atmosphere. There would be no life on Earth without this greenhouse effect.

Most of the energy in the Sun's radiation is concentrated in the visible and ultraviolet regions of the electromagnetic spectrum. When this radiation reaches the Earth's atmosphere, about 30% is reflected back into space, 20% is absorbed by gases in the air and about 50% reaches the surface of the Earth.

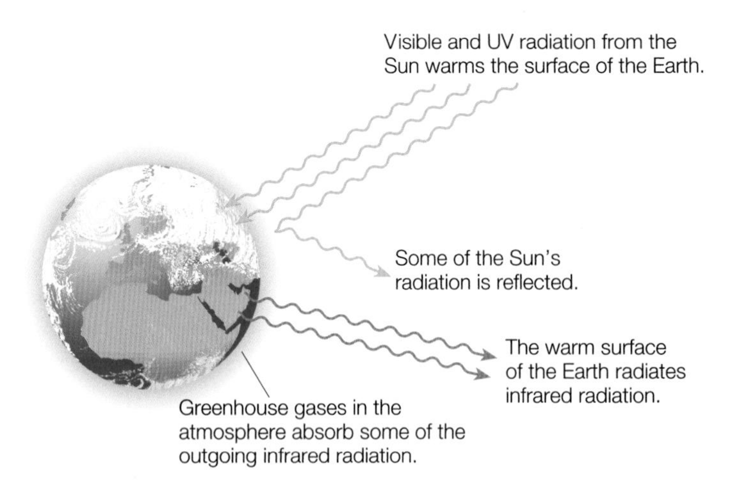

Figure 19.5 ▶ The greenhouse effect

The surface of the Earth is very much cooler than the Sun. This means that it radiates most of its energy back into space at longer, infrared wavelengths. Some of this infrared radiation is absorbed by molecules in the air, warming the atmosphere – this is the greenhouse effect.

The Earth's average temperature stays the same if the energy input from the Sun is balanced by the re-radiation of energy back into space. Anything that changes this steady state gradually leads to global warming or global cooling.

Nitrogen and oxygen make up most of the air but they do not absorb infrared radiation. This is because they are not polar. They are never polar, even when their molecules vibrate, and so they cannot absorb infrared radiation (Section 17.2).

The gases in the air which do absorb infrared radiation are called greenhouse gases. The natural greenhouse gases which help to keep to Earth warm are water vapour, carbon dioxide and methane. Water vapour makes the biggest contribution.

Since the Industrial Revolution, human activity has led to a marked increase in the concentrations of greenhouse gases in the atmosphere. More recently, new synthetic chemicals such as CFCs have joined the mix. CFCs and their related compounds are very powerful greenhouse gases.

Adding greenhouse gases to the air means that more infrared radiation from the surface of the Earth is trapped – and this enhances the greenhouse effect. As a result, the Earth warms up more than it would naturally. As the Earth warms, its climate changes.

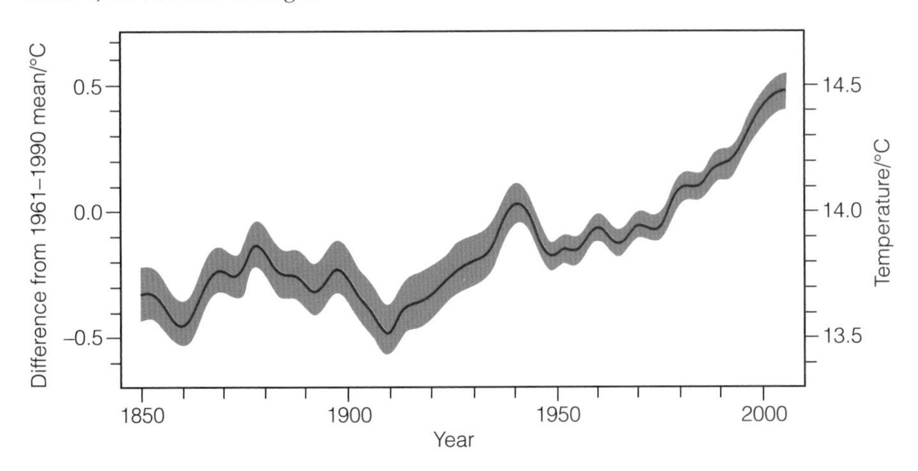

Figure 19.6 ◀
Variations in the global mean temperature since 1850 compared to the average value from 1961 to 1990. The blue area indicates the range of uncertainty around the red line showing the mean value

The contribution that a greenhouse gas makes to the enhanced greenhouse effect is determined by several factors. These include how strongly the gas absorbs infrared energy, how much of it is already in the atmosphere, how much is being added to the atmosphere and how long it survives in the atmosphere.

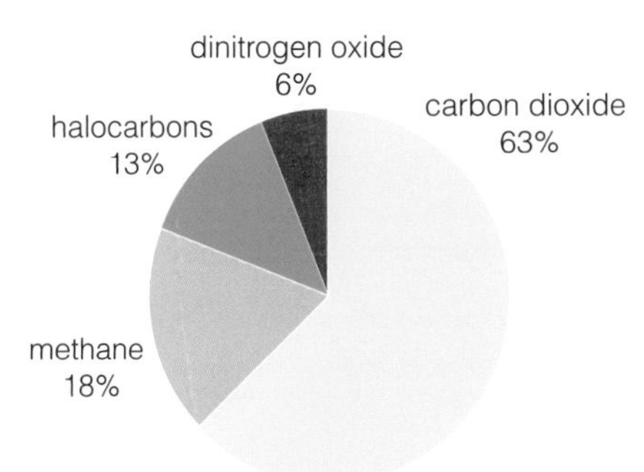

Figure 19.7 ▲
Contributions of the main greenhouse gases to the enhanced greenhouse effect

Test yourself

6 Methane has non-polar molecules yet it can absorb infrared radiation.
   a) Draw the structure of methane and show the polarity of the bonds.
   b) Why is methane non-polar overall?
   c) Using a simple diagram, suggest a pattern of vibrations for C–H bonds in methane that can lead to a varying dipole.

7 Human activities produce a lot of water vapour. Why does this have little effect on the concentration of water in the atmosphere?

8 Give three examples of human activities that have led to a marked increase in the concentration of carbon dioxide in the atmosphere.

9 The concentration of all the CFCs in the atmosphere is about 1 000 000 times less than the concentration of CO_2, yet these compounds make a significant contribution to the enhanced greenhouse effect. Suggest an explanation for this.

10 Why do you think most public debate about climate change and greenhouse gases focuses on carbon dioxide?

11 The blue area on either side of the graph line in Figure 19.6 is narrower for the year 2000 than for the year 1900. Suggest reasons for this.

12 How does the data in Figure 19.6 suggest that recent global warming is the result of human activities and not just the result of natural climate variations?

19.4 Fuels for transport and their carbon footprint

In 2007, the Intergovernmental Panel for Climate Change (IPCC) issued its third assessment of scientific evidence. The IPCC report showed a strong scientific consensus that climate change is now caused by human activity, and that it is a serious and urgent issue. Most governments accept that we must make a huge change in the way we supply and use energy, and in all other activities that release greenhouse gases.

The UK government has a strategy for dealing with climate change. Each year, by law, it must report to Parliament on progress towards meeting the targets aimed at reducing carbon dioxide emissions. Organisations and individuals are being encouraged to consider their own carbon footprint and to find ways of reducing it.

A carbon footprint measures the total amount of carbon dioxide, and other greenhouse gases, released into the air over the full life cycle of a process or product. On average, the carbon footprint for one person in the UK is just under 11 tonnes of carbon dioxide per year from all activities. Of this amount over 40% comes directly from our individual activities, such as heating and lighting our homes and driving vehicles. The average driver in the UK produces 4 tonnes of carbon dioxide per year just from motoring.

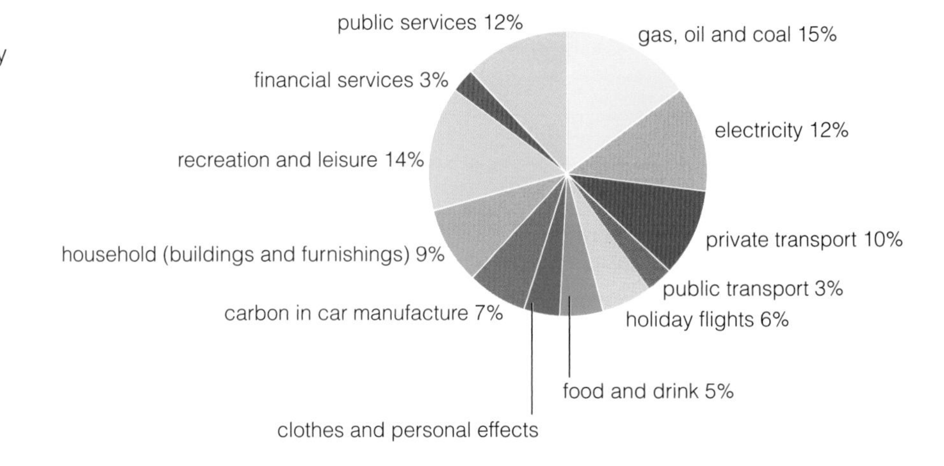

Figure 19.8 ▶
Part of a carbon footprint comes directly from activities that involve burning fuels, such as home heating and travel. Other parts of the footprint are indirect and arise from the whole life cycle of the products and services we use.

Test yourself

13 Identify possible changes to your lifestyle that could significantly reduce your carbon footprint.

Definitions

A **carbon footprint** measures the total amount of greenhouse gases produced by a process or product in tonnes of carbon dioxide.

Carbon dioxide equivalent, CO_2eq, is a measure of greenhouse gases in terms of the quantity of carbon dioxide that would have the same global warming effect.

Bioethanol is ethanol made from plant crops or other kinds of biomass.

A process is **carbon neutral** if the carbon dioxide (or other greenhouse gases) released is balanced by actions which remove an equivalent amount of carbon dioxide from the atmosphere.

Alternative fuels

Almost all forms of transport are based on burning fossil fuels directly or indirectly. Transport makes a big contribution to people's carbon footprint. For this reason, there is growing interest in alternative fuels.

In recent years, there has been a rapid expansion in the production of bioethanol as a fuel throughout the world, particularly in Brazil and the USA. The apparent advantage of bioethanol is that it is carbon neutral. The crops grown to make the ethanol take in carbon dioxide during photosynthesis as they grow; the same quantity of carbon dioxide is then given out as the bioethanol burns.

The problem with bioethanol in the USA is that its production uses a lot of energy from oil. This includes the fuel to make and run farm machinery, energy and chemicals to manufacture pesticides and fertilisers, as well as energy to process the crop and produce ethanol.

Figure 19.9 ▲
A refinery in the USA which converts starch from maize into bioethanol

Studies of the overall impact of biofuels have produced conflicting results. Some have even suggested that more energy is expended in making the bioethanol than is available from burning the fuel. A more optimistic study has suggested that ethanol from maize in the USA does reduce greenhouse gas emissions, but only by about 12% compared to petrol.

Reduction in CO_2 emissions is more favourable for biodiesel from vegetable oil. Biodiesel from soyabeans can reduce emissions by 41% compared with conventional diesel fuel. This is mainly because energy is not needed for distillation during the production of the fuel. In addition, far fewer fertilisers and pesticides are used in growing soyabeans.

Instead of making bioethanol from maize, a better solution is to produce ethanol from non-food sources, such as woody plants and agricultural wastes including straw. Using micro-organisms to break down cellulose to sugars does not compete with food supplies and makes it possible to process large amounts of waste. However, this approach is still at the research and development stage.

The conditions for producing bioethanol in Brazil are more favourable. The ethanol is manufactured by fermenting sugars from sugar cane. The refineries that make bioethanol in Brazil have the advantage that they can meet all their energy needs for heating and electricity by burning the sugar-cane waste. However, there are several negative aspects of this industry in Brazil where, among other things, large-scale deforestation has been carried out to make way for sugar cane plantations.

Test yourself

14 How can motorists reduce their carbon footprint using vehicles running on fuels such as petrol or diesel?

15 a) Why, in theory, is the use of bioethanol as a fuel considered to be carbon neutral?
b) Why, in practice, is the use of bioethanol as a fuel not carbon neutral?

16 Why does the use of bioethanol in Brazil have a smaller carbon footprint compared to the use of bioethanol in the USA?

Activity

Buses running on hydrogen

Hydrogen has advantages as a fuel because it produces only water when it reacts with oxygen. Transport for London ran a three-year trial of buses fuelled by hydrogen. The hydrogen combined with oxygen in fuel cells to provide electricity for the motors. The three buses used in the trial proved so reliable that the trial period was extended. The main aim was to cut local pollution of the air in London streets by avoiding the use of diesel fuel.

There are no natural reserves of hydrogen. The gas must be produced from compounds such as water or methane. Hydrogen is, in effect, a store of energy.

Hydrogen for the London bus trial was supplied by a petrochemical company which makes the gas from natural gas (methane) and steam in a catalytic process. The first stage involves an endothermic reaction which produces hydrogen and carbon monoxide. In the exothermic second stage, the carbon monoxide reacts with more steam to make hydrogen and carbon dioxide.

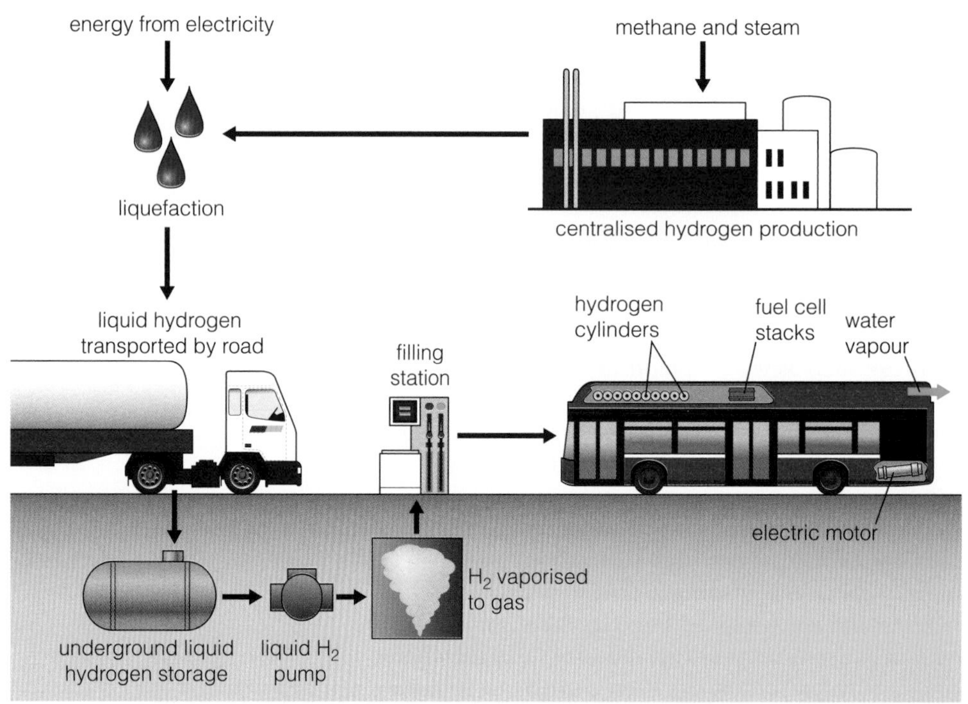

Figure 19.10 ▲ Infrastructure needed to support the trial of fuel-cell buses in London

The carbon dioxide and hydrogen are separated with the help of a molecular sieve which traps carbon dioxide molecules but not the smaller hydrogen molecules. A molecular sieve is usually a zeolite – a crystalline material with channels in the giant structure through which smaller molecules can pass freely. From time to time, the zeolite has to be regenerated by flushing out the carbon dioxide with a stream of hydrogen.

1 a) Write an equation for the first stage – the reversible reaction of methane with steam.

b) What are the conditions that favour the formation of products in this equilibrium?

2 a) Write the equation for the reaction in the second stage, when carbon monoxide reacts with more steam to make hydrogen and carbon dioxide.

b) Overall, in the two stages, how many moles of hydrogen are formed from each mole of methane?

3 Explain why hydrogen is described as a 'store of energy'.

4 Was the use of buses with hydrogen fuel cells in the London trial a carbon-neutral system? Give reasons for your answer.

5 Which of the following methods of making hydrogen have the potential to be carbon neutral, and why?

A Electrolysis of water

B Growing genetically modified algae which release hydrogen during photosynthesis

6 Comment on the extent to which the bus trial in London reduced air pollution.

19.5 Free radicals in the stratosphere

The ozone layer is a concentration of the gas in the stratosphere at about 10–50 km above sea level. At this altitude, ultraviolet (UV) light from the Sun splits oxygen molecules into oxygen atoms.

$$O_2 + \text{UV light} \rightarrow O\bullet + O\bullet$$

These are free radicals which then convert oxygen molecules to ozone.

$$O\bullet + O_2 \rightarrow O_3$$

The ozone formed also absorbs UV radiation. This splits ozone molecules back into oxygen molecules and oxygen atoms, destroying the ozone. This is the reaction that protects us from the worst effects of sunburn.

$$O_3 + \text{UV light} \rightarrow O_2 + O\bullet$$

In the absence of pollutants, a steady state is reached with ozone being formed at the same rate as it is destroyed. Normally the steady-state concentration of ozone is sufficient to absorb most of the dangerous UV light from the Sun.

Reactions with CFCs

The problem with CFCs, and some other halogen compounds, is that they are very unreactive chemically. They escape into the atmosphere where they are so stable that they last for many years, long enough for them to diffuse up to the stratosphere. In the stratosphere, the intense UV light from the Sun splits up CFCs, producing compounds such as hydrogen chloride, HCl.

These chlorine compounds then undergo a series of reactions that only happen in winter in the stratosphere above Antarctica. Once chemists understood these reactions, they could explain why the 'hole' in the ozone layer is most extreme over the South Pole.

In the dark winter over Antarctica, the stratosphere is so cold that clouds of ice crystals form. These clouds circle the pole in a spinning vortex of air that traps the ice crystals. On the surface of the ice crystals, a whole series of reactions take place that cannot happen anywhere else in the atmosphere. The reactions are fast. The end result of the reactions produces chemicals including chlorine, Cl_2, which can destroy ozone in sunlight. When sunlight returns to the Antarctic in the spring of the southern hemisphere, the chlorine molecules are quickly split into chlorine atoms.

Figure 19.11 ▲
Polar stratospheric clouds form between 20 and 30 km above the ground. They only form at very low temperatures. That is why they only appear in winter and mainly over Scandinavia, Scotland, Alaska and Antarctica.

Test yourself

17 Why is the presence of ozone in the stratosphere important to life on Earth?

$$Cl_2 + \text{UV light} \rightarrow Cl\bullet + Cl\bullet$$

Chlorine atoms then react with ozone:

$$Cl\bullet + O_3 \rightarrow ClO\bullet + O_2$$

$$ClO\bullet + O\bullet \rightarrow Cl\bullet + O_2$$

$$\text{Overall: } O\bullet + O_3 \rightarrow 2O_2$$

The second reaction involves oxygen atoms, which are common in the stratosphere, and it reforms the chlorine atom. This means that one chlorine atom can rapidly destroy many ozone molecules. This effect was noticed in the early 1980s when scientists in the Antarctic noticed that the ozone concentration in the stratosphere was much lower than expected. Since then the ozone layer has been monitored by satellites which have confirmed that there is a 'hole' in the ozone layer, not only over the Antarctic but in other regions too.

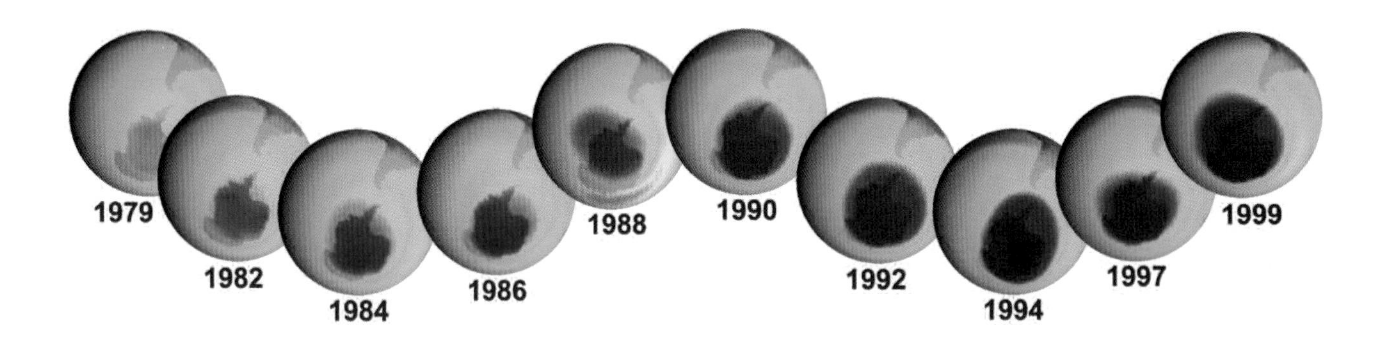

Figure 19.12 ▲
The growth of the ozone hole over Antarctica over 20 years. The dark blue end of the spectrum indicates the region of maximum ozone depletion; the red end of the spectrum shows the area of least depletion.

Test yourself

18 a) Using curly half-arrows, show how chlorine molecules split to form chlorine atoms.
b) Why are chlorine atoms examples of free radicals?

19 a) Identify two reactions that remove ozone from the stratosphere.
b) Why can a single chlorine atom destroy as many as 100 000 ozone molecules?

20 Why is UV spectroscopy a suitable technique for measuring ozone concentrations in the stratosphere?

21 Why is the depletion of the ozone layer most severe over Antarctica?

Worldwide production of CFCs has fallen sharply since their impact on the ozone layer was discovered. At a United Nations conference in 1987, governments agreed on large reductions in the production of CFCs, leading to an eventual ban – this was the Montreal Protocol. The protocol has been revised regularly to keep it up to date. It is often quoted as the most successful international treaty about the environment to date.

Concentrations of the key CFCs in the atmosphere have levelled off, or started to fall, since the Montreal Protocol came into effect. The concentration of halons from fire extinguishers have continued to rise but the rate of increase has slowed. Meanwhile the concentrations of the hydrochlorofluorocarbons (HCFCs) have increased because they were the early replacements for CFCs as solvents and refrigerants (Section 15.8).

The effect of oxides of nitrogen on the ozone layer was discovered before the effects of CFCs were understood. Like CFCs, the nitrogen oxides NO and NO_2 react catalytically with ozone. They upset the steady state in the stratosphere and speed up the rate of breakdown of ozone.

$$O_3 + \text{UV light} \rightarrow O_2 + O\bullet$$

$$NO + O_3 \rightarrow NO_2 + O_2$$

$$NO_2 + O\bullet \rightarrow NO + O_2$$

$$\text{Overall: } 2O_3 \rightarrow 3O_2$$

This happens to some extent naturally. Micro-organisms in the soil and in the oceans reduce nitrogen compounds to dinitrogen oxide, N_2O. Some of this gas gets carried up into the stratosphere, where it reacts with oxygen atoms to form NO. High-flying aircraft can significantly increase the role of oxides of nitrogen in destroying ozone. This is especially so for supersonic aircraft that fly in the stratosphere. They can release nitrogen oxides from their engines directly into the ozone layer at altitudes of 20 km.

Test yourself

22 Explain why concentrations of CFCs in the atmosphere have started to fall, while concentrations of halons and HCFC have increased.

23 Write an equation to show the formation of NO from N_2O.

24 Suggest reasons why the effect of CFCs, and related compounds, on the ozone layer is much more significant than the effect of nitrogen oxides from aircraft.

REVIEW QUESTIONS

1 a) What is the origin of the infrared radiation absorbed by gases in the air that leads to global warming? **(3)**

b) Explain, with examples, how the model of polar covalent bonds allows chemists to understand why some gases in the atmosphere contribute to global warming, while others do not. **(3)**

2 a) Explain, with examples, the meaning of the terms 'carbon neutral' and 'carbon footprint'. **(4)**

b) How is society changing as a result of new scientific information about the environmental effects of using fossil fuels? **(3)**

3 a) Use the example of CFCs to show how an understanding of the mechanisms of reactions can help scientists to understand the causes of environmental problems. **(4)**

b) Why have scientists recommended that CFCs should no longer be used in aerosols, foams and refrigerants? **(2)**

c) How have governments, industry and individuals responded to the scientific information about ozone-depleting chemicals? **(3)**

4 a) Give examples to explain the difference between greenhouse gases and gases that help to deplete the ozone layer. **(4)**

b) Give two examples of compounds that are both greenhouse gases and ozone-depleting chemicals. Explain why this is possible. **(3)**

The periodic table of elements

Key

relative atomic mass **atomic symbol** name atomic (proton) number

1	2											3	4	5	6	7	0(8)
							1.0 **H** hydrogen 1										(18)
(1)	(2)											(13)	(14)	(15)	(16)	(17)	4.0 **He** helium 2
6.9 **Li** lithium 3	9.0 **Be** beryllium 4											10.8 **B** boron 5	12.0 **C** carbon 6	14.0 **N** nitrogen 7	16.0 **O** oxygen 8	19.0 **F** fluorine 9	20.2 **Ne** neon 10
23.0 **Na** sodium 11	24.3 **Mg** magnesium 12	(3)	(4)	(5)	(6)	(7)	(8)	(9)	(10)	(11)	(12)	27.0 **Al** aluminium 13	28.1 **Si** silicon 14	31.0 **P** phosphorus 15	32.1 **S** sulfur 16	35.5 **Cl** chlorine 17	39.9 **Ar** argon 18
39.1 **K** potassium 19	40.1 **Ca** calcium 20	45.0 **Sc** scandium 21	47.9 **Ti** titanium 22	50.9 **V** vanadium 23	52.0 **Cr** chromium 24	54.9 **Mn** manganese 25	55.8 **Fe** iron 26	58.9 **Co** cobalt 27	58.7 **Ni** nickel 28	63.5 **Cu** copper 29	65.4 **Zn** zinc 30	69.7 **Ga** gallium 31	72.6 **Ge** germanium 32	74.9 **As** arsenic 33	79.0 **Se** selenium 34	79.9 **Br** bromine 35	83.8 **Kr** krypton 36
85.5 **Rb** rubidium 37	87.6 **Sr** strontium 38	88.9 **Y** yttrium 39	91.2 **Zr** zirconium 40	92.9 **Nb** niobium 41	95.9 **Mo** molybdenum 42	[98] **Tc** technetium 43	101.1 **Ru** ruthenium 44	102.9 **Rh** rhodium 45	106.4 **Pd** palladium 46	107.9 **Ag** silver 47	112.4 **Cd** cadmium 48	114.8 **In** indium 49	118.7 **Sn** tin 50	121.8 **Sb** antimony 51	127.6 **Te** tellurium 52	126.9 **I** iodine 53	131.3 **Xe** xenon 54
132.9 **Cs** caesium 55	137.3 **Ba** barium 56	138.9 **La*** lanthanum 57	178.5 **Hf** hafnium 72	180.9 **Ta** tantalum 73	183.8 **W** tungsten 74	186.2 **Re** rhenium 75	190.2 **Os** osmium 76	192.2 **Ir** iridium 77	195.1 **Pt** platinum 78	197.0 **Au** gold 79	200.6 **Hg** mercury 80	204.4 **Tl** thallium 81	207.2 **Pb** lead 82	209.0 **Bi** bismuth 83	[209] **Po** polonium 84	[210] **At** astatine 85	[222] **Rn** radon 86
[223] **Fr** francium 87	[226] **Ra** radium 88	[227] **Ac*** actinium 89	[261] **Rf** rutherfordium 104	[262] **Db** dubnium 105	[266] **Sg** seaborgium 106	[264] **Bh** bohrium 107	[277] **Hs** hassium 108	[268] **Mt** meitnerium 109	[271] **Ds** damstadtium 110	[272] **Rg** roentgenium 111							

Elements with atomic numbers 112–116 have been reported but not fully authenticated

*Lanthanide series	140 **Ce** cerium 58	141 **Pr** praseodymium 59	144 **Nd** neodymium 60	[147] **Pm** promethium 61	150 **Sm** samarium 62	152 **Eu** europium 63	157 **Gd** gadolinium 64	159 **Tb** terbium 65	163 **Dy** dysprosium 66	165 **Ho** holmium 67	167 **Er** erbium 68	169 **Tm** thulium 69	173 **Yb** ytterbium 70	175 **Lu** lutetium 71
*Actinide series	232 **Th** thorium 90	[231] **Pa** protactinium 91	238 **U** uranium 92	[237] **Np** neptunium 93	[242] **Pu** plutonium 94	[243] **Am** americium 95	[247] **Cm** curium 96	[245] **Bk** berkelium 97	[251] **Cf** californium 98	[254] **Es** einsteinium 99	[253] **Fm** fermium 100	[256] **Md** mendelevium 101	[254] **No** nobelium 102	[257] **Lr** lawrencium 103

Index

Page numbers in bold refer to illustrations.